Mechatronics and Robotics

Mechatronics and Robotics

Edited by:

Zoran Gacovski

ARCLER
PRESS

www.arclerpress.com

Mechatronics and Robotics

Zoran Gacovski

Arcler Press
224 Shoreacres Road
Burlington, ON L7L 2H2
Canada
www.arclerpress.com
Email: orders@arclereducation.com

ISBN: 978-1-77407-775-7 (Hardcover)

DECLARATION

ABOUT THE EDITOR

Dr. Zoran Gacovski has earned his PhD degree at Faculty of Electrical engineering, Skopje. His research interests include Intelligent systems and Software engineering, fuzzy systems, graphical models (Petri, Neural and Bayesian networks), and IT security. He has published over 50 journal and conference papers, and he has been reviewer of renowned Journals. Currently, he is a professor in Computer Engineering at European University, Skopje, Macedonia.

TABLE OF CONTENTS

LIST OF CONTRIBUTORS

A. Gasparetto
Polytechnic Department of Engineering and Architecture, University of Udine, Udine, Italy

L. Scalera
Polytechnic Department of Engineering and Architecture, University of Udine, Udine, Italy

Theodoros Foradis
Electrical and Computer Engineering, University of Patras, Patras, Greece

Kleanthis Thramboulidis
Electrical and Computer Engineering, University of Patras, Patras, Greece

Kleanthis Thramboulidis
Visiting Professor Helsinki University of Technology, Electrical & Computer Engineering, University of Patras, Patras, Greece.

Mohd Hudzari Razali
Faculty of Bioresources & Food Industry, Universiti Sultan Zainal Abidin, Tembila Campus, Besut, Terengganu Darul Iman, Malaysia

Syazili Roslan
Faculty of Bioresources & Food Industry, Universiti Sultan Zainal Abidin, Tembila Campus, Besut, Terengganu Darul Iman, Malaysia

Abdul Ssomad Mohd Abd Halim
Faculty of Bioresources & Food Industry, Universiti Sultan Zainal Abidin, Tembila Campus, Besut, Terengganu Darul Iman, Malaysia

Hayan Basit
Faculty of Bioresources & Food Industry, Universiti Sultan Zainal Abidin, Tembila Campus, Besut, Terengganu Darul Iman, Malaysia

Ashraf Elfasakhany
Department of Mechanical Engineering, Faculty of Engineering, Taif University, Al-

Haweiah, Saudi Arabia
Tecnológico de Monterrey, Campus Ciudad Juárez, Ciudad Juarez, Mexico

Eduardo Yanez
Tecnológico de Monterrey, Campus Ciudad Juárez, Ciudad Juarez, Mexico

Karen Baylon
Tecnológico de Monterrey, Campus Ciudad Juárez, Ciudad Juarez, Mexico

Ricardo Salgado
Tecnológico de Monterrey, Campus Ciudad Juárez, Ciudad Juarez, Mexico

Pedram Bagheri
Department of Electrical and Computer Engineering, University of Calgary, Calgary, Alberta, Canada

Laleh Behjat
Department of Electrical and Computer Engineering, University of Calgary, Calgary, Alberta, Canada

Qiao Sun
Department of Mechanical and Manufacturing Engineering, University of Calgary, Calgary, Alberta, Canada

Firas A. Raheem
Control and Systems Department, University of Technology, Baghdad, Iraq

Umniah I. Hameed
Control and Systems Department, University of Technology, Baghdad, Iraq

Rogério Rodrigues dos Santos
School of Mechanical Engineering, Federal University of Uberlândia, Uberlândia, Brazil

Valder Steffen, Jr
School of Mechanical Engineering, Federal University of Uberlândia, Uberlândia, Brazil

Sezimária de Fátima Pereira Saramago
Faculty of Mathematics, Federal University of Uberlândia, Uberlândia, Brazil

Ke Wang, Jiping Zhou
School of Mechanical Engineering, University of Yangzhou, Yangzhou, China

Chen Liu
Department of Transportation and Environmental Engineering, Hiroshima University, Higashi-Hiroshima, Japan

Yukio Fujimoto
Department of Transportation and Environmental Engineering, Hiroshima University, Higashi-Hiroshima, Japan

Yoshikazu Tanaka
Department of Transportation and Environmental Engineering, Hiroshima University, Higashi-Hiroshima, Japan

Shuncai Yao
Department of Electric Engineering, North University of China, Taiyuan, China

Jindong Tan
Department of Electrical Computer Engineering, Michigan Technological University, Houghton, USA

Hongxia Pan
Department of Electric Engineering, North University of China, Taiyuan, China

G. Padmavathi
Department of Computer Science, Avinashilingam University for Women, Coimbatore, Tamil Nadu, India

D. Shanmugapriya
Department of Information Technology, Avinashilingam University for Women, Coimbatore, Tamil Nadu, India

M. Kalaivani
Department of Computer Science, Avinashilingam University for Women, Coimbatore, Tamil Nadu, India

Mohammed A. Al-Moniee
Saudi Arabian Oil Company (Saudi Aramco), Dhahran, Saudi Arabia

Xiangyang Zhu
Saudi Arabian Oil Company (Saudi Aramco), Dhahran, Saudi Arabia

Rikke Markfoged
Danish Technological Institute (DTI), Aarhus, Denmark

Aabdullah H. Al-Wadei
Saudi Arabian Oil Company (Saudi Aramco), Dhahran, Saudi Arabia

Poul L. Pedersen
Danish Technological Institute (DTI), Aarhus, Denmark

Anders K. Tuxen
Danish Technological Institute (DTI), Aarhus, Denmark

Fuad I. Al-Nuwaiser
Saudi Arabian Oil Company (Saudi Aramco), Dhahran, Saudi Arabia

Lone Tang
Danish Technological Institute (DTI), Aarhus, Denmark

Tinna Staghøj Roesen
Danish Technological Institute (DTI), Aarhus, Denmark

Thomas Lundgaard
Danish Technological Institute (DTI), Aarhus, Denmark

Jingyang Peng
Department of Mechanical Engineering, University of Saskatchewan, Saskatoon, Canada

Xiongbiao Chen
Department of Mechanical Engineering, University of Saskatchewan, Saskatoon, Canada

Mosfequr Rahman
Georgia Southern University, Statesboro, GA, USA

Mahbub Ahmed
Southern Arkansas University, Magnolia, AR, USA

Masud Nawaz
Georgia Southern University, Statesboro, GA, USA

Gustavo Molina
Georgia Southern University, Statesboro, GA, USA

Abdur Rahman
Edinboro University, Edinboro, PA, USA

Jwo Ming Jou
Department of Mechanical Engineering, Cheng Shiu University, Kaohsiung City, Taiwan

Minoru Yaga
Department of Mechanical Systems, Faculty of Engineering, University of the Ryukyus, Okinawa, Japan

Yusuke Uechi
Mie Metal Industry Co. Ltd., Mie, Japan

Hiroaki Ozono
Graduate School of Engineering and Science, University of the Ryukyus, Okinawa, Japan

Masaaki Ishikawa
Department of Mechanical Systems, Faculty of Engineering, University of the Ryukyus, Okinawa, Japan

Isao Teruya
Department of Mechanical Systems, Faculty of Engineering, University of the Ryukyus, Okinawa, Japan

Y. Cao, X. B. Chen
Department of Mechanical Engineering, University of Saskatchewan, Saskatoon, Canada

LIST OF ABBREVIATIONS

AMR	Anisotropic Magneto Resistive
AFM	Atomic Force Microscope
AMS	Autonomous microbe sensor
ARX	Auto-regressive
ARX	Autoregressive exogenous
COTS	Commercial off-the-shelf
DTI	Danish Technological Institute
DOF	Degrees-of-freedom
EPFL	Ecole Poly-technique Fédérale de Lausanne
EDM	Electrical Discharge Machining
TEKES	Finnish Funding Agency for Technology and Innovation
AMS	Fluorescence count
FAO	Food and Agriculture Organization
4 DOF	Four degree of freedom
FCS	Free Cartesian space
IAS	Industrial automation systems
IAT	Industrial Automation Thing
IC	Integrated circuits
IoMT	Internet of manufacturing things
IoT	Internet of Things
MRC	Maxwell resistive capacitor
MIC	Microbiologically influenced corrosion
MEMS	Micro-Electro-Mechanical-Systems
MIM	Model Integrated Mechatronics
MTSs	Model-driven development of complex mechatronic systems
NC	Normally closed
NO	Normally open

OO	Object- oriented
PSO	Particle Swarm Optimization
PFE	Photo-ferroelectric
PEA	Piezoelectric actuator
PVDF	Piezoelectric film
PIM	Platform Independent Model
PI	PrandtlIshlinskii
PLC	Programmable Logic Controllers
PD	Proportional-Derivative
PID	Proportional-integral- derive
QoS	Quality of Service
RRT	Rapidly-Exploring Random Tree
SOA	Service-oriented architecture
SMC	Sliding mode control
SMOs	Sliding mode-based observers
SMC	Sliding model control
SRB	Sulfate-reducing bacteria
SysML	Systems Modeling Language
TBD	Track Before Detect
UV	Ultraviolet

PREFACE

Mechatronics is a synergy of mechanical engineering, electrical engineering and information and communication technologies (computer science, informatics).

Highly automated systems are all around us, from music and video devices to factory facilities. Due to the continuous advancement of the technology, where there are no longer any sharp boundaries between individual fields of expertise, there is a need for experts who will be able to design, operate and maintain such complex systems.

Mechatronics is not a new technical branch, but a new approach in the field of technical sciences in an effort to educate engineers who are able to fully understand the functioning and design of modern sophisticated products (devices) and systems.

Mechatronics is presented - not as a new engineering branch, but as a new concept, a new approach to engineering that requires interaction and integration between different engineering disciplines. The afore-mentioned technical areas are taught as much as is necessary for understanding and servicing individual systems, which require further specialization. The engineer should be able to understand the principle of system operation, the role of individual components and diagnose system operation. He is the link between the technicians who are specialized, (he is the team leader).

Mechatronics is the process of embedding intelligence into physical systems.

The characteristics of mechatronic products and systems are:

- functional interaction between mechanical, electronic and information technologies;
- spatial integration of subsystems into one unit; intelligence related to the control functions of the mechatronic system;
- adaptability, the convenience with which - mechatronic products can be adapted to changing tasks and situations;
- multifunctionality related to microprocessor functions determined by a computer program;
- hidden functions performed by microelectronics, difficult to see and understand by the end consumers.

Mechatronics is an interdisciplinary field that integrates knowledge in electronics, mechanical engineering and computer science. Mechatronics specialists will be the most flexible, the most versatile and the most modern trained professionals of the future.

This edition covers different topics from mechatronics and robotics, including mechatronics basics, robotics arms and manipulators, sensors and actuators in mechatronics.

Section 1 focuses on mechatronics basics, describing a brief history of industrial robotics in the 20th century; an IoT model for cyber-physical manufacturing systems; 3+1 SysML view-model in model integrated mechatronics; design and development of mechatronic application in agricultural irrigation device.

Section 2 focuses on robotics arms and manipulators, describing design and development of a competitive low-cost robot arm with four degrees of freedom; an adaptive robust approach to modeling and control of flexible arm robots; interactive heuristic d* path planning solution based on PSO for two-link robotic arm in dynamic environment; optimal task placement of a serial robot manipulator for manipulability and mechanical power optimization; kinematical analysis and simulation of high-speed plate carrying manipulator based on Matlab.

Section 3 focuses on sensors in mechatronics, describing flexible impact force sensor; a sensing and robot navigation of hybrid sensor network; a study on vehicle detection and tracking using wireless sensor networks; deployment of pre-industrial autonomous microbe sensor in Saudi Arabia's injection seawater system.

Section 4 focuses on actuators, describing a survey of modeling and control of piezoelectric actuators; experimental investigation of photostrictive materials for MEMS application; theory and simulation analysis of the mode shape and normal shape actuators and sensors; experimental study of the response of transonic diffuser flow to a piezoceramic actuator at diffuser throat; an ARX-based PID-sliding mode control on velocity tracking control of a stick-slip Piezoelectric-driven actuator.

CHAPTER

1

A Brief History of Industrial Robotics in the 20th Century

A. Gasparetto, L. Scalera

Polytechnic Department of Engineering and Architecture, University of Udine, Udine, Italy

ABSTRACT

Industrial robotics is a branch of robotics that gained paramount importance in the last century. The presence of robots totally revolutionized the industrial environment in just a few decades. In this paper, a brief history of industrial robotics in the 20th century will be presented, and a proposal for classifying the evolution of industrial robots into four generations is set forward. The characteristics of the robots belonging to each generation are mentioned, and the evolution of their features is described. The most

Citation:- Gasparetto, A. and Scalera, L. (2019), "A Brief History of Industrial Robotics in the 20th Century". *Advances in Historical Studies*, **8**, 24-35. doi: 10.4236/ahs.2019.81002.

significant milestones of the history of industrial robots, from the 1950's to the end of the century, are mentioned, together with a description of the most representative industrial robots that were designed and manufactured in those decades.

Keywords: Industrial Robots, History, Unimate, Stanford Arm, Delta Robot

INTRODUCTION

The idea to design and build some sort of beings, or devices that could carry out repetitive or heavy tasks, thus relieving men from this burden, dates back to ancient times. Since the Greek-Hellenistic age some of these devices, which were named automata, have been designed and created by ingenious inventors, belonging to several different civilizations through the centuries. The term "automata" mainly refers to human-like devices, while the term "robot" has a more general meaning.

The origin of the term "robot" is placed in more recent times: namely, it comes from the Czech word "robota", meaning "heavy work" or "forced labor". The introduction of this term is due to the Czech writer Karel Čapek (1890-1938), who used it for the first time in 1920 in his novel "R.U.R.: Rossum's Universal Robots" (Čapek, 2004) .

On the other hand, the word "Robotics" was employed for the first time by Isaac Asimov (1920-1992) in his novel "Runaround" (1942), contained in the famous series "I, Robot". In that novel he defined three rules concerning the behavior of robots and the interaction with humans: these rules would later be named the three Laws of Robotics (Asimov, 1942).

In the literature, there are not so many works dealing with Robotics from a historical perspective. For instance, Ceccarelli dealt with this topic in (Ceccarelli, 2001) and (Ceccarelli, 2004), while Gasparetto (2016) presents a historical outline of robotics from ancient times until the Industrial Revolution.

Industrial applications of Robotics gained a paramount importance in the last century. The beginning of "Industrial Robotics", as we currently define it, can be dated back to the 1950's, although some kinds of automatization in the industrial environment started to appear since the times of the Industrial Revolution.

In this paper, the main milestones of the history of industrial robotics, from its beginning (in the 1950's and even earlier) to the end of the 20th century, will be mentioned and described. The evolution of industrial robots can be subdivided in four categories, as in (Zamalloa, 2017), the first three covering the timespan from the 1950's to the end of the 1990's. The robots of the fourth generation (which ranges from 2000 to nowadays), that are characterized by high-level "intelligent" features (such as the capability of performing advanced computations, logical reasoning, deep learning, complex strategies, collaborative behavior), are not analyzed in this paper.

In the scientific literature, not so many works on the history of industrial robots are present. In some books, such as in (Ceccarelli, 2004), (Mathia, 2010), (Siciliano & Khatib, 2008), and (Westerlund, 2000), there are some sketches on this topic, which could also be found in some internal reports, e.g. in (Lagerberg & Jonson, 2016) and (Wallen, 2008) .

THE FIRST GENERATION OF INDUSTRIAL ROBOTS (1950-1967)

Some Authors (Zamalloa, 2017) proposed a chronological categorization of industrial robots, by defining four "generations".

The first generation of industrial robot spans from 1950 to 1967. The robots of this generation were basically programmable machines that did not have the ability to really control the modality of task execution; moreover, they had no communication with the external environment. With respect to the hardware, the first generation robots were provided with low-tech equipment, and servo-controllers were not present (Wallen, 2008). A peculiar feature of these robots is the strong noise they produced, when their arms collided with the mechanical stops built to limit the movement of the axes. With respect to the actuators, almost all the first generation robots employed pneumatic actuators and were controlled by a sort of "logic gates" acting as automatic regulators. Such "logic gates" were basically cams which activated pneumatic valves, or relays which controlled solenoid valves. Finally, the tasks that the first generation robots were capable to perform were necessarily quite trivial, such as loading-unloading or simple material handling operations.

The history of industrial robotics is conventionally set in the 1950's, although some developments in automation had taken place before: namely, a "programmable" paint-sprayer device invented by Pollard and Roselund in 1938 (Koetsier, 2019), and a tele-operated "manipulator" invented by Goertz in 1949, for instance. However, the turning point for industrial robotics was due to the genius of George Devol, who designed in 1954 a "Programmable Article Transfer" (this was the name given when the patent request (Devol, 1954) was filed). Such a device was the base for the development of Unimate that is considered the first "true" industrial robot in history.

Devol (Malone, 2011) , who was basically a scientist, needed an entrepreneurial mate with whom set up in a more concrete way his idea of a robotic manipulator which could be used to automatize industrial processes. In 1954 he met Joseph Engel Berger, an engineer employed in the spatial industry, and they agreed on setting up a company which could manufacture robots for industrial applications. This led to the foundation of a company named Unimaton, which produced in 1961 the first Unimate robot (Figure 1). Unimate, which was hydraulically actuated, was immediately installed in an automotive company, namely in the General Motors factory located in Trenton (New Jersey, USA), where it was employed to extract parts from a die-casting machine (as mentioned earlier, it was used for a single task, because it was very complicated to reprogram it). In the following years, several other Unimates were installed in automotive factories, where were mainly employed for spot-welding of cars and for handling of workpieces (Wallen, 2008).

In the same years, several robot manufacturing companies were born, due to the fact that many entrepreneurs understood that this kind of devices had great potentialities, especially in the automotive sector. Companies like Ford and General Motors started to consider the automatization of their productive plants and needed device such as the new robot to achieve this goal. Thus, there was a sudden increase in the orders of robotic devices, which gave the opportunity to thrive to many robot manufacturers.

One of such companies was, for instance, AMF Corporation: in 1962 they manufactured a new robot that was called Versatran (i.e. "versatile transfer"). It was a cylindrical robot (Figure 2) that was ordered by Ford for its production plants in Canton (Ohio, USA), thus enjoying a good popularity (Birnie, 1974). The Versatran was also the first robot to be installed in a productive site in Japan (1967). In 1969, the presence of robots in Japanese companies had a significant increase, also due to the fact that Unimation allowed to Kawasaki Heavy Industries Ltd. the license to produce robots for the Japanese and the Asian market. In particular, this led to the development of the Kawasaki-Unimate 2000, the first industrial robot ever built in Japan.

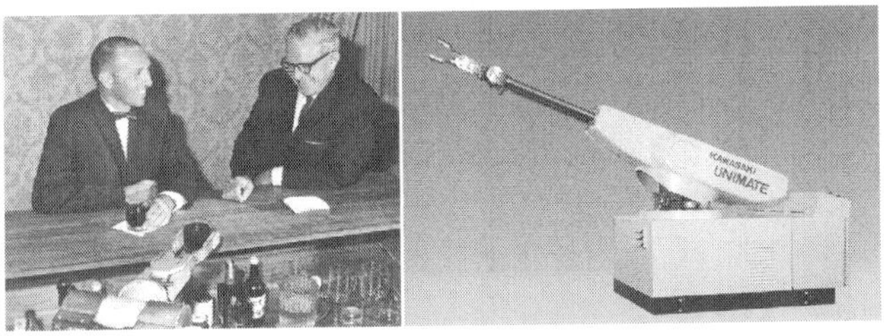

Figure 1. Joseph Engelberger and George Devol (left); the Unimate robot (right).

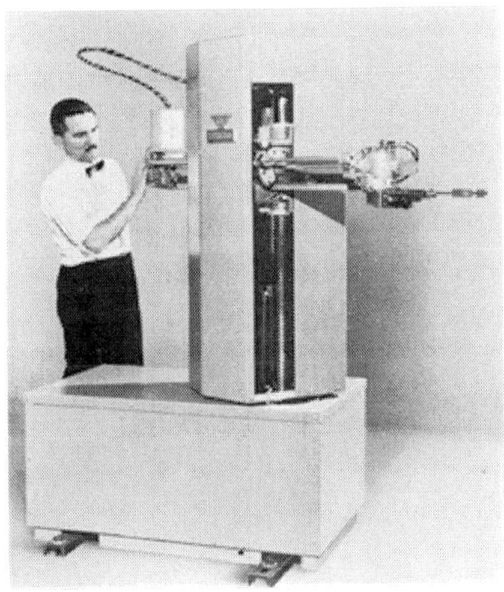

Figure 2. The Versatran robot.

Since then, Japan became one of the leading countries in the world in the field of industrial robotics. It was in Japan that the first national robotic association in the world was founded in 1971, namely JIRA (Japanese Robotics Association).

The diffusion of robots in Europe took place more or less at the same time as in Japan, namely in the second half of the 1960's. The first robotic installation in Europe appeared in 1967 in the company Svenska Metallverken located in Upplands Väsby (Sweden): it consisted of robots carrying out simple pick-and-place tasks (Wallen, 2008). Two years later, in Norway, the first painting robot was manufactured by the Tralffa Company and installed in a plant for wheelbarrow painting.

The appearance of robots allowed to automatize important production tasks such as welding. Unimation was the first company to manufacture welding robots, while General Motors was the first company to install this kind of robotic devices at their automotive plants in Lordstown (Ohio, USA) in 1969. Three years after, FIAT installed the first welding robots in Europe at their plants in Turin (Italy).

THE SECOND GENERATION OF INDUSTRIAL ROBOTS (1968-1977)

The industrial robots of the second generation (conventionally ranging from 1968 to 1977) were basic programmable machines with limited possibilities of self-adaptive behavior and elementary capabilities to recognize the external environment (Zamalloa, 2017). These robots used servo-controllers, which enabled them to perform both point-to-point motion, and continuous paths as well. Their control system consisted of microprocessors or of Programmable Logic Controllers (PLC), and they could be also programmed by an operator by means of a teach box. With respect to those belonging to the first generation, these robots could carry out more complex tasks (e.g. control of work centers). However, their level of versatility was not very high, because each robot had its own software, which was dedicated to a specific task. Hence, these robots turned out to be application-specific devices, meaning that it was very difficult to employ the same robot for different tasks, since this would require a substantial modification of the controller, and a thorough reprogramming of the operating software. With respect to diagnostics, the robots of the second generation were not very per formant, since the only diagnostic reports they could produce were those related to failures, which were reported by means of indicator lights, without any hint related to the cause of the failure that was left to the operator to trace.

At the beginning of the history of industrial robotics, the robots had hydraulic actuators. The shift from hydraulic to electric actuators took place in the 1970's, when the electronic components needed to govern a robot reached the full technical maturity. As a matter of fact, microprocessors and other components started to be widely used at that time: this allowed the robot manufacturers to dispose of powerful and cost-effective devices which could be employed to implement control systems able to deal with a complex and computationally expensive task such as the control of a robot.

The economic and geopolitical situation, at the international level, also pushed the trend towards electrically driven robots: for instance, the price of crude oil suddenly grew after the oil crisis following the Kippur war (October 1973). The companies were thus forced to find more efficient ways of production: robot (and in particular electrically driven robots) was consistent with the goals of reducing the costs and increasing the productivity. This gave a boost to the installations of industrial robots, which increased more than 30% per year in the second half of the 1970's (Karlsson, 1991).

From the scientific point of view, a significant base for the development of electrically driven robots was the merit of Victor Scheinman (Scheinman, 1973). Scheinman was a mechanical engineering student at Stanford University who in 1969 designed and built the famous Stanford Arm (Figure 3). This robot was the first prototype of a robot actuated by electric motors (6 DC motors) and controlled by a PDP-6 microprocessor. The Stanford Arm had 5 revolute and one prismatic joint, for a total of 6 DOFs, and its kinematic chain was made of harmonic drives and spur gear reducers. Its inverse kinematics could be analytically solved in a closed form, which allowed a fast trajectory execution. Moreover, some sensors (tachometers and potentiometers) were mounted on the robot, in order to measure position and velocity of the robot joints.

Figure 3. The Stanford Arm.

Four years later (1973), Scheinman designed another electrical robot, named Vicarm that was smaller and lighter than the industrial robots of that time. This made Vicarm particularly suitable for use in tasks, such as assembly of parts, where the robot was not required to lift and carry heavy loads. These features of Vicarm were so appreciated, that Unimation bought the company that produced Vicarm and exploited its know-how to design and manufacture (in 1978) the famous PUMA robot (Figure 4).

The PUMA (an acronym for Programmable Universal Machine for Assembly) was considered for many decades the archetype of the anthropomorphic robots, and its kinematics is still taken as an example in several robotic books at the undergraduate as well at the graduate level.

In the meanwhile, other companies developed and manufactured other types of industrial robots. For instance, KUKA developed in 1973 the Famulus robot, whose name in Latin means "servant". In 1974 Cincinnati Milacron, a major machine tool manufacturer, developed a robot called T3 (an acronym for "The Tomorrow Tool"), which was installed in several automotive plants, and especially in the Volvo plants in Sweden. T3 was the first commercially available minicomputer-controlled industrial robot (Figure 5).

In 1974, the Swedish company ASEA (now ABB) started the production of the robots of the famous and successful IRB series, well known worldwide also for their typical orange color. The first robot of this series, that was issued for more than 20 years, was the IRB-6, which was largely employed in productive sites for complex tasks (machining, arc-welding), for its ability to move smoothly along continuous paths (Figure 6).

In the same year, the Japanese company Hitachi developed the robot HI-T-HAND Expert, that is relevant in the history of industrial robotics for the precision it reached in insertion operations (it could insert mechanical parts with a clearance of about 10 micrometers). It was also provided with a force feed-back control system and a flexible wrist mechanism.

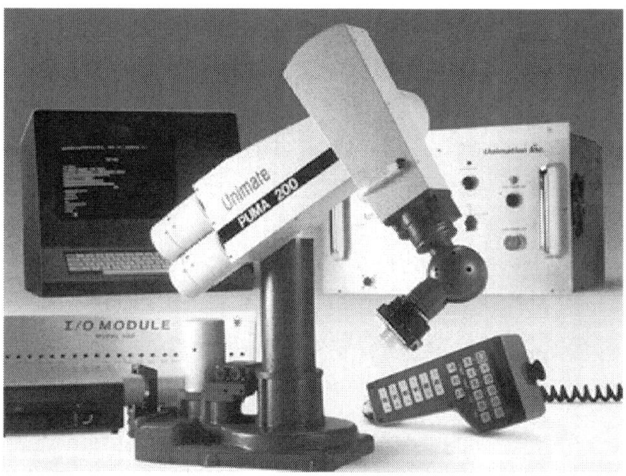

Figure 4. The PUMA robot.

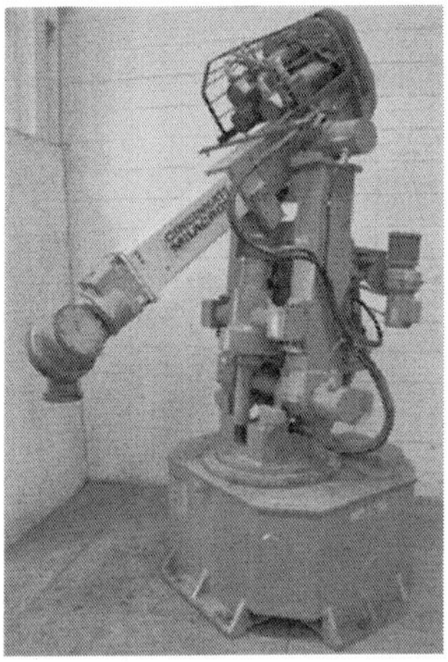

Figure 5. The Cincinnati Milacron T3 robot.

Figure 6. The "legendary" ABB IRB robot.

THE THIRD GENERATION OF INDUSTRIAL ROBOTS (1978-1999)

The industrial robots of the third generation (conventionally ranging from 1978 to 1999) were characterized by a larger extent of interaction with both the operator and the environment, through some kind of complex interfaces (such as vision or voice). They also had some self-programming capabilities, and could reprogram themselves, although by a little amount, in order to execute different tasks (Zamalloa, 2017). These robots were provided with servo controls, and could execute complex tasks, by moving either from point to point or along continuous paths. They could be programmed either on-line (the operator could use a teach box with a keyboard) or off-line, being connected to a PLC or a PC, which allowed to use a high-level language for motion programming and enabled the robots to be interfaced with a CAD or a database. The possibility of high-level, off-line programming enlarged the operational potential of the robots: for in-stance, they could elaborate data from sensor reading, in order to adjust the robot movements taking into account changes in the environment (e.g. changes in position and orientation of the work pieces). Moreover, the diagnostic capabilities could be greatly enhanced: these robots could produce not only an indication of failure detection, but also a report on the location and on the type of the failure. In addition, some sort of "intelligence" was present in the robots of the third generation, with some (although limited) adaptive capabilities. These capabilities could be employed in some more complex tasks (such as tactile inspection, assembly operations, arc welding), by using the data coming from vision or perception systems to locate the objects and the work pieces and guide the joint movements according to the task to be performed, taking into account the possibility of small changes in the position of the objects.

Between the end of the 1970's and the beginning of the 1980's, other scientific and technical improvements contributed to the diffusion of robots.

In 1978, a novel kinematic structure was proposed by the Japanese scientist Hiroshi Makino from Yamanashi University. Such a structure was made of three revolute joints with parallel axes and a prismatic joint lying at the end of the kinematic chain (Figure 7). The robot with this structure was named SCARA (an acronym from "Selective Compliance Assembly Robot Arm"), since its compliance in the horizontal direction resulted lower than the compliance in the vertical direction. For this reason, as well as for the lightness of the kinematic chain (that allowed a simpler and faster controller), this robot was suitable to be employed in tasks such as the assembly of small objects (Makino & Furuya, 1980).

Another relevant technical improvement in industrial robotics was the appearance of direct drive actuated robot. The first prototype of this kind was the CMU Direct Drive Arm (Asada & Kanade, 1983), developed in 1981 by Kanade and Asaka at Carnegie Mellon University (Pittsburgh, USA). This kind of robot featured higher accuracy and faster operations because the motors connected directly to the arms eliminating the need for intermediate gear or chain systems.

Both the aforementioned findings were employed in the AdeptOne (Figure 8), the first commercially available direct-driven SCARA robot (1984).

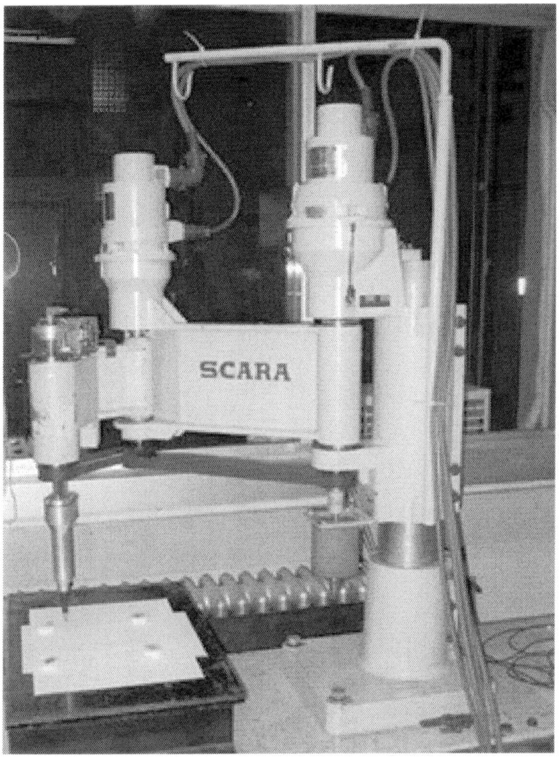

Figure 7. One of the first prototypes of SCARA robot, designed by Hiroshi Makino.

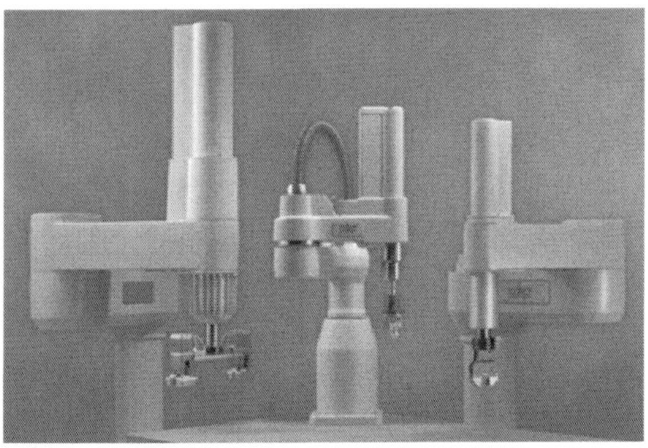

Figure 8. Examples of AdeptOne SCARA robots.

These scientific and technical findings contributed to the diffusion of robots in applications different from those in the classical automotive sector, in particular in the production of consumer goods, especially those in the electronic market. SCARA robots were employed in these types of assembly lines, especially in Japan. For this reason, Japan became in the 1980's not only the world leader in robot manufacturing (with more than 40 companies that built and sold robots worldwide), but also in the production of electronic consumer goods.

Robotics in the 1980's was a rising star, not only in Japan but in all the developed countries. It appeared as a promising field that drew the interest of journalist, scientists, policy makers and also common people.

From the scientific and technical point of view, this is the time when the robots became even more versatile, by exploiting important improvements both with respect to the hardware and the software. Concerning the former aspect, robots started to be provided with advanced sensors (e.g. cameras, force sensors, laser scanners); concerning the latter aspects, the control software became more "intelligent" by introducing come techniques related to Artificial Intelligence. Both these aspects increased the versatility and the flexibility of the robots, which could be employed in more and more complex tasks.

Despite the significant progress undergone in the 1980's, the need for robots that could carry out task at high speed pushed the scientific research to design innovative kinematic structures. The idea of employing parallel kinematic chains instead of the classical serial kinematic chains was put forward and led to a type of lightweight robot featuring the capability of moving at high speed. The archetype of this kind of robots was the Delta robot (that appeared in 1992), conceived by the Swiss scientist Reymond Clavel at the Ecole Poly-technique Fédérale de Lausanne (EPFL). This type of robot, designed by Clavel in his PhD thesis, had three translational DOFs and one rotational DOF (Clavel, 1991). With respect to serial robots, parallel robots featured a smaller work-space, but the capability of operating at much higher speed. The kinematic architecture of the Delta robot was copied in many parallel manipulators, devoted to high speed pick-and-place operations.

The first application of Delta robots was developed by the Swiss company Demaurex in 1992: six Delta robots were operating inside a work cell for loading pretzels into trays (Figure 9). Some years afterwards (1998) ABB developed the Flex-Picker, the world's fastest picking robot, based on the structure of the Delta robot (Figure 10).

In this decade, the first attempts to synchronize robots took place: in 1994 Motoman developed MRC, the first robot control system which provided the synchronized control of two robots (up to 21 axes).

The end of the third generation is conventionally set to the end of the century; beginning from the year 2000, the industrial robots are considered to belong to the fourth generation (which extends up to the current days). Such robots feature high-level "intelligent" capabilities (such as performing advanced computations, logical reasoning, deep learning, complex strategies, collaborative behavior).

Figure 9. Delta robots in the Demaurex work cells for the packaging of pretzels.

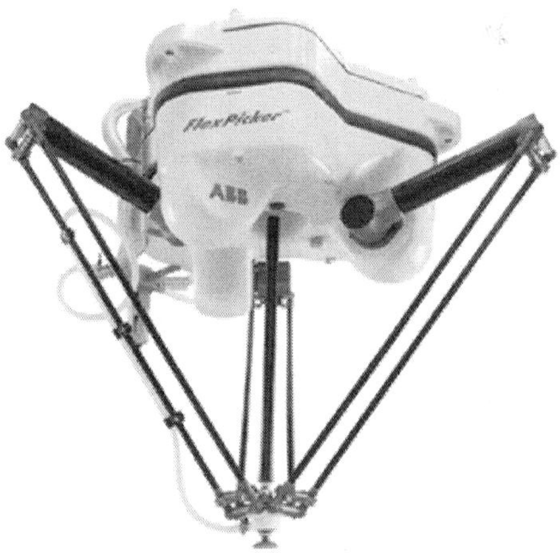

Figure 10. The ABB Flex-Pitcher robot.

CONCLUSION

In this paper, a brief history of industrial robotics in the 20th century was presented. The evolution of the industrial robots was conventionally categorized into four generations, of which the first three cover the timespan from the 1950's to the end of the century.

In this historical sketch, not only the scientific and technical evolution was taken into account, but some considerations about the economic and geopolitical issues that determined the diffusion of industrial robots, were also done.

The evolution of industrial robotics is not over, but is still developing in the current days: innovative ideas and novel hardware devices, together with some new programming techniques connected with Artificial Intelligence, are revolutionizing the concept of industrial automation and giving a new youth to the factory environment.

CONFLICTS OF INTEREST

The authors declare no conflicts of interest regarding the publication of this paper.

REFERENCES

1. Asada, H., & Kanade, T. (1983). Design of Direct-Drive Mechanical Arms. Journal of Vibration, Acoustics, Stress, and Reliability, 105, 312-316. https://doi.org/10.1115/1.3269106
2. Asimov, I. (1942). I, Robot. In Runaround. New York: Spectra Books.
3. Birnie, J. V. (1974). Practical Implications of Programmable Manipulators. Industrial Robot: An International Journal, 1, 122-128. https://doi.org/10.1108/eb004720
4. Capek, K. (2004). R.U.R. Rossum's Universal Robots. Fayetteville: Penguin Ed.
5. Ceccarelli, M. (2001). A Historical Perspective of Robotics toward the Future. Fuji International Journal of Robotics and Mechatronics, 13, 299-313. https://doi.org/10.20965/jrm.2001.p0299
6. Ceccarelli, M. (2004). Fundamentals of Mechanics of Robotic Manipulation. Dordrecht: Kluwer/Springer. https://doi.org/10.1007/978-1-4020-2110-7
7. Clavel, R., (1991). Conception d'un robot parallèle rapide à 4 degrés de liberté. Ph.D. Thesis, Lausanne, Switzerland: EPFL.
8. Devol, G. (1954). Programmable Article Transfer. https://patents.google.com/patent/US2988237
9. Gasparetto, A. (2016). Robots in History: Legends and Prototypes from Ancient Times to the Industrial Revolution. In Explorations in the History of Machines and Mechanisms. History of Mechanism and Machine Science (vol. 32, pp. 39-49). Berlin/Heidelberg: Springer. https://doi.org/10.1007/978-3-319-31184-5_5
10. Karlsson, J. M. (1991). A Decade of Robotics; Analysis of the Diffusion of Industrial Robots in the 1980s by Countries, Application Areas, Industrial Branches and Types of Robots. Stockholm, Sweden: Mekanforbundets Forlag.
11. Koetsier, T. (2019). Towards the Global Intelligent Machine. In The Ascent of GIM, the Global Intelligent Machine. History of Mechanism and Machine Science (vol 36, pp. 293-317). Berlin/Heidelberg: Springer. https://doi.org/10.1007/978-3-319-96547-5
12. Lagerberg, T., & Jonson, J. (2016). Une bio du robot. Petite histoire de la robotique industrielle. ABB Review 3/16. http://new.abb.com/about/technology/abb-review

13. Makino, H., & Furuya, N. (1980). Selective Compliance Assembly Robot Arm. Proceedings of 1st International Conference on Assembly Automation (ICAA), Brighton, March 25-27, 77-86.
14. Malone, B. (2011). George Devol: A Life Devoted to Invention, and Robots. IEEE Spectrum Magazine. https://spectrum.ieee.org/automaton/robotics/industrial-robots/george-devol-a-life-devoted-to-invention-and-robots
15. Mathia, K. (2010). Robotics for Electronics Manufacturing. Principles and Applications in Clean-room Automation. Cambridge: Cambridge University Press. https://doi.org/10.1017/CBO9780511712173
16. Scheinman, V. (1973). Design of a Computer Controlled Manipulator. PhD Thesis, Stanford, CA: Stanford University. http://www.dtic.mil/docs/citations/AD0708069
17. Siciliano, B., & Khatib, O. (2008). Springer Handbook of Robotics. Berlin Heidelberg: Springer-Verlag.
18. Wallen, J. (2008). The History of the Industrial Robot. Technical Report, Linkopings: Automatic Control at Linkopings Universitet. http://www.control.isy.liu.se/publications
19. Westerlund, L. (2000). The Extended Arm of Man—A History of the Industrial Robot. Stockholm: Informationsforlaget.
20. Zamalloa, I., Kojcev, R., Hernandez, A., Muguruza, I., Usategui, L., Bilbao, A., & Mayoral, V. (2017). Dissecting Robotics—Historical Overview and Future Perspectives.

CHAPTER 2

From Mechatronic Components to Industrial Automation Things: An IoT Model for Cyber-Physical Manufacturing Systems

Theodoros Foradis, Kleanthis Thramboulidis

Electrical and Computer Engineering, University of Patras, Patras, Greece

ABSTRACT

IoT is considered as one of the key enabling technologies for the fourth industrial revolution that is known as Industry 4.0. In this paper, we consider the mechatronic component as the lowest level in the system composition hierarchy that tightly integrates mechanics with the electronics and software required to convert the mechanics to intelligent (smart) object offering well defined services to its environment. For this mechatronic component to be

Citation:- Foradis, T. and Thramboulidis, K. (2017), "From Mechatronic Components to Industrial Automation Things: An IoT Model for Cyber-Physical Manufacturing Systems". Journal of Software Engineering and Applications, 10, 734-753. doi: 10.4236/jsea.2017.108040.

integrated in the IoT-based industrial automation environment, a software layer is required on top of it to convert its conventional interface to an IoT compliant one. This layer, which we call IoT wrapper, transforms the conventional mechatronic component to an Industrial Automation Thing (IAT). The IAT is the key element of an IoT model specifically developed in the context of this work for the manufacturing domain. The model is compared to existing IoT models and its main differences are discussed. A model-to-model transformer is presented to automatically transform the legacy mechatronic component to an IAT ready to be integrated in the IoT-based industrial automation environment. The UML4IoT profile is used in the form of a Domain Specific Modelling Language to automate this transformation. A prototype implementation of an Industrial Automation Thing using C and the Contiki operating system demonstrates the effectiveness of the proposed approach.

Keywords: Mechatronics, Cyber-Physical Systems, Internet of Things (IoT), Contiki, UML4IoT Profile

INTRODUCTION

Based on one of the most commonly used definitions, the term Mechatronics emphasizes on the synergistic integration of the three discipline areas, i.e., mechanical engineering, electronics and intelligent computer control, in the design and manufacture of products and processes [1] , i.e., it emphasizes on synergy. What is not clear by this definition is the level at which this integration should be performed, i.e., at the system level, which is the traditional approach, or at the subsystem or even at the mechanical unit (component) level.

The latter is proposed in Model Integrated Mechatronics [2] and refined with the 3+1SysML-view model [3] [4]. This approach defines the Mechatronic component as the main building block that abstracts the mechanical object to the software level, and transforms it to a smart object by adding additional functionality to the one offered by the mechanical part.

The so constructed mechatronic components are integrated with cyber components and humans to construct the industrial automation system. This approach slightly finds its road to production in the context of Industry 4.0, e.g., [5], since it greatly reduces the coupling between the system components compared to the traditional one, which considers the integration of the three disciplines at the system integration level.

A substantial number of communication mechanisms and middleware are used for the integration of the constituent components of industrial automation systems (IAS). DPWS [6], one of the most recent, is the result of extensive research based on the service-oriented architecture (SOA). It is based on SOAP which was the dominating technology used in SOA implementations. However, the REST paradigm is gaining more attention in manufacturing the past few years [7] [8]. A comparison of REST with SOAP in the context of the manufacturing domain can be found in [7]. Cloud computing has also attracted the interest of researchers. Cloud manufacturing defines a new service-oriented manufacturing model where cloud computing and IoT deeply influence the development process of manufacturing systems [9]. IoT technologies greatly reduce the time for decision-making that is very critical in modern manufacturing environments [10]. Both, cloud computing and IoT, have been moving from buzzwords and hypes to tangible practical technologies that subtly influence and change our world [11]. A state-of-the-art survey in cloud manufacturing is given in [9], which considers IoT as a technology that deeply influences the development of cloud manufacturing.

IoT is aligned well with the architecture of a manufacturing enterprise and it is able to provide "vital solutions to planning, scheduling, and controlling of manufacturing systems at all levels" [12]. Several approaches consider IoT as a technology that can be utilized as an integration mechanism to be used down to the sensor and actuator level of the industrial automation system. Others consider IoT as the new logical transition from the automation and connectivity concepts that exist in the IAS domain for many years. Bradley et al. [13] , in an article with title "The Internet of Things—The future or the end of mechatronics", argue that many of the smart components associated with the IoT will be essentially mechatronic in nature, and will be constructed as far as it regards their interaction with the physical world on the conventional hierarchical model. This model considers the controller of the mechatronic component in the loop with the controlled physical unit through sensors and actuators.

For the conventional mechatronic component to be integrated in the IoT- based industrial environment a software layer is required on top of it to convert its conventional interface to an IoT-compliant one. Thus, the adoption of the IoT as integration technology for the system, transforms the conventional mechatronic component to an Industrial Automation Thing (IAT). This transformation is more likely, as authors also argue in [13] , to bring significant changes to the way mechatronic, and related, systems are designed and configured. There is already an increasing complexity in the job of the industrial engineer in the task of transferring the functionality of the physical world in the software world in the level of the IAT. To this, the complexity of adding an extra layer to transform the conventional mechatronic component to an IAT is added. New protocols, languages, environments and architectural paradigms should be used and successfully integrated with the already used conventional architectures and this complicates the job of the industrial engineer. To this direction, authors in [14] present UML4IoT with focus on the modeling of the IAT. UML4IoT is a UML-based approach that realizes the model driven engineering paradigm to exploit IoT in the manufacturing domain.

In this paper, 1) we extend the model of the IoT introduced in [14] , and 2) define a model-to-model transformer to automate the construction of IATs based on the Contiki OS [15] and the C language. In the extended IoT model, the IAT is still the key artifact for the adoption of the IoT infrastructure in the manufacturing domain. Based on this model the manufacturing system is considered as a composition of cyber-physical and cyber components along with humans [16]. All these components are considered as Things, either permanent or on demand that collaborate exploiting an IoT communication infrastructure to realize a higher level of behavior, i.e., the one of the system level.

The presented approach is discussed in comparison with other approaches and mainly the IoT-A reference architecture that has been adopted by the Papyrus for IoT project [17], which is building a platform for the design of IoT systems in general. This project has many similarities with our project; both projects use UML and SysML as modeling languages, and Papyrus as tool to provide a modeling solution for IoT. Our approach focuses on manufacturing systems. More specifically, it focuses on the case that a high-level design specification for the mechatronic component is not available. Specific annotations were defined to annotate the C source code specification of the mechatronic component so as to automatically transform the mechatronic component to an IAT. The approach is presented using as case study the Liqueur production laboratory system. The LWM2M IoT application protocol [18] running on top of CoAP [19] is used as the IoT protocol stack.

The remainder of this paper is structured as follows. Section 2 positions this research against related work. In Section 3, the proposed extension to the IoT model which is used in the UML4IoT approach is presented and discussed in comparison with existing IoT models. In Section 4, the

Contiki based Industrial Automation Thing is presented along with the case study. The model-to-model transformer for the C language and the Contiki operating system is presented in section 5 and the paper is concluded in the last section.

RELATED WORK

IoT offers new levels of connectivity in the industrial domain that may lead to higher efficiency, flexibility, and interoperability among industries [20]. However, not only many definitions exist for the IoT but also several models. These models, e.g., ETSI, IETF, SENSEI, have been developed to capture the key concepts of the domain and provide the infrastructure to develop frameworks and architectures for the systems based on IoT. Fei et al. [21] claim that even though IoT has been used in various application domains there is still no clear and uniform definition and architecture about it. Therefore, the IoT domain suffers from an inconsistent usage and understanding of the meaning of several key terms, as also claimed in [22]. The definition of an IoT domain model is a prerequisite for defining IoT Architectures. A detailed discussion on the IoT models can be found in [23], where the IoT-A reference model for IoT is presented and validated. Authors in [24] present and discuss applications of IoT in Manufacturing and describe a five-layer architecture for manufacturing based on IoT.

Authors in [25] focus on the device nature of Thing and consider sensors, actuators and controllers as IoT devices, i.e., things. They focus on the data field structures and evaluate the benefit of using an IP smart gateway as the decentralized peripheral to integrated sensors, actuators and the controller and claim that this may improve the performance of IoT devices. We do not agree with the use of sensors and actuators as first-class model elements in the high-level design specification of the system. Sensors and actuators are just technology artefacts used to integrate the physical with the cyber world so they have no place in the high-level design spec of the system. In our approach, the IAT, which encapsulates sensors, actuators and the controller, plays the role of the Thing and represents the key construct in the IoT manufacturing environment.

MDD becomes more and more popular in the development of embedded software systems and various reports refer efficiency gains, from up to 50%, for example, in the development in the car industry [26], with high error reductions and a rapid increase of the maturity level of developed products. MDD is considered as a promising solution to address the complexity of software development in IoT [27] and improve quality characteristics of the produced software. Malavolta and Muccini [28] argue that MDD is the right tools to address the complexity of wireless Sensor Networks development exploiting abstraction, reuse, separation of concerns and automation. They present a framework to systematically study, classify and compare existing MDA approaches in this area [28]. Several works publish results that exploit the MDD paradigm in IoT based systems to improve their quality characteristics but also the ones of their development process.

Thang et al. [27] present FRASAD, a framework based on MDA to manage the complexity of IoT applications. They present a rule based model and a domain specific language to describe the application, using the sensor node as key concept. The primary objective is to model the sensor node software. Contiki is also supported among other OSs by this framework. Authors assume that the application logic of the sensor node program is captured in a Platform Independent Model (PIM), which is constructed using a set of rules they have defined to describe behavior of the sensor node programs. This PIM is next mapped through a Domain Specific Language to the specific platform where it is indented to be executed. However, the approach focuses more on the message dissemination compared to the processing which is considered as an optional part of a sensor node. Our approach focuses on the interface of the Industrial Automation Thing; the behaviour which is very complicated compared to one of a sensor node is defined using another DSML we have defined for structuring the cyber-physical component.

Authors in [29] present the software architecture of a platform developed to address issues, among which the lack of development toolkits, that limit the diffusion of IoT within industrial environments. They also describe an innovative, IoT oriented, model-driven development toolkit that focuses on the seamless integration of heterogeneous industrial devices and sensors, into existing legacy systems by transforming them into web services. The proposed toolkit allows inexperienced developers to discover and compose distributed devices and services into mashups using a modelling tool. Thus, the use of the MDD approach is mainly on the generation of the mashups and does not focus on the modelling of a mechatronic component as is the case of our approach. Furthermore, authors do not refer or describe the domain modelling language that they use in their MDD approach.

A very early approach to model complex IoT systems with UML and then generate RESTful interfaces from these models is presented by Prehofer

[30]. Authors do not define any DSML but they construct class diagrams and state charts using only primitive UML model elements. Authors in [31] describe an approach to define a visual DSML for the IoT based on UML. They model the Thing, which they consider as key construct for building an IoT system, using the UML component construct and its interface using provided and required interfaces. Authors do not address the mapping of the conventional object- oriented (OO) interfaces of the Thing with the ones of the REST paradigm.

Yingfeng Zhang et al. [32] present a real-time information capturing and integration architecture of the internet of manufacturing things (IoMT) to provide a new paradigm by extending the techniques of IoT to the manufacturing field.

They use the term manufacturing thing but they do not focus on its structure and its development process. Instead, they focus on the overall architecture for a manufacturing system and describe a framework with focus on real-time tracking and tracing for the dynamic monitoring of the manufacturing process.

To the best of our knowledge there is no other work that focuses on the automation of the transformation process of the conventional mechatronic component to an IoT compliant one, i.e., to an Industrial Automation Thing ready to be integrated into the IoT-based industrial environment.

TOWARDS AN IOT MODEL FOR MANUFACTURING

The IoT-A Reference Model

De et al. [33] describe the key concepts of the IoT-A reference architecture that is a result of an EU funded IoT project. These concepts and their interrelations are depicted in Figure 1.

Based on this, device is attached to entity, which is associated with resource that is accessed through service. In more detail, authors consider the entity as the "thing" in the IoT, i.e., the focus of interactions by humans and/or software agents.

The device represents the hardware component that is either attached to an entity or it exists in its environment and monitors it. The resource is the actual software component that provides information on the entity or enables the controlling of the device. A service exposes the functionality of a device by accessing its hosted resources.

The Proposed IoT Model

Figure 2 captures the high-level key concepts of our IoT model. Based on this, the IoT is defined as a composition of Things and a processing and communication (IPV6-based) infrastructure. Any artifact that is able to communicate with other Things using the processing and IPV6-based communication infrastructure (Proc & Comn Infr) is considered as Thing. Things collaborate to achieve higher level of behavior compared to the one offered by each one of the collaborating Things. Collaborating Things form a new Thing of type System As Thing that represents a system of Things.

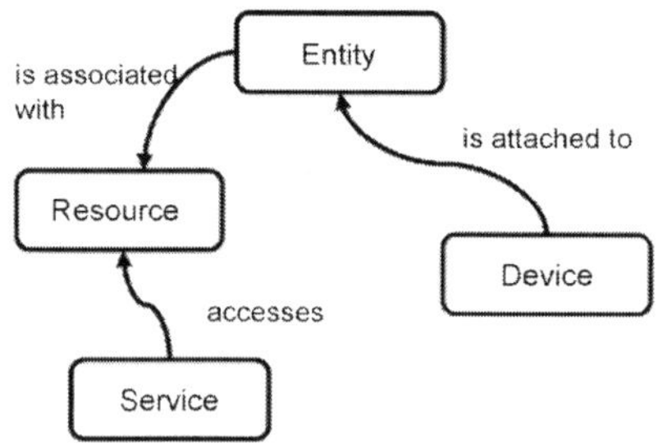

Figure 1. Example key concepts and interaction in the IoT-A model (De et al. 2011).

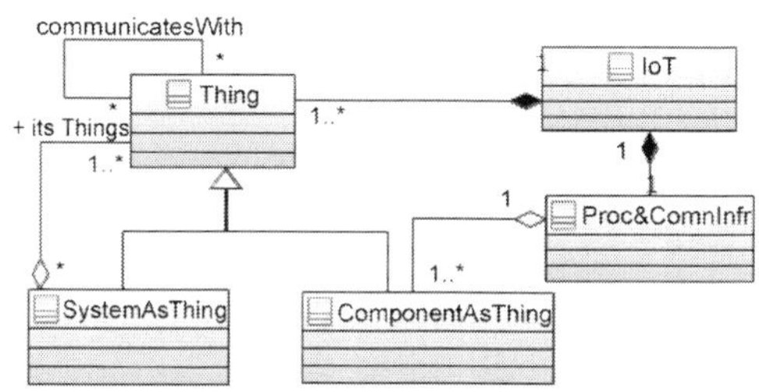

Figure 2. High level key concepts in IoT.

A Thing may be either a system of Things (System As Thing) or a component (Component As Thing). A Component As Thing is a Thing that does not utilize IoT for the integration of its constituent components. This type of Thing is used to represent in the IoT world conventional systems or components that have been transformed to Things (Component As Thing) by adding on top of their conventional interface an IoT-compliant one. The smallest Thing of this type is a sensor or actuator. The Proc&ComnInfr is a composition of: 1) processing nodes and 2) communication devices, i.e., gateways, bridges, switches, routers, etc. Proc&ComnInfr is modelled as a composition of Things (Component As Thing) assuming that these devices are IoT enabled; this will be the case in the near future. The Cloud is considered part of the Proc&Comn In fr.

The Model of Thing in Manufacturing

Figure 3 presents the proposed model for the Thing. Based on this, a Thing is either real, cyber or virtual. A Real Thing is either permanent or on demand Thing. A permanent Thing is a composition of a cyber-physical object (Cp Object) and a cyber IoT enabler (Cyber Iot Enabler). An On Demand Thing is an aggregation of a Physical Object and a cyber-physical IoT enabler (Cp Iot Enabler). As cyber-physical IoT enabler we model any device such as laptop, tablet, mobile phone, wearable, RFID reader that provides an IoT like interface and is able to interact with a physical object. As physical object we consider a human, an inanimate object or even animal with an embedded or attached tag. A human interacts with an app, i.e., application specific IoT wrapper and is temporarily transformed to a Thing.

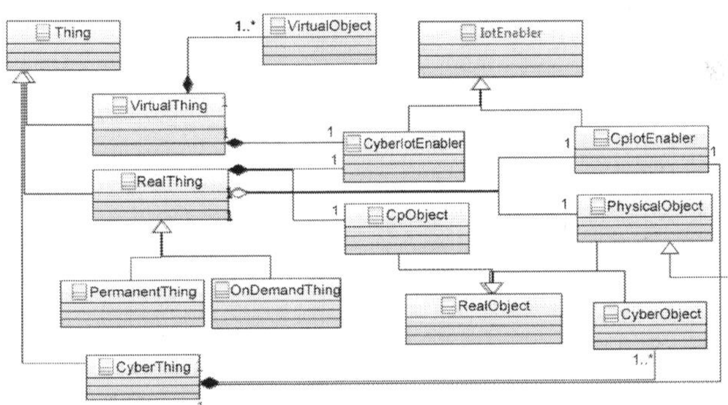

Figure 3. The model of thing in IoT.

Physical objects of type animals or inanimate objects with an embedded or attached tag interact with an RFID reader with IoT like interface for the same reason. A cyber-physical object is any physical object that: 1) implements a kind of functionality, i.e., material and/or energy transformations and 2) has been transformed to a smart object by appending on it information processing functionality. A mechatronic component is an example of cyber-physical object. A cyber IoT enabler integrated with virtual objects and deployed on a processing unit constructs a virtual Thing (virtual Thing). The Io Twrapper of the UML4IoT approach is an example of cyber IoT enabler.

A cyber Thing is defined as a composition of one Cp Iot Enabler and one-to-many cyber Objects. This allows the developer to optionally group cyber objects of the system design model and map these to one cyber Thing.

A Thing may expose to its environment: 1) part of its structure in terms of properties and/or 2) part of its processing or storage functionalities. These functionalities are exposed as services. Services are discriminated to: a) IoT infrastructure services (Iot Service), b) application domain services and c) application specific services. All types of services may be managed (activated, configured, updated, etc.) through the IoT communication infrastructure. Communication infrastructure services may be considered similar to the Industrial Automation Thing services with the remark that Industrial Automation Things perform material, energy and information processing while communication infrastructure Things perform only energy and information processing. For example, device management services, defined by the OMA LWM2M [18] are Iot Services. Native services are services that would be defined for a specific application domain, e.g., home automation, manufacturing, or system specific services, such as the generate Liqueur Type A service of the Liqueur Plant laboratory production system [14]. Device management is implemented by the management interface of the LWM2M. On the other side, Thing management is domain or application specific and should be implemented by specific cyber components on top of the device management and service interface, e.g., the one of LWM2M.

The IoT processing and communication infrastructure (Proc&ComnInfr) should provide an environment for a service-based collaboration of Things. Each Thing implements functionalities offered to the environment as services with negotiated QoS that should be discovered and exploited by other Things. It may also utilize services of other Things to realize its behavior. The cyber components of the manufacturing system model are deployed during the deployment time on Things of type Component As Thing (see Figure 2). UML/SysML design models of the system are marked with the IoT model elements using the UML4IoT profile to automatically transform the system to an IoT-compliant one.

Discussion on the Proposed IoT Model

Our definition for the Thing is different from a widely accepted one, described by De et al. [33], which defines the thing and its relations to devices, resources and services. Device, Resource, Service and Thing are also model elements in the IoT model presented by Stephan [34]. In our approach, the

Component As Thing encapsulates and hides how sensing of its physical part as well as actuation are realized, since this is an implementation issue. Based on this, we do not consider sensors and actuators first class model elements in our model and we do not capture these artifacts in the design model of the system. Thus, actors of the OO approach or terminators of the SA approach are modeled either as Industrial Automation Things or as human Things.

This definition of Thing satisfies the requirement set by Zhuming et al. [12] according to which all interactions among the system constituent components, which may be humans, machines and products should be performed under the same umbrella. This allows the developers to focus on the system's functionality and not worry about interactions, thus increasing the productivity.

In the design model of the system we do not capture resources. A resource is a technology artifact used: 1) to represent the exposed properties and services of a Thing, and 2) to access these through a well-defined set of operations to achieve low coupling between collaborating Things. The LWM2M defines a set of such operations, i.e., READ, EXECUTE, WRITE, etc., implemented on top of the http operations.

The RESTful as well as the SOAP paradigms can be utilized for accessing the services offered by Things. Thus, we adopt a different meaning for service from the widely accepted and described, e.g., [18], where access to resources from the outside world finally happens through services. IoT-A [23] which is a result of the IoT-A EU project, consider the service as an entity that accesses a Resource which is associated with an Entity that has attached a Device. It should also be noted that while a resource is defined in IoT-A as the core software component that represents an entity in the digital world, a Device is attached to a Resource. We do not adopt this model because it is technology driven. Our model focuses on the system modeling level and its objective is to offer a platform independent modeling of the target system. In our model, a Thing has structural (attributes) and behavioral properties (functions/methods). Those properties that are accessible from its environment are represented as resources. The REST ful paradigm is adopted for accessing the resources. In this context and to exploit the benefits of IoT, the networking entities of the IoT Proc&ComnInfr are also considered as Things (IoT-Thing Networking Thing) that provide their own set of information processing services required to establish the communication infrastructure of the IoT.

Plant processes as well as other functionalities of the plant are assigned to cyber objects of the system's design model. These cyber objects may be marked as cyber Things. Alternatively, cyber objects may be deployed on other cyberThings or on Things of the Proc&ComunInfr. In both cases the corresponding services are mapped to resources of the corresponding Thing.

Our approach differs from the one of the ebb its platform [20] that identifies the following four layers which are considered required to bind the physical world with software services:

1) Physical-world layer, where devices, sensors and physical-objects are captured,
2) The IoT layer,
3) The internet-of-services layer and,
4) The business system mediation and product life cycle layer.

Pramudianto et al. [35] capture sensors and actuators as first-class model elements in their IoT meta model and use the term virtual object to refer to the software entity that acts as a proxy of the real-world object. In our model, the meaning of the virtual Thing is completely different. We use the term software representative (SR) to refer to what authors in [35] call virtual object.

A CONTIKI BASED INDUSTRIAL AUTOMATION THING

The Liqueur Production Laboratory System

The Liqueur production mechatronic system, used as case study in this work, is composed of the following mechatronic components: smartSilo1, smartSilo2, smartSilo3, smartSilo4 and smart Pipe (see Figure 4). The system is based on the case study initially used by Basile et al. [36] and then extended by Thram boulidis [16] to be compliant with the mechatronic component concept. The smart Silo mechatronic components are reserved in couples to produce specific types of liqueurs. Smart Silos 1 and 4 form one couple; smartSilos 2 and 3 form the other couple. A mechatronic component has a well-defined interface through which exposes its behavior to be used by the liqueur production processes. This interface exposes the functionalities offered by the silo such as fill, empty, mix and heat. Using the common pipe at the same time for liquid transfer among the silos is not allowed. Moreover, mixing the liquid in two silos at the same time is not permitted due to a constraint in power consumption. Implementation issues regarding the physical silo are encapsulated and hidden from the mechatronic component's environment.

Our intention is to integrate the components of this conventional mechatronic system using IoT and gain from the low coupling that this technology introduces among the interacting components. The use of the IoT will also enable the system to exploit the benefits of this technology regarding the user interaction by allowing end users to produce custom types of liqueur. The end user would be able to define, through an app (my Liqueur App), the production parameters of the desired type of liqueur, as shown in Figure 4.

The Cyber-Physical Component

The legacy smart Silo Mechatronic/cyber-physical component is composed of the physical silo (physical part), a processing, storage and communication unit and the low-level control software (cyber part) required for the smart Silo to provide a higher level of abstraction functionality compared to the one provided by the physical silo.

As shown in Figure 5, which presents the high-level architecture of the cyber-physical component, the software part is composed of two main parts. The first part is the software representation of the physical object, i.e., the mechanical unit, into the software domain.

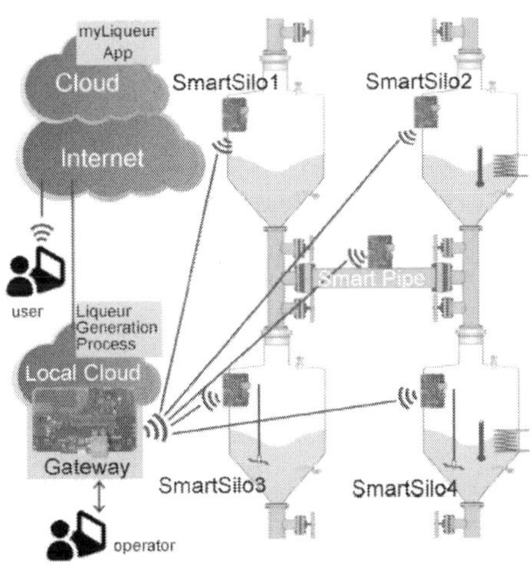

Figure 4. The Liqueur production system used as case study.

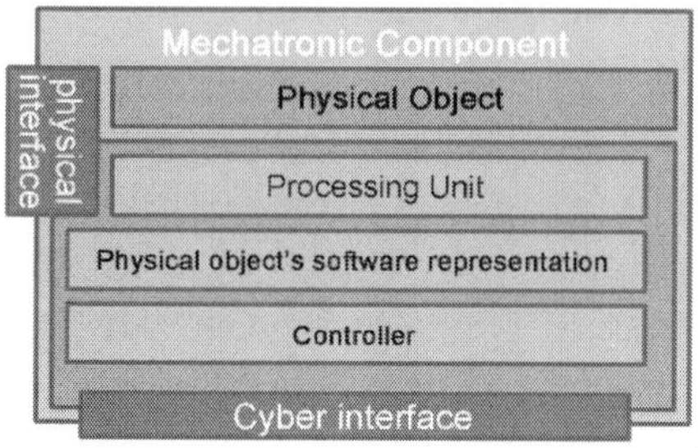

Figure 5. The architecture of the cyber- physical component.

This part does not add any extra functionality; it only encapsulates the details of the integration of the physical world with the cyber world. On top of this, another part, the controller in Figure 5, transforms the physical object to a smart one adding extra functionality. This part encapsulates the low-level control of the physical object required to transform the physical world object into a smart cyber-physical component that provides its functionality through a well-defined interface.

We use the Interface construct of UML to specify the cyber interface of a cyber-physical component. The Interface is used in UML to declare a set of public features and obligations that together constitute a coherent service [37]. In this sense, an Interface specifies a contract that any instance of the mechatronic component shall fulfil. The UML class diagram of Figure 6 presents the interface of the smart Silo Mechatronic component in terms of provided and required interfaces. The Smart Silo Usage If represents the provided interface while the Smart Silo User If represents the required one. In the required interface, we show how to model the interaction between SmartSilo and its client with the Signal and Reception constructs of UML in order to represent the possibly asynchronous nature of this interaction.

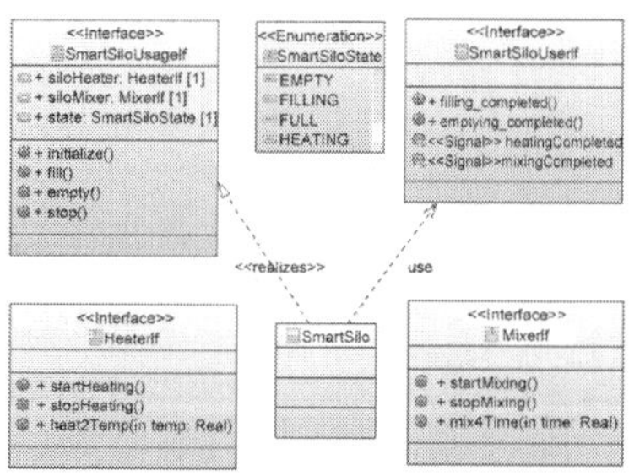

Figure 6. The cyber interface of the SmartSilo mechatronic component.

Thus, the heating Completed and mixing Completed signals sent by the Smart Silo will trigger an asynchronous without a reply reaction to the Smart Silo client, e.g., the type A liqueur generation process, through the corresponding Receptions captured in the Smart Silo User If.

Towards a Contiki-Based Industrial Automation Thing

The 6LoWPAN IoT gateway of Weptech electronic Gmbh running the Contiki operating system is used to host the controller of the smartSilo mechatronic component. The 6LoWPAN IoT gateway, which is based on an ARM®Cortex®? M3 SoC with 512 kB Flash and 32 kB RAM, functions as a border router in a 6LoWPAN network. It connects a wireless IPv6 network, over an 802.15.4 compliant radio interface in the 2.4 GHz band, to the Internet via a 10BASE-T Ethernet interface.

Contiki is a lightweight operating system ported to various microcontroller architectures on resource constraint devices. It was selected mainly for its event-driven kernel that guaranties fast response times to events and to its ability for dynamic loading and replacement of individual programs and services that leads to very flexible Mechatronic components whose behavior may be modified during run-time. The interfacing of the cyber part with the physical one, has been developed using the event-driven handling mechanism of Contiki to get a better response time compared to the traditional scan cycle approach mainly used in industry. Sensor signals generate interrupts which are handled by Contiki and transformed to asynchronous software events. These Events are broadcasted and captured by the corresponding event handling routines. Thus, the high-level sensor signal is transformed to the high Level Reached asynchronous event. This event is handled by the corresponding event handling routine, which is responsible to implement the sensor data handling algorithm. The data handling algorithm sends among others a close signal to the in Valve and activates the sending of a filling Completed event to the client of the mechatronic component, i.e., the liqueur generation process.

The response of the system from the time that the sensor generates the signal to the time that the signal arrives to the in Valve actuator has an average value of 39.20 μs. Listing 1 presents a part of the object-based C implementation of the smart silo cyber part that is related with the interface of the component with its environment. This implementation is for the case that the required interface will be modelled by call back functions instead of signals and receptions. It is evident that both alternatives, i.e., signals or call back functions, imply a tight coupling among the smartSilo and the components that use its behaviour, in the sense that these interfaces should be known in advance for the development of the component's clients.

A model-to-model transformer is required to automate the process of generating the IAT. This transformer will use as input the properly annotated with the DSML conventional mechatronic component. One approach is to mark the UML design specification of the mechatronic component with the stereotypes defined by the UML4IoT profile. If a UML design is not available then the source code of the cyber part of the mechatronic component is properly annotated with specific annotations that have been defined based on the UML profile. An example of annotated code with Java-like annotations is given in Listing 2, where objects and object types of the cyber-physical component as well as their properties that should be exported to the IoT environment are properly annotated.

```
struct silo{
  enum silo_state state;
  struct level_sensor *high_level_sensor,*low_level_sensor;
  struct valve *in_valve,*out_valve;
  ...
  int32_t target_temperature;
  // provided If
  void (*initialize)(void);
  void (*fill)(void);
  void (*empty)(void);
  void  (*stop)(void);
  void (*heat)(void);
  void (*mix)(void);
  // required If
  void (*filling_completed)(void);
  void (*emptying_completed)(void);
```

Listing 1. Part of the C object-based implementation of the cyber part of the mechatronic component.

```
struct valve//@ObjectType (name="valve",id=1004,instanceType=multiple,mandatory=true)
{
  int state;//@ResourceDef (id=5850,name="state",operations=R,type=string,instanceType=single,mandatory=true)
  int id;
  void (*valve_open)(void);//@ResourceDef (id=5851,name="valve_open",operations=E,type=void,instanceType=single,mandatory=true)
  void (*valve_close)(void);//@ResourceDef (id=5852,name="valve_close",operations=E,type=void,instanceType=single,mandatory=true)
};

struct level_sensor//@ObjectType (name="LevelSensor",id=1005,instanceType=multiple,mandatory=true)
{
  int state;//@ResourceDef (id=5550,name="state",operations=R,type=void,instanceType=single,mandatory=true)
};

struct silo{//@ObjectType (name="SmartSilo",id=1003,instanceType=single,mandatory=true)
  enum silo_enum state;//@ResourceDef (id=0,name="state",operations=R,type=string,instanceType=single,mandatory=true)
  int (*fill_op)(void);//@ResourceDef (id=1,name="fill",operations=E,type=void,instanceType=single,mandatory=false)
  int (*empty_op)(void);//@ResourceDef (id=2,name="empty",operations=E,type=void,instanceType=single,mandatory=false)
  int (*initialize_op)(void);//@ResourceDef (id=4,name="initialize",operations=E,type=void,instanceType=single,mandatory=false)
  struct level_sensor high_level_sensor,low_level_sensor;
  struct valve in_valve,out_valve;
  int filling_completed;//@ResourceDef (id=7,name="filling_completed",operations=R,type=boolean,instanceType=single,mandatory=false)
  int emptying_completed;//@ResourceDef (id=8,name="emptying_completed",operations=R,type=boolean,instanceType=single,mandatory=false)
};
```

Listing 2. Part of the C object-based implementation of the cyber part of the mechatronic component annotated with the UML4IoT java-like annotations.

In Figure 7, the Contiki-based silo industrial automation thing developed with the proposed approach is shown. In the current implementation, the Weptech embedded board is used as processing unit while a hardware simulator is used for the silo. The Raspberry Pi and the XDK of Robert Bosch are alternative supported platforms.

The Interfaces of the Industrial Automation Thing

Adopting the OMA LWM2M application protocol, the interface of the Industrial Automation Thing is well defined and independent of the behavior that is implemented by the component. This interface is defined using UML provided and required interfaces as shown in Figure 8. Based on this figure the IAT has three provided interfaces and three required that are independent of the nature of the component. This feature combined with the ability of Contiki to dynamically load and replace individual services, results to a completely flexible component regarding its behavior. New or replaced behavior can be activated by the same well-defined REST interface of the Industrial Automation Thing.

Figure 7. The silo industrial automation thing based on a hardware silo simulator.

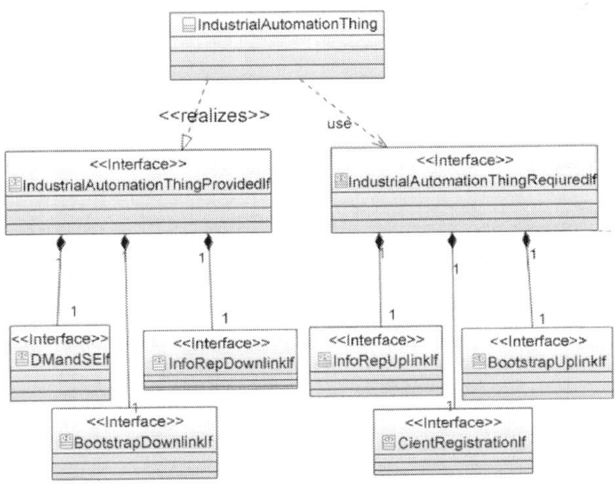

Figure 8. Industrial automation thing interfaces (provided and required).

A comparison with Figure 6 that captures the conventional mechatronic component interface points out the flexibility of the AIT compared to the conventional one. Through this REST interface, resources may be created and used on demand based on requirements assuming that the physical part supports the requested new behavior. Resources, Resource Instances, Objects, Object Instances which are exposed by the IAT as well as their attributes, are accessed by the clients of the IAT through the device management and service enablement interface (DM&SE If). IPSO smart objects [38] have been adopted in this work to satisfy the requirement for interoperability.

AUTOMATING THE GENERATION PROCESS OF IAT

For the automation of the generation process of the IAT a model-to-model transformer has been implemented. This transformer accepts as input the annotated legacy source code and applies a set of transformation rules properly defined to get the source code of the IAT.

The Model-to-Model Transformer

A prototype model-to-model transformer was developed based on Autogen and GNU Guile. Autogen (https://www.gnu.org/software/autogen/) is a GNU tool that supports creation and maintenance of source code. As shown in Figure 9, Autogen takes as input the annotated legacy code of the smart object, a template file and the definitions file. The template file implements the set of transformation rules. It actually defines the structure of the textual output of Autogen, i.e., the structure of the IAT source code, making use of autogen's macro-type format and embedded scheme code. The definitions file provides the information required to instantiate the template file. More specifically, it includes the LWM2M object/resource/instance properties and other source-level information. The definitions file can be generated by the appropriate scheme procedures of the c-parser Guile module that uses as input the annotated legacy source code, as shown in Figure 9.

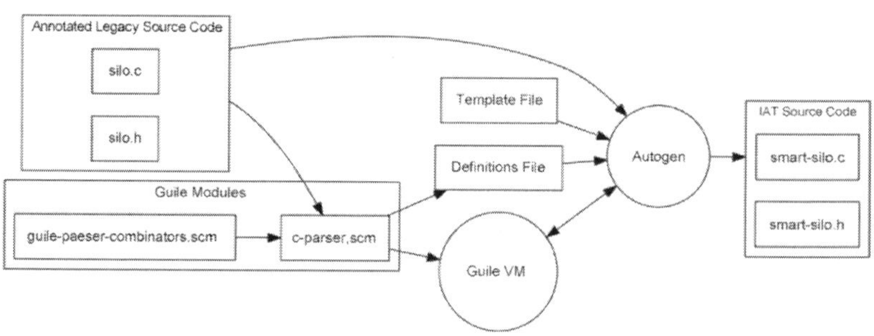

Figure 9. Transformation process of the IAT.

The Guile module that implements the c-parser scheme procedures was developed as a high-level recursive descent parser, based on the guile-parser-combinators module (https://git.dthompson.us/guile-parser-combinators.git). GNU Guile is an implementation of the Scheme programming language (https://www.gnu.org/software/guile/). Autogen collaborates with the Guile VM through guile procedure calls embedded in the template file. The output of the transformation process is the source code of the IAT in a format ready to be compiled with the Contiki operating system and then deployed.

Transformation Rules

The following rules that apply for each object of the lwm2m client have been defined.

Rule 1: Create wrapper functions for annotated behaviors.

For each function with the BehaviorResource annotation create a wrapper function with input parameters:

lwm2m_context_t *ctx, const uint8_t *arg, size_targsize, uint8_t *outbuf, size_t outsize

E.g., Source:static int fill(void);

Target: static intfill(lwm2m_context_t *ctx, const uint8_t *arg, size_targsize, uint8_t *outbuf, size_t outsize);

Rule 2: Create getters/setters functions for each property annotated with the PrimitiveRes annotation.

For each attribute annotated with the PrimitiveRes annotation create the corresponding read and write function depending on the applied operations on the attribute defined in the annotation.

E.g., for the silo_state property

staticintget_silo_state(lwm2m_context_t *ctx, uint8_t *outbuf, size_t outsize) {

char *value;

value = get_silo_state_inString(silo->state);

returnctx->writer->write_string(ctx, outbuf, outsize, value, strlen(value));}

Rule 3: Construct the Resource model.

For each annotated attribute or function create an entry using the LWM2M_RESOURCE_CALLBACK macro of the lwm2m implementation which is integrated into the Contiki OS. E.g., for the silo_state

LWM2M_RESOURCE_CALLBACK(0,{get_silo_state, NULL, NULL}),

Append this entry to a list of resources, i.e., silo_resources, using the LWM2M_RESOURCES macro.

Create the corresponding object instance using the LWM2M_INSTANCE macro and register the resources. E.g., LWM2M_INSTANCE(0, silo_resources)

Append it to the list of silo instances using the LWM2M_INSTANCES.

E.g.,LWM2M_INSTANCES(silo_instances, ….);

Create the Object and registers its instances using the LWM2M_OBJECT macro. E.g.,

LWM2M_OBJECT(silo_obj, 1663, silo_instances);

Rule 4: Modify setter functions.

For each attribute annotated with the ObservableResource annotation, modify its setter function (set_ ())by appending a call to the lwm2m_object_notify_observersfunction.

Assumption: For each observable attribute a setter function exists. E..g.,

Source: void set_filling_completed(){

silo->filling_completed = 1;}

Target: void set_filling_completed(){

silo->filling_completed = 1;

lwm2m_object_notify_observers(&silo_obj, "/0/7"); }

Rule 5: Generate and handle the initialize function for the object.

5.1 Generate an initialize function to initialize the legacy object and register it to lwm2m by a call to thelwm2m_ engine_register_object function.

The legacy initialize function of the object should be properly annotated. E.g.,

void ipso_ silo_init(void) {

silo_init(); // legacy object initialization function

lwm2m_engine_register_object(&silo_obj); }

5.2 Append the initialize function prototype to theipso-objects.hfile. E.g.

void ipso_silo_init(void);

5.3 Append a call statement to the initialize function of each object to theipso_objects_init()function body of theipso-objects.cfile. E.g.,

```
void ipso_objects_init(void) {
ipso_silo_init();
```

CONCLUSION

In this paper, we consider the tight integration of the physical world with the cyber one at the mechatronic component level. A mechatronic component offers its functionality through well-defined mechanical, electrical and software interfaces. In this sense the industrial automation system is a composition of mechatronic components along with cyber components and humans. IoT is adopted for the integration of these components to exploit the benefit of this technology and UML4IoT is utilized to automatically transform the conventional mechatronic component into an IoT compliant cyber-physical one, i.e., to an Industrial Automation Thing. The IoT model used in the UML4IoT approach is extended towards a complete IoT model for the manufacturing domain. The transformation rules required for the development of the model-to-model transformer have been developed and validated through a prototype implementation of the liqueur production laboratory system. The prototype implementation of the silo industrial automation thing based on a Contiki enabled embedded board and the C language is used to demonstrate the applicability and the effectiveness of the proposed approach.

REFERENCES

1. UNESCO Chair (2016) Home Page. On Mechatronics and Mechatronics Research and Application Center. http://mecha.ee.boun.edu.tr/
2. Thramboulidis, K. (2005) Model Integrated Mechatronics—Towards a new Paradigm in the Development of Manufacturing Systems. IEEE Transactions on Industrial Informatics, 1, 54-61. https://doi.org/10.1109/TII.2005.844427
3. Thramboulidis, K. and Scholz, S. (2010) Integrating the 3+1 SysML View Model with Safety Engineering. 2010 IEEE Conference on Emerging Technologies and Factory Automation (ETFA), Bilbao, Spain, 13-16 September 2010, 1-8.https://doi.org/10.1109/etfa.2010.5641353
4. Thramboulidis, K. (2012) Overcoming Mechatronic Design Challenges: The 3+1 SysML-View Model. The Computing Science and Technology International Journal, 2, 6-14.
5. FESTO (2016) Industry 4.0: Efficient Engineering Processes with "OPAK". https://www.festo.com/net/en_corp/SupportPortal/MobilePressDetails.aspx?documentId=368146&q=
6. OASIS (2016) Devices Profile for Web Services (DPWS) Specification. http://docs.oasis-open.org/ws-dd/ns/dpws/2009/01
7. Angulo, P., Guzmán, C.C., Jiménez, G. and Romero, D. (2016) A Service-Oriented Architecture and Its ICT-Infrastructure to Support Eco-Efficiency Performance Monitoring in Manufacturing Enterprises. International Journal of Computer Integrated Manufacturing, 30, 202-214. https://doi.org/10.1080/0951192X.2016.1145810
8. Morgan, J. and O'Donnell, G.E. (2015) Enabling a Ubiquitous and Cloud Manufacturing Foundation with Field-Level Service-Oriented Architecture. International Journal of Computer Integrated Manufacturing, 30, 442-458.
9. He, W. and Xu, L. (2015) A State-of-the-Art Survey of Cloud Manufacturing. International Journal of Computer Integrated Manufacturing, 28, 239-250.https://doi.org/10.1080/0951192X.2013.874595
10. Ghimire, S., Luis-Ferreira, F., Nodehi, T. and Jardim-Goncalves, R. (2016) IoT Based Situational Awareness Framework for Real-Time Project Management. International Journal of Computer Integrated Manufacturing, 30, 74-83.https://doi.org/10.1080/0951192X.2015.1130242

11. Ren, L., Zhang, L., Wang, L., Tao, F. and Chai, X. (2014) Cloud Manufacturing: Key Characteristics and Applications. International Journal of Computer Integrated Manufacturing, 30, 501-515. https://doi.org/10.1080/0951192X.2014.902105
12. Bi, Z., Xu, L.D. and Wang, C. (2014) Internet of Things for Enterprise Systems of Modern Manufacturing. IEEE Transactions on Industrial Informatics, 10, 1537-1546. https://doi.org/10.1109/TII.2014.2300338
13. Bradley, D., Russell, D., Ferguson, I., Isaacs, J., MacLeod, A. and White, R. (2015) The Internet of Things—The Future or the End of Mechatronics. Mechatronics, 27, 57-74.
14. Thramboulidis, K. and Christoulakis, F. (2016) UML4IoT—A UML Profile to Exploit IoT in Cyber-Physical Manufacturing Systems. Computers in Industry, 82, 259-272.
15. Dunkels, A., Gronvall, B. and Voigt, T. (2004) Contiki—A Lightweight and Flexible Operating System for Tiny Networked Sensors. 29th Annual IEEE International Conference on Local Computer Networks, 16-18 November 2004, 455-462. https://doi.org/10.1109/lcn.2004.38
16. Thramboulidis, K. (2015) A Cyber-Physical System-Based Approach for Industrial Automation Systems. Computers in Industry, 72, 92-102.
17. Eclipse (2016) Papyrus for IoT—A Modeling Solution for IoT. https://www.eclipse.org/community/eclipse_newsletter/2016/april/article3.php
18. Open Mobile Alliance (OMA) (2015) Lightweight Machine to Machine Technical Specification. OMA-TS-LightweightM2M-V1_0-20151214-C, Candidate Version 1.0. 14 December 2015.
19. Kovatsch, M., Duquennoy, S. and Dunkels, A. (2011) A Low-Power CoAP for Contiki. 8th IEEE International Conference on Mobile Ad-Hoc and Sensor Systems, Valencia, Spain, 17-21 October 2011, 855-860. https://doi.org/10.1109/MASS.2011.100
20. Khaleel, H., Conzon, D., Kasinathan, P., Brizzi, P., Pastrone, C., et al. (2015) Heterogeneous Applications, Tools, and Methodologies in the Car Manufacturing Industry through an IoT Approach. IEEE Systems Journal, PP, 1. https://doi.org/10.1109/JSYST.2015.2469681
21. Tao, F., Cheng, Y., Xu, L.D., Zhang, L. and Li, B.H. (2014) CCIoT-CMfg: Cloud Computing and Internet of Things-Based Cloud Manufacturing Service System. IEEE Transactions on Industrial Informatics, 10, 1435-1442.https://doi.org/10.1109/TII.2014.2306383

22. Bauer, M., Bui, N., De Loof, J., Magerkurth, C., Nettstrter, A., Stefa, J. and Walewski, J. (2013) Iot Reference Model. In: Bassi, A., et al., Eds., Enabling Things to Talk, Springer, Berlin, Heidelberg, 113-162.https://doi.org/10.1007/978-3-642-40403-0_7
23. IoT-A (2016) Final Architectural Reference Model for the IoT. http://www.meet-iot.eu/deliverables-IOTA/D1_5.pdf
24. Tao, F., Zuo, Y., Xu, L.D. and Zhang, L. (2014) IoT-Based Intelligent Perception and Access of Manufacturing Resource toward Cloud Manufacturing. IEEE Transactions on Industrial Informatics, 10, 1547-1557.https://doi.org/10.1109/TII.2014.2306397
25. Diaz-Cacho, M., Delgado, E., Falcon, P. and Barreiro, A. (2015) IoT Integration on Industrial Environments. 2015 IEEE World Conference on Factory Communication Systems (WFCS), Palma de Mallorca, 27-29 May 2015, 1-7. https://doi.org/10.1109/wfcs.2015.7160553
26. Broy, M., Kirstan, S., Krcmar, H., Schatz, B. and Zimmermann, J. (2013) What Is the Benefit of a Model-Based Design of Embedded Software Systems in the Car Industry? In: Software Design and Development: Concepts, Methodologies, Tools, and Applications, IGI Global, 310-334.
27. Nguyen, X.T., Tran, T., Baraki, H. and Geihs, K. (2015) FRASAD: A Framework for Model-Driven IoT Application Development. 2015 IEEE 2nd World Forum on Internet of Things (WF-IoT), Milan, 14-16 December 2015, 387-392. https://doi.org/10.1109/WF-IoT.2015.7389085
28. Malavolta, I. and Muccini, H. (2014) A Study on MDE Approaches for Engineering Wireless Sensor Networks. 40th EUROMICRO Conference on Software Engineering and Advanced Applications (SEAA), Verona, 27-29 August 2014, 149-157.https://doi.org/10.1109/seaa.2014.61
29. Conzon, D., Brizzi, P., Kasinathan, P., Pastrone, C., Pramudianto, F. and Cultrona, P. (2015) Industrial Application Development Exploiting IoT Vision and Model Driven Programming. 2015 18th International Conference on Intelligence in Next Generation Networks (ICIN), Paris, 17-19 February 2015, 168-175. https://doi.org/10.1109/icin.2015.7073828
30. Prehofer, C. (2015) Models at REST or Modelling RESTful Interfaces for the Internet of Things. IEEE 2nd World Forum on Internet of Things (WF-IoT), Milan, 14-16 December 2015, 251-255.

31. Eterovic, T., Kaljic, E., Donko, D., Salihbegovic, A. and Ribic, S. (2015) An Internet of Things Visual Domain Specific Modeling Language Based on UML. 2015 XXV International Conference on Information, Communication and Automation Technologies (ICAT), Sarajevo, 29-31 October 2015, 1-5. https://doi.org/10.1109/icat.2015.7340537

32. Zhang, Y., Zhang, G., Wang, J., Sun, S., Si, S. and Yang, T. (2015) Real-Time Information Capturing and Integration Framework of the Internet of Manufacturing Things. International Journal of Computer Integrated Manufacturing, 28, 811-822. https://doi.org/10.1080/0951192X.2014.900874

33. De, S., Barnaghi, P., Bauer, M. and Meissner, S. (2011) Service Modelling for the Internet of Things. 2011 Federated Conference on Computer Science and Information Systems (FedCSIS), Szczecin, 18-21 September 2011, 949-955.

34. Haller, S. (2010) The Things in the Internet of Things. Internet of Things Conference 2010, Tokyo, 29 November-1 December 2010.

35. Ferry, P., et al. (2014) IoTLink: An Internet of Things Prototyping Toolkit. IEEE International Conference on Ubiquitous Intelligence and Computing, Bali, 9-12 December 2014, 1-9.

36. Basile, F., Chiacchio, P. and Gerbasio, D. (2013) On the Implementation of Industrial Automation Systems Based on PLC. IEEE Transactions on Automation Science and Engineering, 10, 990-1003. https://doi.org/10.1109/TASE.2012.2226578

37. OMG (2015) Object Management Group, OMG Unified Modeling Language (OMG UML), Version 2.5, OMG Document Number Formal/2015-03-01. http://www.omg.org/spec/UML/2.5

38. Internet Protocol for Smart Objects (IPSO) Alliance (2014) IPSO Smart Object Guideline. IPSO Smart Object Committee, 21 September. https://www.ipso-alliance.org/ipso-community/resources/smart-objects-interoperability/

CHAPTER 3

The 3+1 SysML View-Model in Model Integrated Mechatronics

Kleanthis Thramboulidis

Visiting Professor Helsinki University of Technology, Electrical & Computer Engineering, University of Patras, Patras, Greece.

ABSTRACT

Software is becoming the driving force in today's mechatronic systems. It does not only realize a significant part of their functionality but it is also used to realize their most competitive advantages. However, the traditional development process is wholly inappropriate for the development of these systems that impose a tighter coupling of software with electronics and

Citation:- K. Thramboulidis, "The 3+1 SysML View-Model in Model Integrated Mechatronics," Journal of Software Engineering and Applications, Vol. 3 No. 2, 2010, pp. 109-118. doi: 10.4236/jsea.2010.32014.

mechanics. In this paper, a synergistic integration of the constituent parts of mechatronic systems, i.e. mechanical, electronic and software is proposed though the 3+1 SysML view-model. SysML is used to specify the central view-model of the mechatronic system while the other three views are for the different disciplines involved. The widely used in software engineering V-model is extended to address the requirements set by the 3+1 SysML view-model and the Model Integrated Mechatronics (MIM) paradigm. A SysML profile is described to facilitate the application of the proposed view-model in the development of mechatronic systems.

Keywords: Systems Engineering, System Modeling, Mechatronic Component, Model Driven Development, Model Integrated Mechatronics, SysML Profile, V-Model, IEC61499

INTRODUCTION

Software does not only implement a significant part of the functionality of today's mechatronic systems, but it is also used to realize their most competitive advantages. It is the evolving driver for innovations in many mechatronic systems and in general it is considered as the driving force in improving this kind of systems. However, the traditional development process is wholly inappropriate for the development of systems characterized by complexity, dynamics and uncertainty as is the case with today's mechatronic systems [1]. According to the traditional development process the constituent parts of the mechatronic system, i.e. the mechanical, electronic and software, that constitute the system are developed independently and then are integrated to compose the final system.

The software development starts when the development of electronic and mechanical is already at a stage where any change in these parts is expensive and time consuming. This is why the mechanical and electronic properties impose several constrains and narrow the solution space for software development. Moreover, as claimed in [2] "the actual cooperation during the construction is less developed. There is no joint development process, no joint tool usage, no joint modeling formalism and no joint analysis. Every discipline has its own approaches". As a result of this, the current process that is traditionally divided into software, electronics and mechanics, emphasizes on domain-spanning design methods and tools and is unable to address the demand for synergetic mechatronic dependability predictions; this is why many products suffer from severe dependability problems [3].

An integrated framework for the construction of mechatronic systems is missing [2]. Such a framework should address current challenges in the development of mechatronic systems that include among others, synergistic modeling and integration, design synchronization, as well as model execution and analysis. It should provide the infrastructure required for applying a tight integration of mechanics with electronics and software in order to replace conventionally designed mechanical and electromechanical systems into smart ones where significant part of functionality will be implemented by software. It is also expected to result in massive improvements in system's Quality of Service (QoS) characteristics and allow a smooth integration of dependability predictions during the early development phases.

Model Integrated Mechatronics (MIM) [4] is a paradigm that was proposed to address the need for an integrated development in mechatronic systems. MIM supports the model-driven development of complex mechatronic systems (MTSs) through the evolution of models that have as primary construct the mechatronic component (MTC). The concept of component has already been adopted in the development process of manufacturing systems by other researchers too. However, they mostly focus on the software part of the component; they do not address the whole development process; and they do not provide an architecture for the concurrent engineering of all constituent components, i.e., mechanical, electronic and software.

In this paper, the Systems Modeling Language (SysML) [5] is adopted for the system's modeling process in the MIM paradigm. An architectural view-model, the one called 3+1 SysML is proposed to address the synergistic integration of the constituent parts of mechatronic systems. The main view is the SysML view that corresponds to the mechatronic layer of the MIM Architecture. This view captures the system model that is the one constructed by the MTS developer. Each of the three views is used to describe the system from the perspective of the corresponding discipline. Specific tools of every discipline may be exploited for the model execution and analysis of the SysML models.

SysML is used to represent the models of the mechatronic system that are proposed by the MIM paradigm. A SysML profile was defined using Papyrus, an open source tool for graphical UML 2/SysML modeling. This profile supports not only the modeling of mechatronic systems using the concept of mechatronic component but also a hybrid development process that integrates the traditional approach with the MTC-based one, to allow the reuse of legacy systems. It provides a SysML based implementation of the MIM architecture that will allow the MIM paradigm to be exploited in the development process of real world mechatronic systems. An effective development process that exploits the 3+1 SysML view-model and the SysML profile is described extending the well-known in software engineering V-model [6].

The remainder of this paper is organized as follows. In the next section, a brief introduction to the MIM architecture and the SysML is given and the related work is briefly discussed. Section 3 presents the proposed 3+1 SysML view-model that emphasizes the importance of the common system-level model. In Section 4, a SysML profile that represents the artifacts used in the MIM architecture is described. Section 5 presents the proposed modifications to the widely used in software systems V-model to address the needs of the MTS development process. Future developments and research challenges are discussed in Section 6 and the paper is concluded in the last section.

BACKGROUND AND RELATED WORK

A Brief Introduction to MIM

The upper layer of the MIM architecture that is shown in

Figure 1, i.e. the mechatronic layer, was defined to systematically address complexities in the model-driven development process of component-based mechatronic systems. This layer is projected into three dimensions representing the application, the resource, and the mechanical process respectively. The controlling application software is modeled in the application layer, while the hardware, i.e., computing and communication, as well as the software resources that constitute the infrastructure required for the execution of the controlling application software, are modeled in the resource layer. Mechanical, hydraulic and pneumatic parts are modeled in the mechanical layer.

Mechatronic system integrators work horizontally in the model evolution dimension of the MIM architecture. They interactively compose the MTS using already defined MTCs without worrying on lower layers' implementation details. They go through a model-driven development process to build the MTS using descriptions of already existing MTCs. They only have to capture the application logic in application layer components, as well as to identify their required QoS characteristics from the resource layer infrastructure.

MTC builders work in the model integration dimension and apply an information integration process that crosses the boundaries between mechanical, electronic, and computer science fields. They work horizontally and vertically, either top-down or bottom-up, in the lower three layers of the architecture in a concurrent way. Constructed MTCs are stored in MTC repositories to be discovered and used by mechatronic system integrators [7].

MIM is a new paradigm that promotes model integration not only of implementation space artifacts but also of early analysis and design phase ones. It promotes reuse at the mechatronic level and significantly decreases development and validation time of the system.

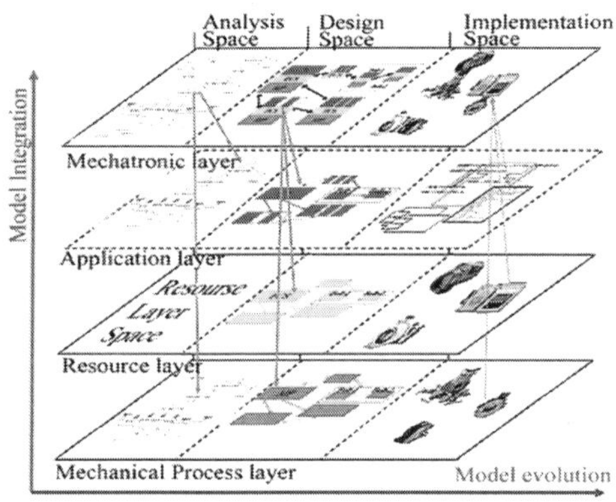

Figure 1. The Model Integrated Mechatronics (MIM) architecture [4].

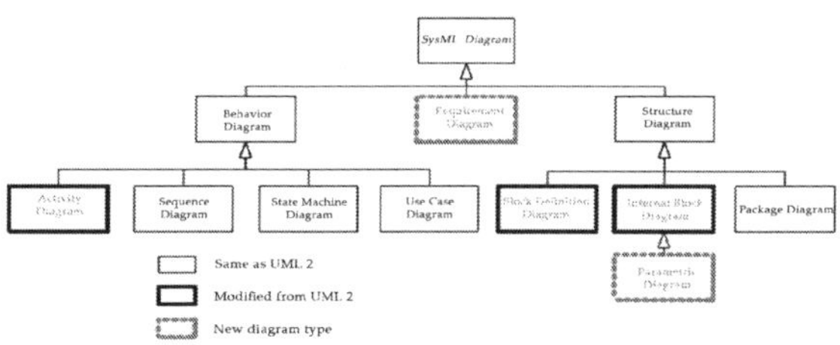

Figure 2. SysML diagrams and its comparison with UML 2 [5].

MIM addresses the need for synergistic integration at the model and process levels; it facilitates the integration between the design processes of the different disciplines which is the approach considered as the most effective to improve the development process of mechatronic systems. The other two approaches being: implementation time integration and design time integration at data and model level [8].

Systems Modeling Language (SysML)

SysML was developed to support the specification, analysis, design, verification and validation of a broad range of complex systems [5]. These systems may include hardware, software, information, processes, personnel, and facilities. The objective of SysML is to unify the diverse modeling languages currently used by system engineers. SysML reuses a subset of UML 2 [9] and provides additional extensions needed to address system engineering aspects not covered by UML 2. It includes diagrams that can be used to specify system requirements, behavior, structure and parametric relationships. Requirements diagram and parametric diagrams are the new diagram types proposed by SysML, as shown in Figure 2 which presents the diagrams that are used by SysML.

SysML provides modeling constructs to represent text-based requirements and relate them to other modeling elements. The requirement diagram can be used to represent many of the relationships that exist between requirements and visualize them. It provides a bridge between traditional requirements management tools and the other SysML models. It can be used to depict the requirements in graphical, tabular, or tree structure format and highlight their relationships, as well as to capture the relationships between requirements and other model elements that satisfy or verify them.

The other new diagram, i.e. the parametric diagram, is used to describe the constraints among the properties associated with blocks. It allows the specification of continuous components by parametric constraints on class attribute values expressing corresponding differential equations. However, the syntax and the semantics of behavioral descriptions captured in parametric diagrams have not been defined to allow the integration with other simulation and analysis modeling techniques for the proper execution of the models. So, the parametric diagram is used to integrate the system descriptive behavior and structure models expressed in SysML with other simulation and engineering analysis models such as performance, reliability, and mass property models.

The fact that SysML is based on UML 2, will allow system engineers modeling with SysML and software engineers modeling with UML 2 to collaborate on models of the mechatronic system. This will improve communication among the various stakeholders who participate in the mechatronic systems development process and will promote interoperability among modeling tools in different disciplines. All the above mentioned characteristics make SysML ideal for the representation of models used in the MIM paradigm.

Related Work

Several researchers are already working in the direction of improving the effectiveness of the development process of mechatronic systems. Schafer and Wehrheim [2] survey on current developments in mechatronics and present the architecture of their mechatronic rail system that seems to provide an excellent platform for studding and analyzing future developments and research chalenges in mechatronic systems. They identify the need of an integrated framework for the construction of mechatronic systems and they discuss future trends in mechatronics especially from the software engineering point of view. Habib [10] argues on the urgent need for theories, models, and tools that should facilitate modeling, analysis, synthesis, simulation, and prototyping of mechatronic systems. He emphasizes the argument that the approach based on optimization within each domain separately will not result in the optimum system design, and he proposes a data and model integration approach to address the integration problem. Burmester *et al.* [11] claim that in today's mechatronic systems most of the control and reconfiguration functionality is realized in software. They present mechatronic UML to exploit the Model Driven Architecture approach for the design of hybrid mechatronic real-time systems that have to fulfill safety-critical requirements. "Mechatronic UML" is defined as an extension of UML to built platform independent models for mechatronic systems. Various UML models have been extended to cover the requirements of modeling the structural view as well as the behavioral view of the system. However, the proposed extension is used to model only the software part of the mechatronic system. Authors in [12] briefly refer to a process model of Robert Bosch GmbH for the development of mechatronic systems in Motor Vehicles to support aspects such as reuse, exchangeability, scalability and distributed development. They use the concept of mechatronic component, even though not well defined, as the basic construct of their process. They argue: a) on the need of a clear specification of component interfaces; and b) the great contribution of re-use to increase the quality properties of mechatronic systems and decrease development time. Nordmann [13] is using the concept of mechatronic component and presents an example of using Active Magnetic Bearings to increase performance, reliability, reusability and longer lifetime. Authors in [14] propose for the development of multidisciplinary systems, such as mechatronics, the integration of the various domain-specific tools. They mainly focus on the integration of used data and models and not on a process level integration. Moreover, none of these approaches provide a high level architecture for an integrated, synergistic development process for mechatronic systems and they do not describe a systems level development process based on the mechatronic component and the emerging standard in the domain that is SysML.

THE 3+1 ARCHITECTURAL VIEW MODEL

Each of the three lower layers of the MIM architecture provides a specific view of the central models that are captured in the upper layer, i.e. the mechatronic layer. Each view is used to describe the system from the perspective of the corresponding discipline. The software view (s-view), for example, provides the models of the software part of the MTS and allows for software specific tools to be used to elaborate and further refine these models. The IEC61499 function block model is an example of such a domain specific model that can be used to further refine the s-view [15]. Figure 3 depicts the 3+1 SysML view-model that is proposed for the development of mechatronic systems. The MTS model is the heart of this architecture and is depicted in the center of the picture. It is surrounded by 4 views which correspond to the roles that engineers play during the development process of mechatronic systems.

The main view is the MTS-view that corresponds to the mechatronic layer of the MIM Architecture. This is the view that is used by the MTS developer. The other 3 views correspond to the 3 lower layers of the MIM architecture. The m-View, for example, corresponds to the Mechanical layer of the MIM architecture and captures all the mechanics, hydraulics and pneumatics of the MTS model. These models are generated by projecting the MTS models to the mechatronic layer and are fully synchronized with the MTS models. Any modifications imposed by the mechanical engineer to the models of this view directly affect the corresponding central MTS models. Moreover, modifications done by the MTS developer on the central models directly affect the corresponding m-view models. The m-view is mainly used during the primitive MTC development process where a concurrent, synergistic engineering on the three views is adopted at the primitive MTC level as a more effective process. It is also used to have a whole view of the mechanical system model and perform optimization and analysis activities on this.Figure 4 presents the structure of the primitive MTC and its interfaces to the environment. A primitive MTC may expose to the environment mechanical, electronic and software interfaces through the corresponding ports. Interfaces between its constituent parts are also shown. Sensors and actuators are used to realize the interactions between the mechanical part and the electronic one.

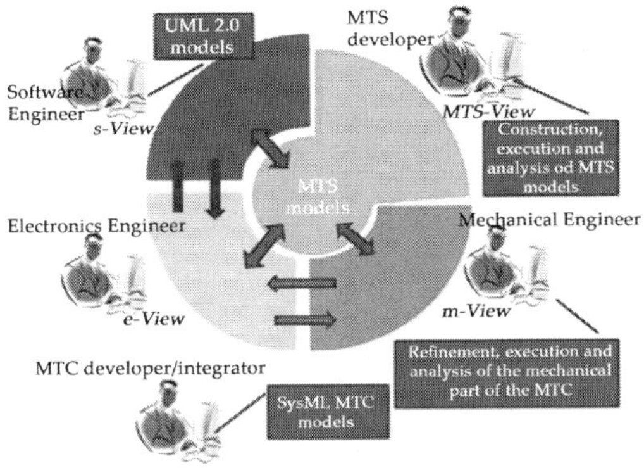

Figure 3. The 3+1 SysML view-model for mechatronic systems development.

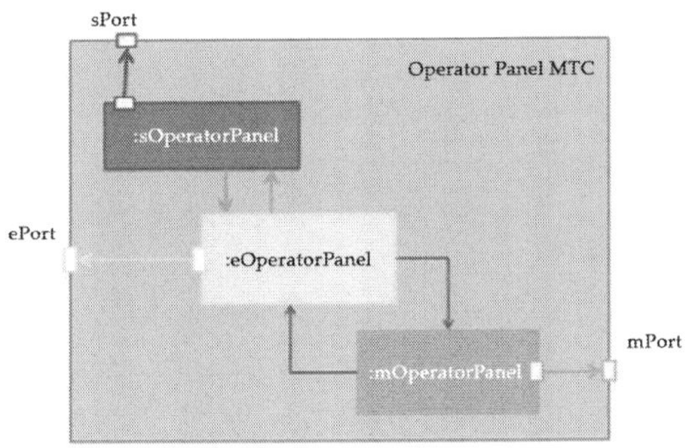

Figure 4. The structure of the primitive mechatronic component.

This means that sensors and actuators are fully encapsulated by the MTC construct. The mechanical part interacts with the environment only through mechanical ports that are mainly in the form of mounting ports or ports for flow of material or energy. A detailed description of mechanical connections that can be discriminated into fixed and moveable arrangements is given in [16]. The sOperator Panel, i.e. the software part of the Operator Panel primitive MTC, exposes its functionality along with the corresponding QoS through a provided interface of the sPort. A hosting functionality for application specific components will be optionally provided by the ePort.

Currently there is no tool to execute the MTS models, not even to analyze their behavior. Discipline specific tools may be exploited for the execution and analysis of MTS models. This is obtained by the proper integration and coordination of specific model execution and analysis tools of the three views. The tool of each view is used to execute the primitive MTC model of its perspective so it has to provide specific interfaces to the tools of the other views, in order to implement the interactions of its own part to the other parts of the primitive MTC. The arrows that cross the boundaries between the three views in Figure 3 represent the interactions of the corresponding models and have to be implemented by specific interfaces of the model refinement, analysis and execution tools of the three disciplines. The AP233 or more formally the ISO 10303-233 standard for systems engineering [17], that provides a data exchange format for the reliable interchange of data between software tools may be exploited to effectively implement these interactions. The execution and analysis of the primitive MTC is obtained through a collaboration of the corresponding tools of the three views. It is clear that the contribution of the three views' specific tools is restricted at the primitive MTC internal level while the execution and analysis of MTC models is done at the MTS level with the coordination and synchronization between MTCs carried by this level. This makes the tool integration a major challenge in the domain of mechatronic systems.

The 3+1 SysML view-model when used with the MTS V-Model that is described in one of the following sections, promote the synergistic integration of the three constituent parts of the mechatronic system and emphasizes the importance of a common model for the system. However, this model can also be used with the traditional development process that is based on the independent development of constituent parts of the mechatronic system and their subsequent integration. Even in this case the existence of a common model for the system greatly improves the effectiveness of the development process.

It should be noted that in each view the corresponding discipline's specific architectures and tools may be exploited, as for example the 4+1 architectural view [18] that may be exploited in the context of the e-view by the software engineer or the MTC developer/integrator.

USING SYSML TO MODEL THE MIM ARTIFACTS

The Systems Modeling Language can be used to represent the artifacts of the mechatronic systems development process that correspond to the system level activities. These include requirements specifications for the MTS and MTC levels, as well as architectural specifications for the MTS and MTC levels. UML 2.0 will be used for the modeling of the software part of the primitive MTC and corresponding tools from the electronics and mechanics domain will be used for the modeling of the other two constituent parts of the MTC. In this section the modeling of the MTS and MTC levels using SysML is considered. It is evident that specific interfaces have to be defined for the integration of the different views and these interfaces have to be realized by the tools used in the various disciplines to create a completely integrated tool chain to support the MTS development process. The SysML to AP233 mapping [19] is towards this direction.

Modeling of the Mechatronic Component

The MTS stereotype that is shown in Figure 5, which presents part of the SysML4MIM profile, is considered as a composition of MTS Components and MTS Connectors (not shown in the figure). The abstract stereotype MTS Componet was defined to provide more flexibility in system modeling. It allows the definition of the different disciplines' components in any level of the system's decomposition hierarchy; it also allows the application of the traditional approach where the system is considered as consisting of mechanical, electronic and software components. An MTS Component that is abstract is specialized to the MTC abstract stereotype and the m Compoment, eComponent and s Component stereotypes. The m Component stereotype is used to represent in the model any mechanical component of the mechatronic system.

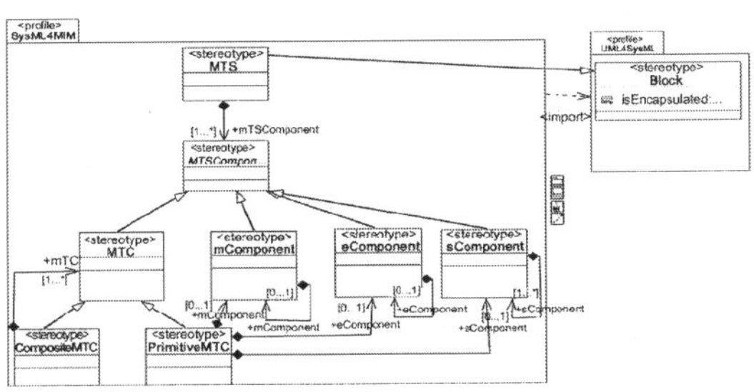

Figure 5. SysML4MIM profile (part); the MTS stereotype.

This way of modeling allows the profile to be used also in the traditional development process of mechatronic systems since the MTS may be considered as a composition of m, e and sComponents. The MTC stereotype which is also abstract is specialized to the Composite MTC and the Primitive MTC. This allows a hierarchical decomposition scheme for the MTS up to the level of primitive MTC that is considered as composition of m, e and sComponents. An m, e and sComponent may be further decomposed in corresponding Components allowing a component based synthesis in each one of the three disciplines. The SysML4UML profile that was created using Papyrus, an open source tool for graphical UML 2 modeling (http://www.papyrusum-l.org/), imports the UML4SysML profile that is already supported by Papyrus. This allows the MTS stereotype to extend the Block stereotype of the UML4SysML profile. All the other components also extend the Block stereotype even not shown in figure. The proposed SysML4MIM profile allows the synergistic integration in the development of mechatronic systems to any level of granularity down to the primitive MTC component which is the one that is not decided or it is not possible to be decomposed into lower layer MTC components.

Modeling of the Mechatronic Port

Allowable inputs and outputs of an MTC are defined using the concept of the port. This allows the design of modular reusable MTCs, with clearly defined interaction points and interfaces with the environment. The construct of Mechatronic port (MTPort) was defined as an extension of the UML port to fulfill this requirement. SysML provides standard ports which support client-server communication and FlowPorts that define flows in and/or out of a block. An MTC may own MTPorts, as shown in Figure 6, which allows the MTC to declare the items it may exchange with its environment and the interaction points through which this exchange is made. Furthermore, MTPorts allow the MTC to declare the provided to the environment services but also the services that the MTC expects from it. An MTPort is defined as an aggregation of mPort, ePort and sPort. Each port is used to represent the interaction point of the corresponding part of the primitive MTC with the environment (see Figure 4). All these ports extend the SysML port; mPort and ePort extend it through the SysML flow port while sPort extends it directly. So, a sPort is characterized by provided and required interfaces. The specification of what can flow in or out of an mPort or ePort is achieved by typing them with a specification of the things that flow in and/or out. It should be noted that an mPort may accept or transmit energy or material but may also accept or transmit information that has been decided to be transferred by mechanical means. Of course the same information may be transferred by electronic signals using an ePort or by software messages using a sPort. The support of several alternatives through configuration, results in increased reuse potential for the MTC. The specification of the services of the sPort is achieved by typing it with the provided and/or required interfaces.

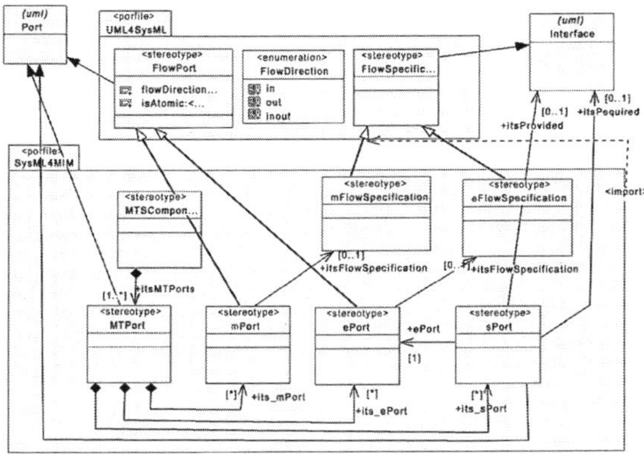

Figure 6. The MTPort stereotype in the SysML4MIM profile.

Flows of mPorts and ePorts may be atomic or non-atomic; an atomic flow is specified with a single type representing the items that flow in or out. A non-atomic flow is specified with a flow specification which lists the items that constitute the flow. A sPort accepts software signals, *i.e.* packages of information, which usually need a more complicated specification supported by UML 2.

THE MTS DEVELOPMENT PROCESS

The MIM development process adopts the V-model as basis and updates it to address the needs of the mechatronic systems domain. Figure 7 presents the proposed MTS V-Model. A system modeling process is applied down to the primitive MTC level, as shown in the left-hand part of the V-model. For primitive MTCs that have to be constructed, a concurrent engineering process of all three constituent parts, i.e., mechanics, electronics and software is adopted, as depicted in the bottom of the V-Model. The system integration and verification process is depicted by the right-hand side of the V-model.

MTS-level requirements are captured using the SysML requirements diagram. Essential use cases, which are used to capture the functional requirements at this level, are defined in abstract, simplified, and independent of technology or implementation way. They are written as "an abstract dialog representing user intentions and system responsibilities, and they are typically small and focused on a highly specific user goal, yielding a finegrained model of user activity" [20]. After the definition of the essential use cases there are two alternatives to proceed in the system's architecture definition phase:

1. Use cases are decomposed in sub-use cases.
2. Responsibilities of the system are identified.

In the first case the decomposition of use cases to sub-use cases allows: a) the reuse of existing components on the basis of their requirement specifications that should have been defined in terms of use cases, and b) modularity and reuse in requirements specification artifacts. In the second case, activity diagrams are defined for each use case in order to identify the activities/responsibilities of the actors and the ones that are required by the system in the context of the specific use case. After this step the list of abstract activities (functions) that have to be performed by the system is available. In other words the responsibilities of the system in the context of the specific use case are defined.

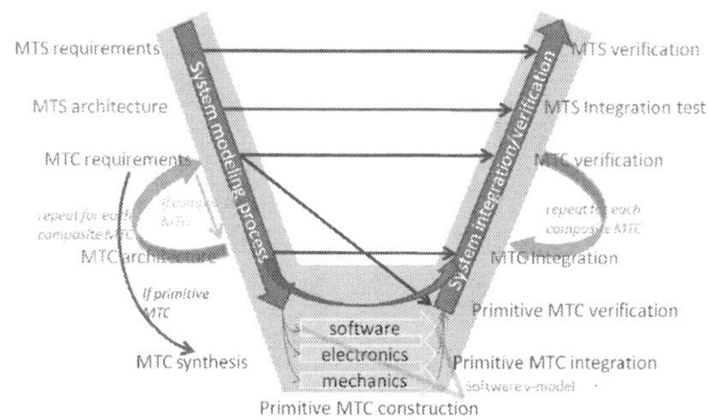

Figure 7. The mechatronic system V-Model (MTS V-Model).

The term responsibility is used to emphasize the fact that only the abstract definition of the activity/function is provided at this time and not its implementation.

As a next step, for both alternatives, a system composition should be proposed to satisfy either the system use-cases or the system responsibilities. Some use cases or system responsibilities may be directly supported by existing mechatronic components. However, it is common that for a required system level use case or responsibility, a collaboration of MTCs has to be defined in order to achieve it. The system level use cases and sub-use cases or the system responsibilities/functions correspondingly are mapped to system's components. An analogous mapping applies also for the non-functional requirements.

The definition of system's structure in terms of MTCs is a design process and results in the selection of system components and the definition of their collaboration. Furthermore each MTC has assigned responsibilities that are handled in the subsequent phases as its required responsibilities. The result in both cases is a system architecture that is comprised of:

- Class or component diagrams to specify the structure of the system;
- Sequence or activity diagrams (or even state charts) to specify the components' collaborations to provide the higher layer functionality.

Domain analysis is used to capture the domain key concepts and provide the information required to create the first architectural model of the system. SysML diagrams are used to specify the proposed architecture. Block definition diagrams (bdd) are used to capture the structure of the system and internal block diagrams (ibd) are used to capture the components' interactions, all expressed using the SysML4MIM profile.

During the architecture definition the developer has to assign the system required responsibilities to the system's components. This assignment results to an architectural diagram that represents the system components, their responsibilities and the components interactions.

The allocation relationship of SysML provides an effective means to capture this assignment and allow the navigation between the system models by establishing cross-cutting relationships among them. There are two alternatives to proceed in the definition of the architecture:

1. The bottom-up approach (synthesis).
2. The top-down (decomposition).

According to the bottom-up approach, for every system-level responsibility a set of commercial off-the-shelf (COTS) MTCs is selected. We assume that each COTS MTC has its own provided functions that are well defined by the developer of the MTC [7]. These provided functions are part of the MTC package that specifies the real world MTC. QoS characteristics are also included in the MTC package and can be used to examine if the QoS aspects of the proposed collaboration scheme satisfy the required system-level QoS aspects for the specific responsibility. If the QoS characteristics of this specific collaboration meet the QoS requirements of the system-level required responsibility, the design is accepted. Either wise corrective actions should be proposed and analyzed. Corrective actions may include: a) re-engineering of the collaboration scenario, b) the use of components with better QoS characteristics than the ones used in the previous design, or c) a combination of the above. It is assumed that the MTC developer has already performed a QoS analysis for the MTC. All this information comprises the offered QoS characteristics of the MTC [4]. It should also be stated that the MTC developer does not know during the MTC's development time all the systems that this MTC will be used in the future.

According to the top-down approach, for every system-level responsibility that has not been assigned to a single MTC, a set of abstract MTCs is defined and the required MTC responsibilities are specified along with the required collaboration scheme. Required system-level QoS characteristics are decomposed to derive the component-level required QoS characteristics.

This process results to the definition of the required QoS characteristics at the level of constituent MTCs. The process of deriving MTC-level QoS characteristics from system-level ones is a complex process and has to be defined. At this time the engineer has well defined required specifications (functional and nonfunctional, including QoS characteristics) for every abstract MTC. Using these required QoS characteristics the engineer is able to select from the market or his components repository the ones that their offered QoS characteristics meet the required ones [8]. If such MTC's do not exist they have to be further analyzed in order to be developed.

Advantages and disadvantages for the above approaches that result in the definition of the architecture of the MTS of its composite MTCs may be identified but it is expected that in the real MTS development process a combination of both approaches will be used resulting to a hybrid more efficient approach.

The above process, bottom-up or top-down, is again applied to every composite MTC that has to be developed. It is applied iteratively down to the primitive-MTC level; the identification of primitive MTCs signals the end of this iteration. For each composite MTC the system modeling process as defined by the left-hand side part of the V-Model is followed. Analysis is applied and its architecture is defined in terms of constituent components (composite and/or primitive). Sequence diagrams are defined to realize use cases of the MTC and identify the activities that are involved in the specific use case.

This is not the case for primitive MTCs that have to bypass the system process and follow a synergistic integration of the three constituent parts, i.e. mechanic, electronic and software (MTC synthesis) as shown in the bottom of the V-model in Figure 7.

For each primitive MTC, verification follows its integration as shown in the right-hand side of the MTS V-Model. Each composite MTC is integrated according to its MTC architecture and then it is verified against its requirements. After the integration and verification of the MTCs of the system, the MTS integration test is performed and the MTS is verified against its requirements.

It should be noted that the system analysis phase is followed by a system architecture design phase as shown in the proposed MTS V-model. This is also the case for the V-Model in software engineering. After this point the proposed V-model is completely differentiated from the traditional software engineering V-model.

After the system architectural design, repetitions of analysis followed by architecture design for every composite MTC are applied following the system modeling process up to the primitive MTC level.

This is the point where the system development process is terminated and the synergistic integration of constituent parts of the primitive MTC is performed working independently but in a synergistic way in the three disciplines. For every primitive MTCs that has complex software constituent part, a software V-model can be applied for its development, as shown in the bottom of the MTS V-Model.

Figure 8 presents two real world MTCs and the Parallel Kinematic Machine evolium MTS, that were developed based on the basic principles of the MIM architecture by a high-tech Italian company. Each axis of the Parallel Kinematic Machine has its own intelligence, so there is no need of an external entity to control the motion trajectory.

FUTURE DEVELOPMENTS AND RESEARCH CHALLENGES

Mechatronic systems development is a very complicated process imposing many challenges. In this section we refer to the ones that are of higher priority considering the 3+1 SysML view-model. The identification of the mechanical discipline information that has to be captured in the system level models is one of the challenges.

Figure 8. Real-world MTS and MTCs

There are two possible approaches: a) exploit SysML constructs to represent as much of the mechanical discipline information, including component interfaces and behavior; and b) extend SysML constructs with new ones with the objective of creating a complete SysML model of the mechanical component. In the first case specific tools will be used for further refinement of the mechanical models and their subsequent execution and analysis. In the latter, the SysML models have to be automatically transformed to models of the specific tools for execution and model analysis. The integration of SysML with the Modelica language (www.modelica.org/) is towards this direction. This challenge is greatly related to the one that concerns the model execution, analysis and assessment of models on the MTS-view level.

Fully automated generation of the three views from the MTS view, as well as automatic update of the MTS-view with changes in the discipline views are important challenges that have to be addressed to improve the effectiveness of the process. This also imposes the challenge of integration of existing mechanical and electronic domain design tools.

The identification and definition of reusable MTCs is another major challenge in mechatronic systems. The Workpartner [21], a mobile service robot, is planned to be used in the context of a TEKES (Finnish Funding Agency for Technology and Innovation) funded project as a case study for the application of MIM, but also as a case study for the identification of reusable MTCs.

Since many of the MTSs are from the safety critical domain, the integration of the MTS V-model development process and the 3+1 SysML view-model with safety engineering is another major challenge for the MIM paradigm to be effectively exploited in safety critical mechatronic systems.

CONCLUDING REMARKS

The traditional approach in the development of mechatronic systems is unable to address the needs of today's complex mechatronic systems. An integrated framework for the construction of mechatronic systems is missing. The work presented in this paper attempts to contribute to this direction by: a) using SysML to define the artifacts of the MIM paradigm; b) proposing the 3+1 SysML view-model imposed by the MIM architecture; and c) extending the well accepted and widely used in the software domain V-model to address the demands of the mechatronic system development process. However, the challenges for a fully automated MTS development process crosses the boundaries of the three disciplines of mechatronic systems and impose a joint effort and collaboration between computer science, electronics and mechanics. The current status of discipline isolation imposed in many cases by the existing structure of engineering degree programs makes the task even more complicated.

ACKNOWLEDGMENTS

Part of this work has been funded by TIKOSU, a project belonging to the Digital Product Process -program of the Finnish Funding Agency for Technology and Innovation (TEKES). The author wishes to thank the partners and especially Jarmo Alanen, Kari Koskinen, and Seppo Sierla for fruitful comments on these ideas. Thanks are also due to Piero Larizza, Eric Coatanea and Jussi Suomela for discussions on these concepts.

REFERENCES

1. G. Rzevski, "On conceptual design of intelligent mecha- tronic systems," Mechatronics, 2003.
2. W. Schafer and H. Wehrheim, "The challenges of build- ing advanced mechatronic systems," Future of Software Engineering, International Conference on Software Engi- neering, IEEE Computer Society, 2007.
3. Philipp Limbourg, "Dependability modelling under un- certainty: An imprecise probabilistic approach," Springer, 2008.
4. K. Thramboulidis, "Model integrated mechatronics: To- wards a new paradigm in the development of manufac- turing systems," IEEE Transactions on Industrial Infor- matics, Vol. 1, No. 1, February 2005.
5. OMG, "OMG Systems Modeling Language (OMG SysML™)," V1.0, September 2007.
6. GD250, "Lifecycle process model 'V-Model'," Available online: http://www.informatik.uni-bremen.de/gd-pa/vmodel/ vm1. htm#application
7. K. Thramboulidis, G. Doukas, and G. Koumoutsos, "A SOA-based embedded systems development environment for industrial automation," EURASIP Journal on Embed- ded Systems, Article ID 312671, pp. 15, 2008.
8. K. Thramboulidis, "Challenges in the development of mechatronic systems: The mechatronic component," 13th IEEE International Conference on Emerging Technolo- gies and Factory Automation (ETFA), Hamburg, Ger- many, September 2008.
9. OMG, "Unified modeling language: Superstructure," version 2.1.1, formal/2007-02-03.
10. M. Habib, "Mechatronics," IEEE Industrial Electronics Magazine, Vol. 1, No. 2, Summer 2007.
11. S. Burmester, H. Giese, and M. Tichy, "Model-driven development of reconfigurable mechatronic systems with mechatronic 'UML' in model driven architecture," Springer Berlin/Heidelberg, Vol. 3599, 2005.
12. K. Knorr, A. Lapp, P. Torre Flores, J. Schirmer, D. Kraft
13. J. Petersen, M. Bourhaleb, and T. Bertram "A process model for distributed development of networked mecha- tronic components in motor vehicles," Proceedings of the IEEE Joint International Conference on Requirements Engineering (RE'02), 2002
14. R. Nordmann, "Use of mechatronic components in rotat- ing machinery,"

Book: Vibration Problems ICOVP 2005, Springer Netherlands, Vol. 111, January 20, 2007.

15. J. El-khoury, O. Redell, and M. Torngren, "A tool integra- tion platform for multi-disciplinary development," Pro- ceedings of the 2005 31st EUROMICRO Conference on Software Engineering and Advanced Applications, 2005.
16. G. Doukas and K. Thramboulidis, "A real-time Linux based framework for model-driven engineering in control and automation," IEEE Transactions on Industrial Elec- tronics (forthcoming).
17. G. Pahl, W. Beitz, J. Feldhusen, and K. H. Grote, "Engi- neering design: A systematic approach," Third Edition, Springer-Verlag London, 2007.
18. R. Eckert, W. Mansel, and G. Specht, "Model transfer among CASE tools in systems engineering," Systems En- gineering, Vol. 8, No 1, pp. 41–50, March 2005.
19. P. Kruchten, "The 4+1 view model of architecture," IEEE Software, Vol. 12, No. 6, pp. 42–50, November 1995.
20. OMG, "SysML and AP233 mapping activity," OMG SysML portal, http://www.omgwiki.org/OMGSysML/ doku. php?id=sysml-ap233:mapping_between_sysml_and_ap233.
21. L. Constantine, "Activity modeling: Toward a pragmatic integration of activity theory with usage-centered design," Technical Paper, Available on-line: http://www. foru se. com/ articles/activitymodeling.pdf
22. "The workpartner mobile service robot," http://automa- tion.tkk.fi/

CHAPTER 4

Design and Development of Mechatronic Application in Agricultural Irrigation Device

Mohd Hudzari Razali, Syazili Roslan, Abdul Ssomad Mohd Abd Halim, Hayan Basit

Faculty of Bioresources & Food Industry, Universiti Sultan Zainal Abidin, Tembila Campus, Besut, Terengganu Darul Iman, Malaysia

ABSTRACT

In order to help the small-scale farmer, an automatic irrigation control system was proposed. This system will provide an irrigation system that will ease the burden of the citizen to take care of the plant. This system will run automatically by referring to the time set by the user. As the name itself

Citation:- Razali, M. , Roslan, S. , Halim, A. and Basit, H. (2016), "Design and Development of Mechatronic Application in Agricultural Irrigation Device". World Journal of Engineering and Technology, 4, 450-459. doi: 10.4236/wjet.2016.43045.

is a water control system, this system will only start irrigating when the time set triggered the water control level for the plant to grow healthily. It will automatically stop when the timer is off (1 hour). The brain of the system is the PLC (Programmable Logic Controller). This is the place where all the activities are done. The irrigation will be provided by a pump that is also connected to the microcontroller. The pump will be activated until the timer has reached its time set. This system will continue running until the user presses the OFF button.

Keywords: Irrigation System, Water Control System, PLC (Programmable Logic Controller)

INTRODUCTION

As technology undergoes rapid advancement, complicated task such as control system is accomplished with a highly automated control system. Example of the system may be in the form of Programmable Logic Controller (PLC) and possibly a host computer and accompanied with signal interfacing to the field devices such as opera- tor panel, motors, sensors, switches, and solenoid valves. Network communication is capable of providing a large scale implementation and process coordination besides enabling greater versatility in understanding distributed control system. Every single integral component module in a control system plays a major role regardless of size. Generally, PLC can be explained as a digital electronic device that uses a programmable memory to store direction and to accomplish functions such as counting, logic sequencing, arithmetic and timing in order to control machine and processes. The term logic was used because the programming is basically concerned with implementing logic and switching operations [1]. Among the most important implementation for PLCs in Southeast Asia is the oil & gas industry. For example, Malaysia is world famous for its palm oil, accounting for a large portion of total global production. Here, many mid and high-end PLCs are used in the production of palm oil, fed mostly by government funding. Meanwhile, large PLCs are used in the oil and natural gas industry in Indonesia and Malaysia. Another important for PLCs implementations in Southeast Asia is the automotive manufacturing industry. Thailand is world-famous for its production of automotive electronics; with industry requirements for PLCs multiply after the country was hit by critical flooding in 2011. Malaysia has a big automotive industry markets in the area among the locals, where a very large population will soon requesting more favorable modes of transport. A third application is in agriculture industry. Thailand and Indonesia are the main contributors to the expansion of this sector, and advanced PLCs will be required to produce innovative food processing and packaging ingredient, aimed at local use, consumption and exports alike. The use of PLCs and other similar devices in the agricultural mechanization industry is growing widespread. Example applications include food processing, building environmental control, grain drying, aquaculture production, and tractor and machinery systems. As we know that Malaysia government policy in terms of making industrializing in agriculture activity sector, so in this project, we introduce the application of PLC that can be used in agriculture activity. It includes how to set up the PLC programmer, integration between the software and hardware, construction of the circuit, and run monitoring.

Problem Statement

The world's population is estimated to reach 9.2 billion by 2050.The UN Food and Agriculture Organization (FAO) has projected that farmers will need to increase their production by 70% more food than in 2006 to meet this demand.

Inability to do so will causing in food scarcity and poorer healthcare in developing countries, with damaging impact for development and the potential for conflict within and between nations. While agriculture productivity has been increasing, production capacity is stagnant, and food security remains a major issue in many nations due to hiking prices as well as availability. Continuing increase costs of input of agriculture, commodity speculation and competition with other uses for crops. In poorer nations, this causing the number of people considered to be 'food insecure' increasing because they are not able to afford enough food even if it were available to buy. Both productivity and absolute production need to increase if this issue is to be addressed. In Malaysia, small farms remain at the Centre of agriculture and rural development. However, one of the main causes for the low agricultural productivity in most developing countries in the region is the lack of appropriate machineries that cater to and suit the requirements of small-scale farms. For this reason, many small farms are deemed as unproductive and inefficient. Farm mechanization plays a significant role in every nation's economy. However, it is often misconstrued to mean modernization, beneficial only to industrialized countries with highly mechanized agriculture. Developing countries often have to rely on a variety of imported farm machines, which are seldom appropriate for small farms [2]. One of the issues that need to be address is the lack of usage of automation such as PLC especially in Malaysia. Many small, medium, and even large-scale farm having less production capacity that can be prevented if the owner using an automation for their respective farm. This Project will give me an overview of how PLC works and insight so more advance automation can be carried out in the future.

Significant of Study

Automation and robotic can play a large role in society to meet the agricultural production needs. For decades automation and robots have played a core role in multiplying the efficiency and lowering the cost of industrial production and products. In the past several years, a similar shift has started to take place in agriculture industry, with GPS- and vision-based such as self-guided tractors and harvesters already being available in the market commercially.

More recently, developers have started to test with autonomous systems that combine operations such as thinning, pruning and harvesting, as well as sowing, spraying, mowing and weed removal. In the fruit production industry, for example, robotic platforms rode by workers have shown to be doubling the efficiency compared to workers using ladders. Advancement in sensors technology and control systems allow for optimum resource and integrated pest and disease management.

This will be a revolution in the way that food nowadays is grown, tended, and harvested. The level of input of engineering technologies into agriculture is generally still low [3]. Automation component such as PLC is an important module in agricultural automation.

It can be found used in irrigation systems, food processing, building environmental control, grain drying, aquaculture production, and tractor and machinery systems. Using automation such as PLC enables us to increase the number of production by reducing error such as human and environmental factor (damaged produce).

Objective of Study

- To develop the hardware and software by using PLC (Programmable Logic Controller) as a main controller.
- To interface PLC module with input and output component.
- To design prototype system for motion control.

LITERATURE REVIEW

Programming Logic Control

Programmable Logic Controller is often found in the field close to the processing unit. Tiny and operator interface of a PLC may be simple as button switch. In later generation PLC manufacturer have added analogue to digital conversion proficiency and provided enough logic to adjust simple control loops [4]. Nowadays there are at least two recognized PLC sizes: Small sized PLC, which is basically a relay replacement and provides a reliable control to stand-alone section of process of PLC. Medium sized PLC that can performs all the relay replacement tasks expected of it, and also performs functions like timing, counting, and complex mathematical applications.

There are five core components in a PLC system:

1) The PLC processor or controller.
2) I/O (Input/Output) modules.
3) Chassis or backplane.
4) Power supply.
5) Programming software that runs in a PC.

Several advantages of PLC: Increased reliability; once a selected program has been drafted and installed, other PLCs can download it without any difficulties. Since almost all the logic pattern is contained in the PLC's internal memory, there is lower chance of it making a logic wiring error.

More Flexibility; It is easier to design and alter a program in a PLC than to wire and rewired a circuit board. Equipment manufacturers can provide system upgrade by simply broadcast out a new program.

Lower Cost; PLC were originally made and designed to replace relay control logic, and the cost reduction have been so exceptional that relay control is becoming non-existent and obsolete except used for power application.Communications Capability; PLC can interact with other controllers or computer equipment to perform such functions as monitoring devices, process parameters, supervisory control, data gathering, and also download and upload of programs.

Faster Response Time; PLCs are designed for real-time and high-speed applications. The PLC operates in real time; causing event taking place in the field will result in the carried out of an operation or output.

Easier to troubleshoot; PLCs have resident diagnostics and override functions that allow users to easily trace and correct software and hardware problems [5].

Many types of PLC are available such as Omron, Mitsubishi, Siemen, Nais and many more. In this project, I am using Omron PLC model CP1E-N30DR-D.This PLC is suitable to control the switching of valve and speed of motor pump (Figure 1).

HARDWARE DESIGN OF PLC

Input Device

Intelligence of an automated system is greatly depending on the ability of a PLC to read in the signal from various types of automatic sensing and manually input field devices. Toggle switches, push buttons, and keypad, which from the basic man machine interface, are types of manual input device. On the other hand, for detection of workplace, monitoring of moving mechanism, checking on pressure and or liquid level and many others, the

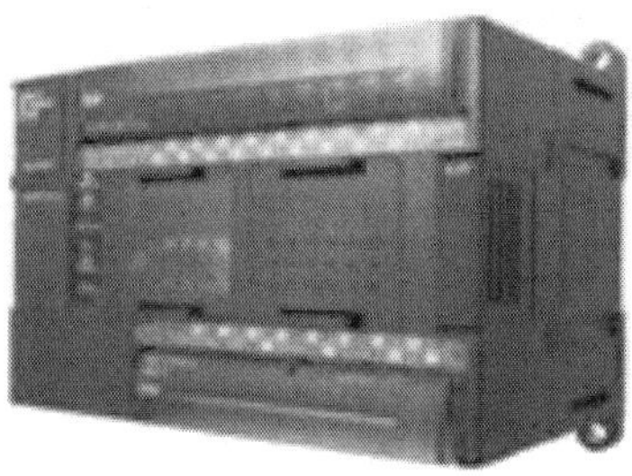

Figure 1. Omron PLC.

PLC will have to tap the signal from the specific automatic sensing device like proximity switch, limit switch, photoelectric sensor, and level sensor and so on (Figure 2). Types of input signal to be connected to the PLC would be analogue or ON/OFF logic. These input signals are interfaced to PLC through various types of PLC input module.

Output Devices

An automatic system is incomplete and the PLC system is virtually paralyzed without means of interface to the field output devices. Some of the most commonly controlled device is relay indicator, motors, solenoids, buzzers and so on (Figure 3). Through activation of motors and solenoids the PLC can control from simple pick and place system to a much complex servo positioning system. These type of output devices are the mechanism of an automated system and so its direct effect on the system performance.

Software Design of PLC Programming

Ladder Logic

To ease the use of PLC's we programmed it using ladder logic format. Ladder logic is a visual representation of a set of inputs and outputs. Ladder Logic format resembles familiar hardware systems and it doesn't require extra training for technicians, engineers and also those who without backgrounds in these hardware systems. The name is derived from the fact that the diagram upon completion resembles a ladder. A Ladder Logic diagram consists of a vertical line on the left hand side, known as the hot rail, and a vertical line on the right hand side, known as the neutral rail [6] . They are connected by lines, known as rungs; with several different symbols each represents an input or output. The logic is following the path and determining if the rung is true. For the rung to be true it is necessary to be able to flow across closed contacts to the opposite rail. If the rung is true the output is then considered true. Figure 4 is sample ladder logic diagram.

Relay

Relay is one of the switch types which are electrically operated. Most of the relay uses an electromagnet to operate a switching mechanism mechanically. It is used to control a circuit by a low-power signal (with complete electrical isolation between control and controlled circuits), where several circuits must be controlled with one signal. A contactor is one of the relay types that can handle a high power that required controlling an electric motor electrical circuit from overload or faults; in modern electric power systems these functions are performed by digital instruments still called "protective relays". In general, components in relay are inductor coil, a spring (not shown in figure), swing terminal, and two high power contacts named as normally closed (NC) and normally open (NO). Relay uses an Electromagnet to move swing terminal between two contacts (NO and NC). When there is no power applied to the inductor coil (Relay is OFF), the spring holds the swing terminal is attached to NC contact. Figure 5 shows the pin diagram for relay:

The relay will start operating when the required power is applied to the inductor coil will generates a magnetic field which will move the swing terminal from normally close contact to normally open contact. When the power is OFF, the spring will move back to normally close contact.

Water Pump Motor

A stepper motor is electromechanical actuators, changing electrical pulses into mechanical actions. When the signals are applied to the system with the right sequence pulses, the shaft of a stepper motor rotates in distinct step.

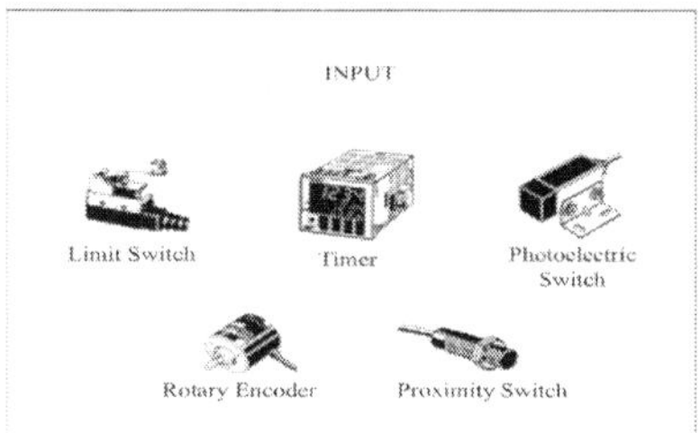

Figure 2. Type of input.

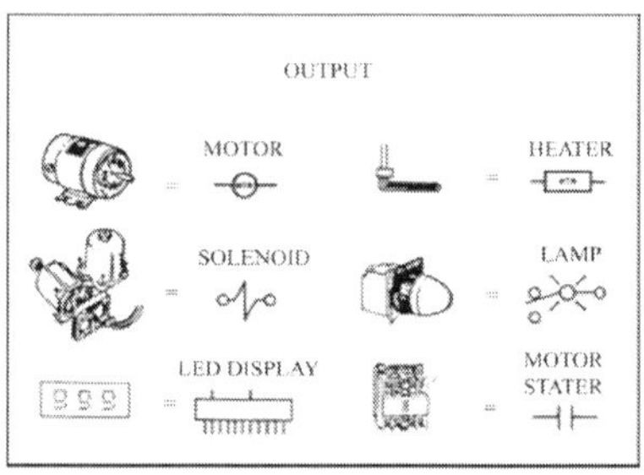

Figure 3. Type of output.

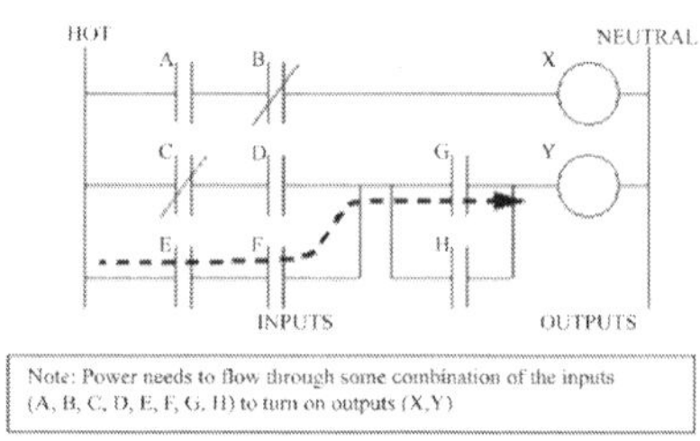

Figure 4. Ladder logic programming.

The stepper motors revolution has various direct relationships to applied input pulses. The speed of the shafts rotation is connected to the frequency of the input pulses and the length of rotation is associated to the amount of input pulses applied. Meanwhile the direction of the shaft rotation is depends on sequences of the input pulses. Not like other type of motor, the stepper motor has no contacts or brushes. It is a synchronous motor in order to rotate armature magnet through the magnetic field switching. The essential function of a stepper motor is to translate switching excitation changes into precisely 15 defined increments of rotor position. Generally the stepper motor capable works as an electric motor when the drive running without commutated. The rotor of the stepper motor can be permanent magnet or variable reluctance motor which has a toothed block of some magnetically soft material.

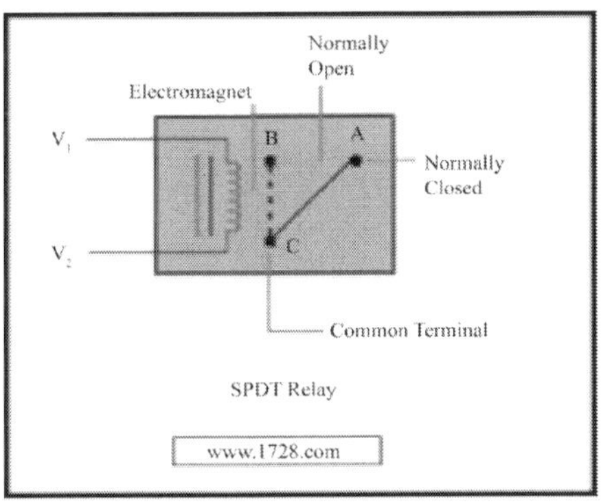

Figure 5. Relay general pin diagram.

Typically, all the windings in the stepper motor are part of the stator. A stepping motor control system consists of three basic elements; controller, driver and motor [7].

MATERIALS AND METHODOLOGY

Materials

1) PLC training kit
   - PLC microcontroller (CP1E-N30DR-D)
   - CX one PLC programmer
   - Stepper motor
   - Touch-screen teach pendant

2) S8VK-G01505 Omron Automation
3) Omron MY2N 24VDC Relay
4) Wire cutter
5) Power source
6) Connecting Wire
7) Water pump motor
8) Push switch

Methodology

Figure 6 showed the methodology or work flow of the project that has been used as the guideline in order to do the project. In started with investigate the topic and objectives with supervisor. After doing literature review, equipment that needed was investigated like relay, water pump motor, PLC and AC/DC converter.

Pilot Experiment

Pilot experiment was done first so we will learn first on how to write a ladder logic system and install it into PLC.

This experiment was done using PLC training kit as a prototype that already had been assembled so it can be used as a learning tool before actual experiment.

The programming that we are going to use is CX one programmer that can be used to write ladder logic system. First, we need to connect PLC to a computer so we will be able to install the programing into our PLC. We put the logic:

When push button is ON, turn the stepper motor ON. After that, we write this logic into a ladder logic system (CX one).

Then we load this program into the PLC. We connect the sensor input to the PLC, the one we specified in our program (push button).

Lastly, we connect the PLC External Output Terminal (specified by our program) to the stepper motor. Now, we execute the logic program on the PLC.

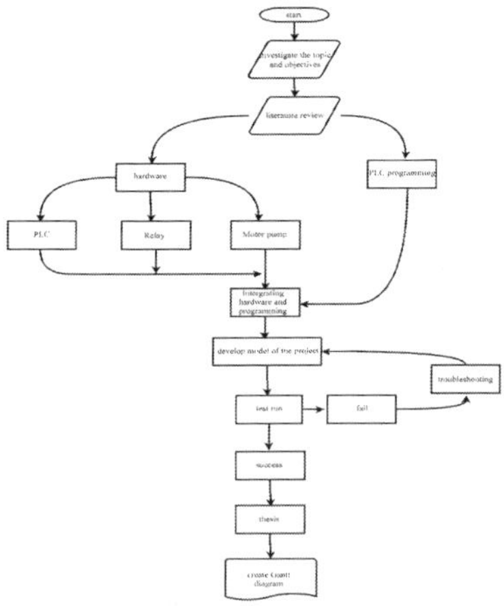

Figure 6. Flowchart of construction of model.

Construction of Model

Firstly, we need to connect our PLC to a computer using an adapter. Then we run the program (CX One programmer shown in Figure 7) and write the intended logic sequence for our experiment that is:

1) When switch is ON, turn the Timer A for 100 seconds.
2) During Timer A ON, the water pump motor is ON until Timer A is OFF.
3) When Timer A is OFF, turn ON Timer B (100 seconds).
4) When Timer B is OFF, switch ON the water pump motor and Timer C (100 seconds).
5) When Timer c is OFF, turn ON Timer D (100 seconds).
6) When Timer D is OFF, the counter resets the main switch and the cycle continue.

Next, we load the program that has been written into the PLC. After that, connect the push switch to the PLC by using connecting wire, the one we specified in our programming earlier. Then, connect the PLC External Output Terminal to the water pump motor (output). Now, we execute the logic program on the PLC. After the model has been completed, we do several test run for our model to make sure the PLC programming work as it intended. Any failure will be observed and documented so we will be able to do troubleshooting for our model. Figure 8 shows the model of construction software for irrigation control.

RESULT

Software Development PLC Programming

Programming instructions:

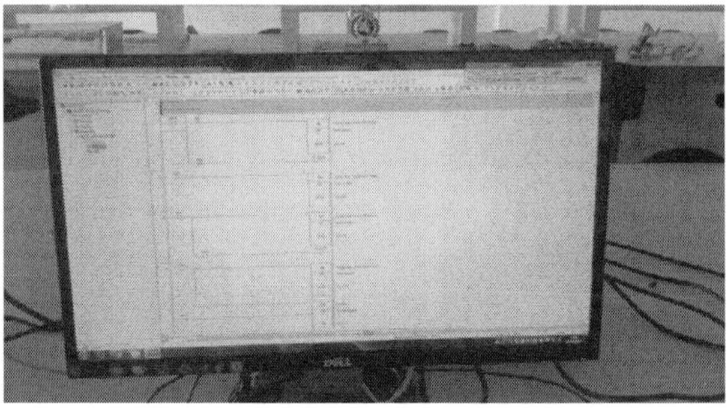

Figure 7. CX one programmer.

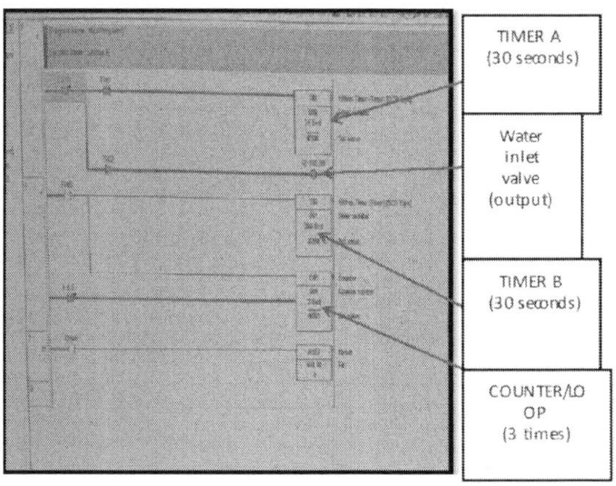

Figure 8. Construction software model for irrigation control.

1) When switch is ON, turn the Timer A for 1 hour.
2) During Timer A ON, the water pump motor is ON until Timer A is OFF.
3) When Timer A is OFF, turn ON Timer B (8 hours).
4) When Timer B is OFF, switch ON the water pump motor and Timer C (1 hour).
5) When Timer C is OFF, turn ON Timer D (14 hours).
6) When Timer D is OFF, the counter RESET the main switch and the cycle continue.

Hardware Development

Figure 9 shows an overview of work table in Farm Mechanization laboratory in Universiti Sultan Zainal Abidin campus. Figure 10 shows constructed model for irrigation control.

Simulation

Figure 11 shows an algorithm for automatic irrigation control.

DISCUSSION

Introduction

Both hardware and software part are developed and built successfully. The system is complete when both of them are connected together. This chapter explains all the results related to both part.

Hardware

The microcontroller is connected directly to AC to DC adaptor that is functioning as power supply to switch on the control unit. The moisture sensor will start measuring and the data obtain will be sent to control unit in voltage form that will be processed by PIC microcontroller. Since the microcontroller has the capability of 10-bit analog to digital converter, it will convert and manipulate the data obtained to get the actual value of humidity. After that, these values are displayed on the LCD. The control unit also will instruct the pump to start working.

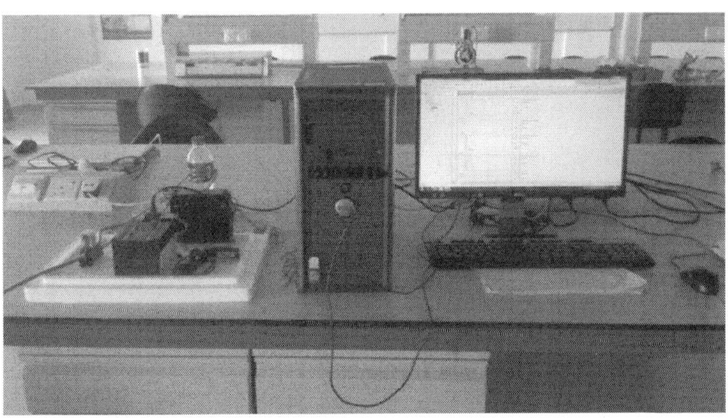

Figure 9. Overview of work table in Farm Mechanization Lab.

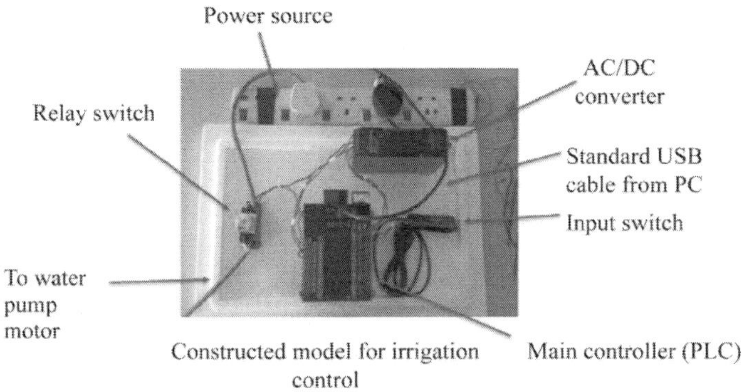

Figure 10. Constructed model for irrigation control.

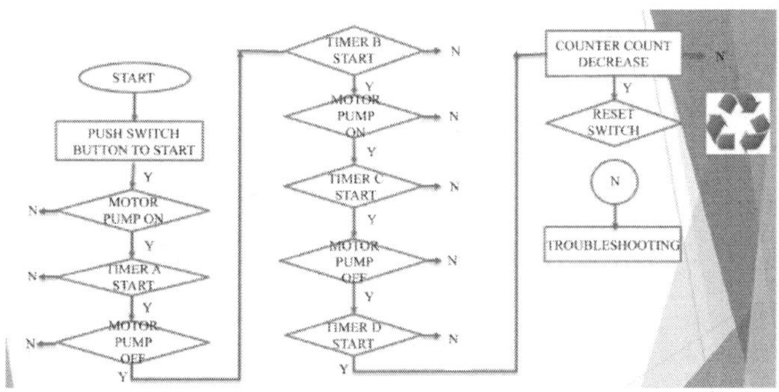

Figure 11. Algorithm for automatic irrigation control.

CONCLUSION

The project involved the design activity on both software and hardware development of PLC in agriculture mini hydroponic system. The design on circuitry system is involved for connecting between input and output system including push button device, microcontroller device, 240 volts power supply and water motor pump. The software development is created from basic concept on timer and counter application. By using this technique, this project was successful to automate basic irrigation system for hydroponic. The system will automatically circulate the fertilizer mix water throughout the system on selected period of morning and afternoon daily with less intervention of the user.

REFERENCES

1. Selvaraj, K. (2010) Development of a "Programmable Logic Controller Circuitry" for Optimal Power Distribution in a Manufacturing Industries. Journal of Computer Science, 6, 250-252.http://dx.doi.org/10.3844/jcssp.2010.250.252
2. (2005) Small Farm Mechanization Systems Development, Adoption and Utilization. Report, FFTC Annual Parts, Sri Lanka, 13-17.
3. Kamaruddin, R., Rukunuddin, I.H. and Seng, O.H. (2007) Research and Development of Agricultural Engineering in Malaysia. Paper, Country Asian, United Nations Centre, Pacific Engineering, Agricultural, 1-11.
4. Halim, M.Z.B. (2007) The Temperature Control System Using PLC.
5. Ismail, M.H. (2008) Rain Water PLC Based Detector and Valve Switcher.
6. Schumann, D., Fietsam, J., Rochel, E., Dixon, K., Imbayan, M. and Ibayan, M. (n.d.) Literature Review for the Design of a SCADA System.
7. Shah, C. (2004) Sensor Less Control of Stepper Motor Using Kalman Filter. Department of Electrical and Computer Engineering, Cleveland State University, Master of Science in Electrical Engineering.

CHAPTER 5

Design and Development of a Competitive Low-Cost Robot Arm with Four Degrees of Freedom

Ashraf Elfasakhany[1,2], Eduardo Yanez[2], Karen Baylon[2], Ricardo Salgado[2]

[1] *Department of Mechanical Engineering, Faculty of Engineering, Taif University, Al-Haweiah, Saudi Arabia*

[2] *Tecnológico de Monterrey, Campus Ciudad Juárez, Ciudad Juarez, Mexico*

ABSTRACT

The main focus of this work was to design, develop and implementation of competitively robot arm with enhanced control and stumpy cost. The robot arm was designed with four degrees of freedom and talented to accomplish

Citation:- A. Elfasakhany, E. Yanez, K. Baylon and R. Salgado, "Design and Development of a Competitive Low-Cost Robot Arm with Four Degrees of Freedom," Modern Mechanical Engineering, Vol. 1 No. 2, 2011, pp. 47-55. doi: 10.4236/mme.2011.12007.

accurately simple tasks, such as light material handling, which will be integrated into a mobile platform that serves as an assistant for industrial workforce. The robot arm is equipped with several servo motors which do links between arms and perform arm movements. The servo motors include encoder so that no controller was implemented. To control the robot we used Labview, which performs inverse kinematic calculations and communicates the proper angles serially to a microcontroller that drives the servo motors with the capability of modifying position, speed and acceleration. Testing and validation of the robot arm was carried out and results shows that it work properly.

Keywords: Robot Arm, Low-Cost, Design, Validation, Four Degrees of Freedom, Servo Motors, Arduino Robot Control, Labview Robot Control

INTRODUCTION

The term robotics is practically defined as the study, design and use of robot systems for manufacturing [1]. Robots are generally used to perform unsafe, hazardous, highly repetitive, and unpleasant tasks. They have many different functions such as material handling, assembly, arc welding, resistance welding, machine tool load and unload functions, painting, spraying, etc.

There are mainly two different kinds of robots: a service robot and an industrial robotic. Service robot is a robot that operates semi or fully autonomously to perform services useful to the well-being of humans and equipment, excluding manufacturing operations [2]. Industrial robot, on the other hand, is officially defined by ISO as an automatically controlled and multipurpose manipulator programmable in three or more axis [1]. Industrial robots are designed to move material, parts, tools, or specialized devices through variable programmed motions to perform a variety of tasks. An industrial robot system includes not only industrial robots but also any devices and/or sensors required for the robot to perform its tasks as well as sequencing or monitoring communication interfaces.

In 2007 the world market grew by 3% with approximately 114,000 new installed industrial robots. At the end of 2007 there were around one million industrial robots in use, compared with an estimated 50,000 service robots for industrial use [3].

Due to increase using of industrial robot arms, an evolution to that topic began trying to imitate human movements in a detail mode. For example a group of students in Korea made a design of innovations that robotic arm take account of dancing hand, weight lifting, Chinese calligraphy writing and color classification [4]. Another group of engineers at USA develop eight degrees of freedom robot arm. This robot is able to grasp many objects with a lot of shapes from a pen to a ball and simulating also the hand of human being [5]. In space, the Space Shuttle Remote Manipulator System, known as SSRMS or Canadarm, and its successor is example of multi degree of freedom robot arms that have been used to perform a variety of tasks such as inspections of the space shuttle using a specially deployed boom with cameras and sensors attached at the end effector and satellite deployment and retrieval manoeuvres from the cargo bay of the space shuttle [6].

In Mexico, Scientists are on track to design and develop many robot arms, and the Mexican government estimates that in Mexico there are about 11,000 robotic arms used in different industrial applications. However, the experts think that the apogee of the robot arms is not only of higher quality, but also accurately, repeatability, and stumpy cost.

Most robots are set up for an operation by the teachand-repeat technique. In this mode, a trained operator (programmer) typically uses a portable control device (a teach pendant) to teach a robot its task manually. Robot speeds during these programming sessions are slow.The present work is part of a two-phase project, which requires a mobile robot to be able to transport the tools from the storage room to the industrial cell. In this phase in the project, which carried out at Monterrey University of Technology, Mexico, the main focus was to design, development and implementation of an industrial robotic arm with stumpy cost, accurate and superior control. This robot arm was designed with four degrees of freedom and talented to accomplish simple tasks, such as light material handling, which will be integrated into a mobile platform that serves as an assistant for industrial workforce.

MECHANICAL DESIGN

The mechanical design of the robot arm is based on a robot manipulator with similar functions to a human arm [6-8]. The links of such a manipulator are connected by joints allowing rotational motion and the links of the manipulator is considered to form a kinematic chain. The business end of the kinematic chain of the manipulator is called the end effector or end-of-arm-tooling and it is analogous to the human hand. Figure 1 shows the Free Body Diagram for mechanical design of the robotic arm. As shown, the end effector is not included in the design because a commercially available gripper is used.

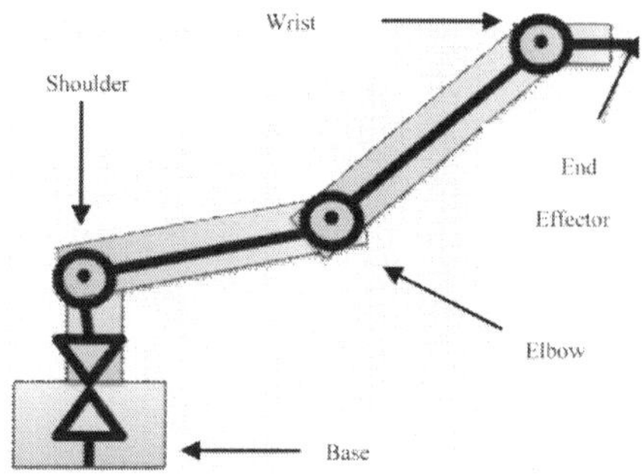

Figure 1. Free body diagram of the robot arm.

This is because that the end effector is one of the most complex parts of the system and, in turn, it is much easier and economical to use a commercial one than build it.

Figure 2 shows the work region of the robotic arm. This is the typical workspace of a robot arm with four degree of freedom (4 DOF). The mechanical design was limited to 4 DOF mainly because that such a design allows most of the necessary movements and keeps the costs and the complexity of the robot competitively. Accordingly, rotational motion of the joints is restricted where rotation is done around two axis in the shoulder and around only one in the elbow and the wrist, see Figure 1.

The robot arm joints are typically actuated by electrical motors. The servo motors were chosen, since they include encoders which automatically provide feedback to the motors and adjust the position accordingly. However, the disadvantage of these motors is that rotation range is less than 180° span, which greatly decreases the region reached by the arm and the possible positions [9]. The qualifications of servo motors were selected based on the maximum torque required by the structure and possible loads. In the current study, the material used for the structure was acrylic.

Figure 3 shows the force diagram used for load calculations. The calculations were carried out only for the joints that have the largest loads, since the other joints would have the same motor, i.e. the motor can move the links without problems. The calculations considered the weight of the motors, about 50 grams, except for the weight of motor at joint B, since it is carried out by link BA. Figure 4 shows the force diagram on link CB, which contains the joints (B and C) with the highest load (carry the links DC and ED) and the calculations are carried out as follows.

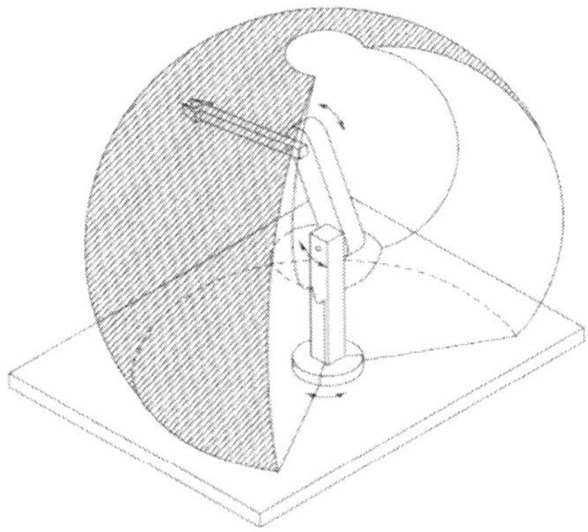

Figure 2. Work region of the robotic arm.

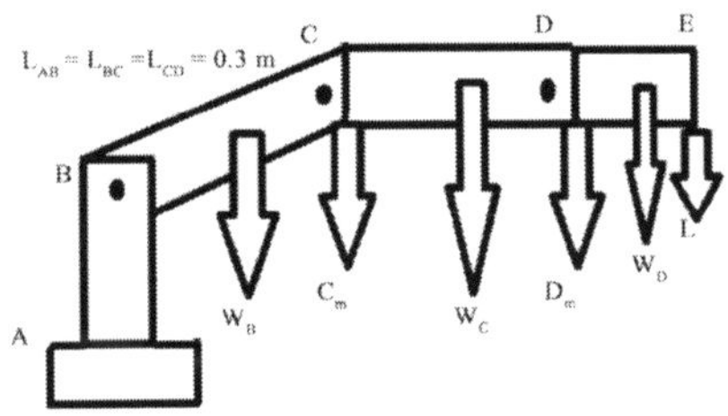

Figure 3. Force diagram of robot arm.

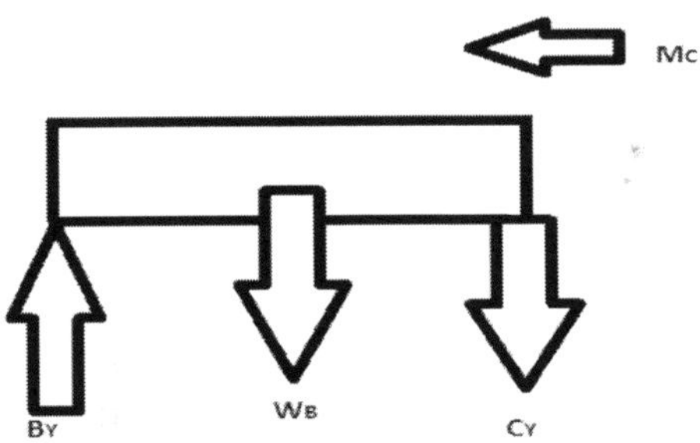

Figure 4. Force diagram of link CB.

The values used for the torque calculations:

$W_d = 0.011$ kg (weight of link DE)

$W_c = 0.030$ kg (weight of link CD)

$W_b = 0.030$ kg (weight of link CB)

L = 1 kg (load)

C_m = Dm = 0.050 kg (weight of motor)

L_{BC} = 0.14 m (length of link BC)

L_{CD} = 0.14 m (length of link CD)

L_{DE} = 0.05 m (length of link DE)

Performing the sum of forces in the Y axis, using the loads as shown in Figure 4, and solving for CY and CB, see Equations (1)-(4). Similarly, performing the sum of moments around point C, Equation (5), and point B, Equation (6), to obtain the torque in C and B, Equations (7) and (8), respectively.

$$\sum F_y = (L + W_d + D_m + W_c + C_m)g - C_Y = 0 \tag{1}$$

$$C_Y = (1.141\,\text{kg})\,9.8\,\text{m/s}^2 = 11.18\,\text{N} \tag{2}$$

$$\sum F_y = (L + W_d + D_m + W_c + C_m + W_B)g - C_B = 0 \tag{3}$$

$$C_B = (1.171\,\text{kg})\,9.8\,\text{m/s}^2 = 11.4758\,\text{N} \tag{4}$$

$$\begin{aligned}\sum M_c = &-\left(\frac{W_c L_{CD}}{2}\right) - W_D\left(L_{CD} + \frac{L_{DE}}{2}\right) \\ &- L(L_{CD} + L_{DE}) - D_m(L_{CD}) + M_c = 0\end{aligned} \tag{5}$$

$$\begin{aligned}\sum M_B = &-L(L_{BC} + L_{CD} + L_{DE}) - W_D\left(L_{BC} + L_{CD} + \frac{L_{DE}}{2}\right) \\ &- D_m(L_{BC} + L_{CD}) - W_c\left(L_{BC} + \frac{L_{CD}}{2}\right) \\ &- C_m(L_{BC}) - W_B\left(\frac{L_{BC}}{2}\right) + M_B = 0\end{aligned} \tag{6}$$

$$M_c = 1.968\,\text{Nm} = 278.6\,\text{oz/in} \tag{7}$$

$$M_B = 3.554\,\text{Nm} = 503.38\,\text{oz/in} \tag{8}$$

The servo motor that was selected, based on the calculations, is the Hextronik HX12K, which has a torque of 280 oz/in. This motor was recommended because it is much cheaper than any other motor with same specifications. Since we need more torque at joint B, see Equation (8), we used two motors at point B to comply with the torque requirements; however, one motor is enough for the other joints. Using two motors at joint B is much cheaper than using one big motor with 560 oz/in. Other relevant characteristics of the motors, which can be shown in Figure 5, are that they can turn 60 degrees in 130 milliseconds and they have a weight of 47.9 grams each.

Once the initial dimensions for the robot arm and the motor were defined, the design were carried out using the SolidWorks platform; design should carefully take into account the thickness of the acrylic sheet and the way that the pieces would be attached to each other. The acrylic sheet used to make the robot is 1/8 thickness and that thin sheet was chosen because it easier for machining and less weight with a good resistance.

Figure 5. Servo motor.

During design, we faced some difficulties due to the way of joining thin acrylic parts strongly. It was needed tools to burn and join the acrylic parts and that weren't available and the team considered that a mechanical junction based on screws and nuts would be much strong than other alternatives, such as glue for example. In order to accomplish this, a small feature was designed which allowed to fasten the bolts with the nuts without having to screw in the thin acrylic layer. The result of this process was the tridimensional design shown in Figure 6.

By end of design, each part was printed in full scale in cardboard paper and then we verified all the dimensions and the interfaces of the assembly. In turn, we built the first prototype of the robot arm. Next, parts of the robot arm were machined from the acrylic sheet using a circular saw and Dermal tools. The detailing on the parts was done in a professional workshop since the parts of robot arm were too small and it is not an easy for accomplishing such small and accurate cuts.

During assembling the robot parts with the motors, few problems pop up.

There were critical points that did not resist the fastening and, in turn, may break down; hence, reinforcements in these points were considered. The final result of the robot arm is shown in Figure 7.

ROBOT ARM INVERSE KINEMATICS

To validate the right positioning of the robotic arm, inverse kinematics calculations are carried out. Such calculations are used to obtain the angle of each motor from a position given by using the Cartesian coordinate system, as shown in Figure 8.

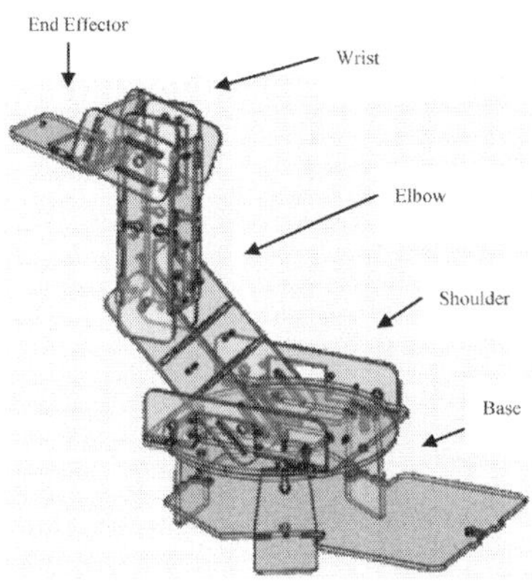

Figure 6. Robot arm 3D model

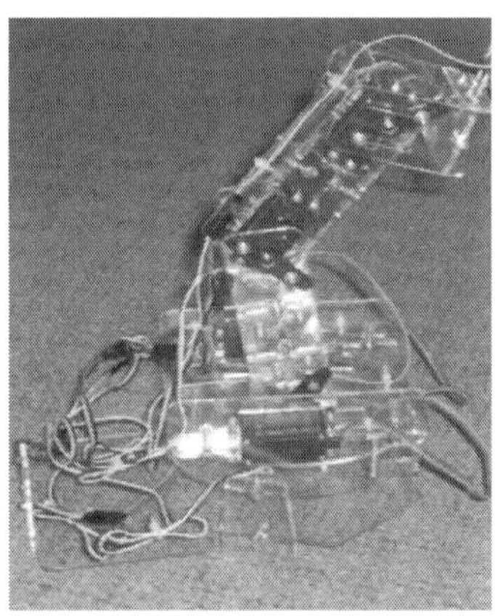

Figure 7. Robot arm complete assembly.

Each motor will have a specific function: the motor located in the A union positions the final element in the y axis, the motors B and C positions the final element in the x and z axis.

The problem was simplified by using the xz plane, as shown in Figure 9. In which the following known values were defined [9]:

L_{AB}: the forearm length.

L_{BC}: the arm length.

z: the position in the z axis.

x: the position in the x axis. y: the position in the y axis.

Using trigonometry relations, as shown in Figure 9, the motor angles θ2 and θ1 are obtained, as seen in Equations (9) and (10).

$$\theta_2 = 180^{\circ} - \text{arcCos}\left(\frac{LAB^2 + LBC^2 - x^2 - z^2}{2 \times LAB \times LBC}\right) \tag{9}$$

$$\theta_1 = \text{arcTan}\left(\frac{z}{x}\right) + \text{arcCos}\left(\frac{LAB^2 - LBC^2 + x^2 + z^2}{2 \times LAB \times \sqrt{x^2 + z^2}}\right) \tag{10}$$

$$\theta_0 = \text{arcTan}\left(\frac{y}{x}\right) \tag{11}$$

The motor B is going to use θ_1 and the motor C is going to use θ_2. The angle for the motor A is calculated as seen in Equation (11).

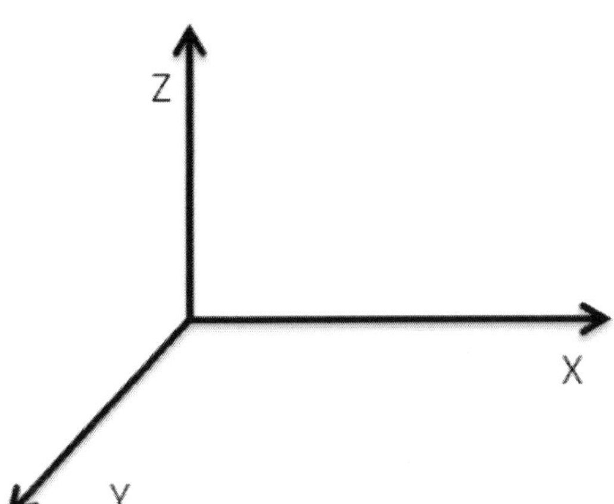

Figure 8. Coordinate system.

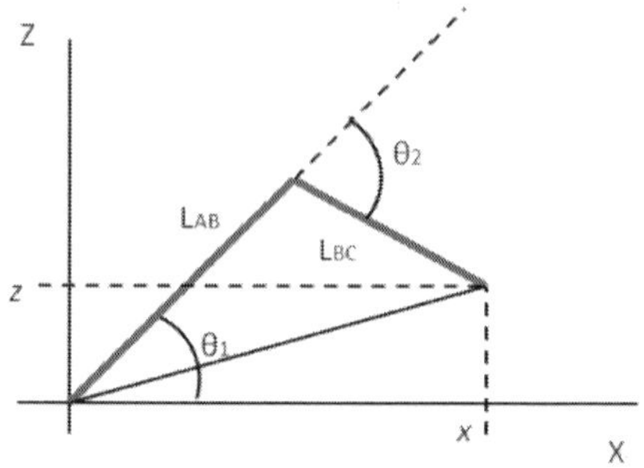

Figure 9. xz Plane.

With these calculations, the angles of servomotors are obtained and in turn they take the action to move the whole structure to the specific position.

END-EFFECTOR SELECTION

The end effector is probably one of the most important and most complex parts of the system. Wisely, it is much easier and economical to use a commercial one than build it. The end effector varies mainly according to the application and the task that the robot arm accomplishes for; it can be pneumatic, electric or hydraulic. Since our robot arm is based in an electric system, we may choose electric basis of end effector. Besides, the main application of our system is handling, accordingly, the recommended type of our end effector is a gripper, as shown in Figure 10. Please note that the end effector is controlled by a servo motor and, in turn, the total servo motors used for our robot arm will be 5 motors that move the structure.

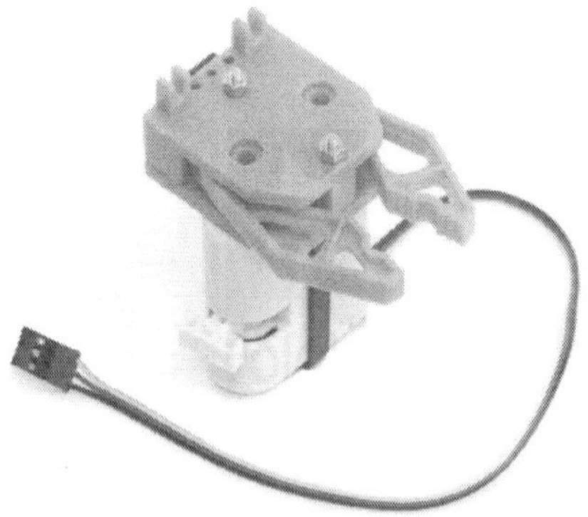

Figure 10. Gripper with servo.

ROBOT ARM CONTROL

The robot arms can be autonomous or controlled manually. In manual mode, a trained operator (programmer) typically uses a portable control device (a teach pendant) to teach a robot to do its task manually. Robot speeds during these programming sessions are slow. In the current work we enclosed the both modes.

The control for the presented robot arm consists basically of three levels: a microcontroller, a driver, and a computer-based user interface. This system has unique characteristics that allow flexibility in programming and controlling method, which was implemented using inverse kinematics; besides it could also be implemented in a full manual mode. The electronic design of control is shown in Figure 11.

The microcontroller used is an Atmega 368 which comes with a development/programming board named "Arduino", as shown in Figure 12. The programming language is very similar to C but includes several libraryies that help in the control of the I/O ports, timers, and serial communication. This microcontroller was chosen because it has a low price, it is very easy to reprogram, the programming language is simple, and interrupts are available for this particular chip.

The driver used is a six-channel Micro Maestro servo controller board. It supports three control methods: USB for direct connection to a computer, TTL serial for use with embedded systems, such as the Arduino microcontroller, and internal scripting for self-contained and host controller-free applications. This controller, as shown in Figure 13, includes a 0.25 μs resolution for position and built-in speed and acceleration control.

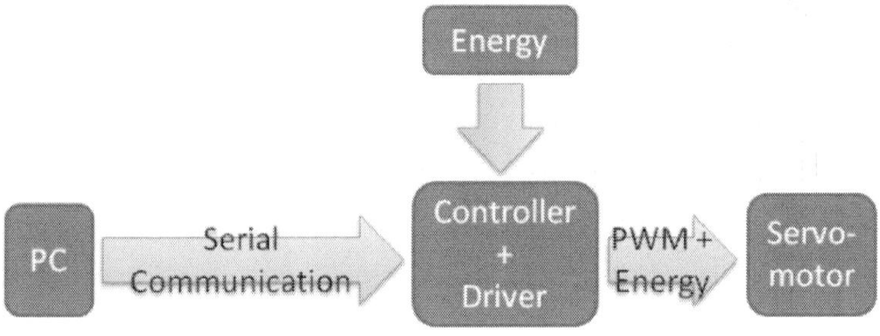

Figure 11. Electronic scheme of control.

Figure 12. Arduino microcontroller board.

Figure 13. Servo controller driver.

The user interface depends on the control method used, i.e., inverse kinematics or a full manual mode. In the following, each interface is described:

Inverse Kinematics Control

In this control method, the user inputs the coordinate system position where the gripper should be. As consequence, interface is generated with Labview through a visual user, as shown in Figure 14. The program automatically performs the inverse kinematics calculations to obtain the angles that each motor should have and then sends a command either to the microcontroller or directly to the driver that will move the robot to the specified position. Communication is performed with the RS-232 protocol. In the following, you may see the Labview user interface inputs and output.

The Labview user interface inputs are:

x axis position.

y axis position.

z axis position.

Gripper opening.

Gripper attack angle.

Serial port.

The Labview user interface outputs are:

Motor A angle.

Motor B1 angle.

Motor B2 angle.

Motor C angle.

Attack angle.

Gripper angle

Such output variables are treated and sent by an appropriate way, so that information can be interpreted in a correct manner. The outputs are sent via the serial port which is communicated with the controller. When the button "Move" is clicked, a process will take place, as shown in Figure 15. With this action, the robotic arm will change its position according to the input values. In addition, it has a standby button that stops the communication controller.

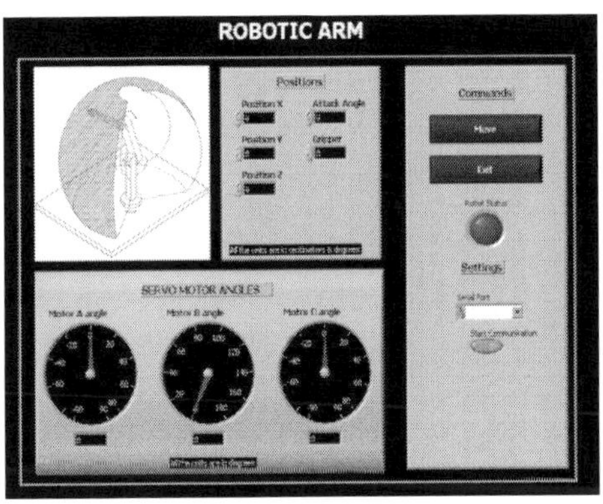

Figure 14. Labview user interface.

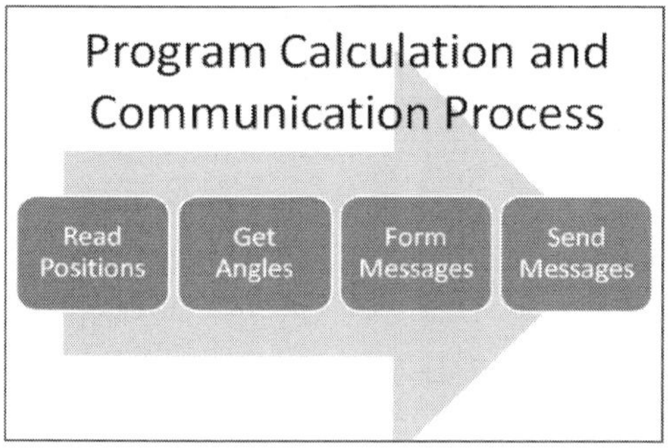

Figure 15. Program process.

The main advantages of this approach are that it uses an efficient way of moving and offers further capabilities that could be implemented, such as position and sequence learning. A disadvantage, on the other hand, is that the possible positions that have valid angles after the inverse kinematics calculations are very limited because the servo motors have a restraint of 180°.

Manual Control

This type of control is an extra option for our system that useful in specific positions. In case of mandatory positions that the inverse kinematics mode cannot calculate their valid angles, we may use the manual control instead. Basically, manual control consists of a series of analog inputs, such as potentiometers, that are connected with the microcontroller which will interpret the values and send a command to the servo driver. In order to implement this, a control board, as shown in Figure 16, should be built to work as an interface with the user. Possible implementation includes a teaching feature where the microcontroller stores positions in memory and by a keypad or a series of switches we may recall these positions.

TESTING AND VALIDATION

Several tests were carried out to validate the robot arm and its components. The testes covered both the particular elements and the overall system, as shown in Figure 17. For the microcontroller, the tests are occurred by sending different commands by the software to the microcontroller and check changes on the output which was connected to a servo motor that turned on or off depending on the command.

The servo motors were tested afterwards by sending different direct pulses to each servomotor and verifying the response of moving to the right position. We used a mark to know where the initial position was and the final position of the motors is determined by sending a signal with the microcontroller and, in turn, it is interpreted by the servo and compared to the signal provided by the encoder, resulting in the rotation to the desired position. During this test, the servo motor was inconsistence with the robot arm system because of an incorrect polarization.

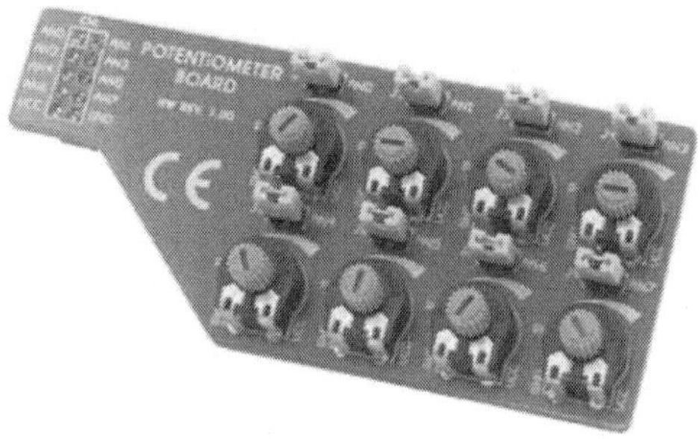

Figure 16. Potentiometer board.

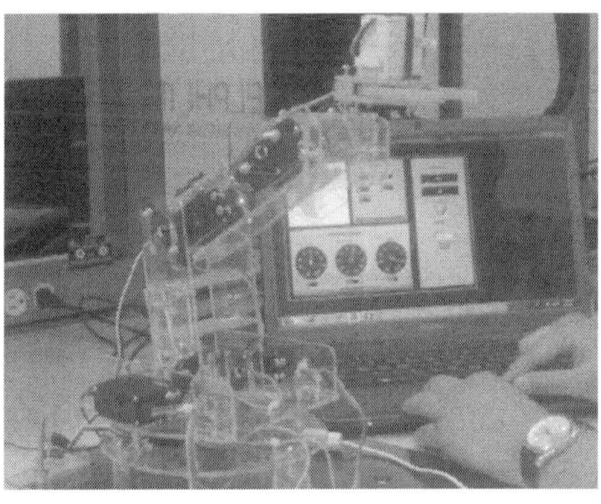

Figure 17. Robot arm tests.

The servo motor driver was also tested using the Lab view software to send commands to the microcontroller which sent the specific commands to the driver which had one motor connected to change the position according to the commend. It is important to notice that at the beginning of the project a different servo motor driver was selected but several problems related to the communication between them and the microcontroller were present. So we choose a driver that allows the data to be sent directly from the computer to it with only a USB wire, so the microcontroller would only be used in case of the implementation of manual control.

Other tests were performed to verify the functionality of the whole system, as shown in Figure 18.

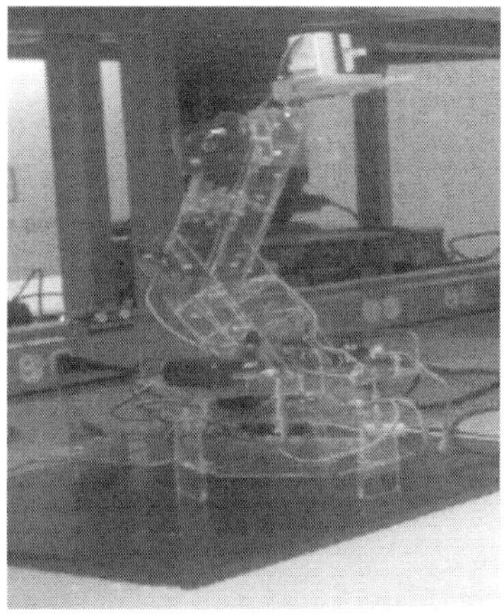

Figure 18. Robot arm in action.

Those tests were occurred by introducing a specific position in the Labview interface and measuring the distance between a reference point and the final point in order to verify: the correct transformation from inverse to direct kinematics, the relationship between the specified angles and the rotation of the motors.

Testing and validation of the robot arm is one of the tasks that require elongated time because several iterations are needed. During our tests, many problems arise as: wrong angle calculations, wrong calibration of the motors, problems with the physical angle and position measurements, and one of the servo motors burned because of an overload that wasn't expected.

RESULTS AND DISCUSSIONS

Results from the robot arm at different operating conditions are presented as follows:

Servo Motors Movement Range

The limits of the servo motors were obtained since specification of this type of motors contains that it has less than a 180 degree span. The real range for all motors was found to be in the range 125 142 degrees, as shown in Table

1. This clearly demonstrated that real operation of robot arm is different from the stander case.

Table 1. Motor angle ranges.

Motor	Angle Range
Motor A	130°
Motor B1	135°
Motor B2	140°
Motor C	142°
Motor Attack Angle	125°

Current Consumption

The current consumption depends on the load and the type of motion of the robotic arm. In the current study, there are 4 levels of current consumptions:

- Low (from 0 to 200 mA). This consumption takes place when the robot is at rest (not motion case).
- Normal (from 200 to 500 mA). This happened when the robot arm is moving with capability to go to the target without needs of great torque.
- High (from 500 mA to 900 mA). This range is reached at the beginning of carrying loads. By overcoming the initial moment of inertia for loads, the normal range takes a place.

Over current (more than 900 mA). The load is too heavy and the motor cannot move at all. For being under this condition for more than one minute, the motor will burn, i.e. it is not possible to be used any more.

Maximum Load

These results were obtained using different weights; a bag of corn was used with a scale to determine bag' weight. Results carried out by using the robot arm to pick up the bag and move it to specific positions. Table 2 presents the current consumption at different weights of bag of corn. From Table 2, it can be seen that the robot can move without problems at loads lower than 50 grams. At loads 60 grams, the robot arm start having difficulties and after passing 80 grams severe condition occurred where irreversible damage could be happened in motors.

Final Position

Results show the precision of the robot arm to move different weight (<50 grams) is presented in Table 3. As shown, the robot arm is able to perform the movement to the position specified. However, this movement is not smooth and sometimes the motors do not have enough force, especially when the load is heavy. In addition, some problems may appear due to synchronizing the two bottom motors. The steps of the two motors were not coincidental and that causes tension in the acrylic parts, which in case of being too much will break the parts.

Table 2. Load vs. current consumption.

Load	Current Consumption
20 grams	Low
40 grams	Normal
50 grams	Normal
60 grams	High
80 grams	Overcurrent
100 grams	Overcurrent

Table 3. Precision on all axis.

Axis	Precision (+/-)
x	1 cm
y	2 cm
z	1 cm

CONCLUSIONS

This paper presents the design, development and implementation of robot arm, which has the talent to accomplish simple tasks, such as light material handling. The robot arm was designed and built from acrylic material where servo motors were used to perform links between arms and execute arm movements. The servo motors include encoder so that no controller was implemented; however, the rotation range of the motor is less than 180° span, which greatly decreases the region reached by the arm and the possible positions. The design of the robot arm was limited to four degrees of freedom since this design allows most of the necessary movements and keeps the costs and the complexity of the robot competitively. The end effector is not included in the design because a commercially available gripper is used since it is much easier and economical to use a commercial one than build it.

During design, we faced some difficulties due to the way of joining thin acrylic parts strongly. A mechanical junction based on screws and nuts is used and in order to accomplish that, a small feature was designed which allowed fastening the bolts with the nuts without having to screw in the thin acrylic layer.

To control the robot arm, three approaches are implemented: a microcontroller, a driver, and a computer-based user interface. This system has unique characteristics that allow flexibility in programming and controlling method, which was implemented using inverse kinematics; besides it could also be implemented in a full manual mode. This robotic arm is contrast with others as being much cheaper than available robot arms, also it can be controlled all of its movements from a computer, using a Labview interface.

Several tests were carried out to validate the robot arm where the testes covered both the particular elements and the overall system; results at different operating conditions show trustful of the robot arm presented.

REFERENCES

1. Manipulating Industrial Robots—Vocabulary, International Organization for Standardization Standard 8373, 1994.
2. Industrial and Service Robots, IFR International Federation of Robotics, 2010. http://www.ifr.org/home
3. Case Studies and Profitability of Robot Investment, The IFR Statistical Department, 2008. http://www.ifrstat.org/downloads/2008_Pressinfo_english .pdf
4. R. J. Wang, J. W. Zhang, et al., "The Multiple-Function Intelligent Robotic Arms," FUZZ-IEEE Journal, Korea, 20-24 August 2009, pp. 1995-2000.
5. L. B. Duc, M. Syaifuddin, et al., "Designing 8 Degrees of Freedom Humanoid Robotic Arm," International Conference on Intelligent and Advanced Systems, Kuala Lumpur, 25-28 November 2007, pp. 1069-1074.
6. C. R. Carignan, G. G. Gefke and B. J. Roberts, "Intro to Space Mission Design: Space Robotics," Seminar of Space Robotics, University of Maryland, Baltimore, 26 March 2002.
7. Occupational Safety and Health Administration Technical Manual, OSHA 3167, United States Department of Labor, 1970.
8. B. Siciliano, L. Sciavicco, L. Villani and G. Oriolo, "Robotics, Modelling, Planning and Control," Springer, London, 2009.
9. M. P. Groover and M. Weiss, "Robotica Industrial, Technologia, Programacion y Aplicaciones," Mc-Graw Hills, Mexico D.F., 1989.

CHAPTER 6

An Adaptive Robust Approach to Modeling and Control of Flexible Arm Robots

Pedram Bagheri[1], Laleh Behjat[1], Qiao Sun[2]

[1]Department of Electrical and Computer Engineering, University of Calgary, Calgary, Alberta, Canada

[2]Department of Mechanical and Manufacturing Engineering, University of Calgary, Calgary, Alberta, Canada

ABSTRACT

In this paper, a novel adaptive robust approach to modeling and control of a class of flexible-arm robots subject to actuators unmodeled dynamics is proposed. It is shown how real-time signals measured from a dynamical

Citation:- Bagheri, P., Behjat, L. and Sun, Q. (2018), "An Adaptive Robust Approach to Modeling and Control of Flexible Arm Robots". World Journal of Engineering and Technology, 6, 282-295. doi: 10.4236/wjet.2018.62017.

system can be utilized to improve the accuracy of the mathematical model of flexible robots. Given the elasticity of the robot's arms, flexible manipulators have both passive and active degrees of freedom. A nonlinear robust controller is designed for the active degrees of freedom to enable the robot to follow desired trajectories in the presence of actuators unmodeled dynamics. Furthermore, it is shown that under some feasible conditions, another nonlinear robust controller is designed for the passive degrees of freedom. Moreover, to use the system response for model extraction, two auxiliary signals are proposed to provide sufficient information for improving the accuracy of the dynamics of the system numerically. Additionally, two adaptive laws are proposed in each case to update the two introduced auxiliary signals. As a result, the controller controls the passive degrees of freedom after the active degrees of freedom converge to their desired trajectories. Simultaneously, the information collected from the system to update the auxiliary signals enhances the model accuracy. In the end, simulation results are presented to verify the performance of the proposed controller.

Keywords: Adaptive Control, Robust Control, Vibration Control

INTRODUCTION

The area of flexible-arm robots has attracted much attention during the last few decades [1] [2]. This interest is due to the advantages that flexible arms offer compared to their rigid counterparts. Weight reduction, lower energy consumption, and faster system response are among several benefits utilized in their numerous applications such as space missions [3].

There has been a great number of studies coping with controlling flexible-arm robots, many of which investigate both theoretical and experimental aspects in this field [4] [5]. On grounds of the flexibility of the arms, along with a trajectory-tracking control problem, vibration control should be also considered for such systems to improve the control system performance [6]. As a result of the vibrations caused by the flexibility of the arms, designing controllers for such systems becomes a challenging task. There exist several research works in the literature addressing the flexibility of the arm. Passive control methods are one way to deal with the vibrations of elastic arms which require modification in physical parameters of the system structure [7]. Due to these structural modifications, absorption properties of the structure can be employed to increase damping properties of the arms. Active control approaches have also been widely used to control flexible systems, in which actuation moments and forces are applied to address the vibrations [8] [9]. However, the mere use of passive control methods to reduce vibrations does not seem to suffice. Merely using active control approaches is not sufficient as well. It is because vibration modes with frequencies near actuators frequencies can lead to instability. Hence, a combination of passive and active controllers can be employed as a suitable solution to vibrations reduction [10] [11]. The boundary feedback scheme is another way to dampen the vibrations of flexible manipulators [12]. Luo et al. [13] controlled the vibrations of a class of flexible robots using a shear feedback control method. Lyapunov-based control has also been vastly utilized to cope with the challenges associated with controlling flexible structures [14]. As another example of Lyapunov-based methods, Dadfarnia et al. [15] used the Lyapunov stability theory to develop a piezoelectric controller for flexible robots.

Infinite dimensionality is one of the most considerable challenges in modeling flexible robots. The existence of flexibility in the system results in dynamics governed by partial differential equations. Thus, techniques such as modal truncation are employed to express the dynamics by a set of ordinary differential equations [16]. As examples of such methods, Arts et al. [17] proposed an adaptive model integration method as a model reduction technique for planar flexible manipulators. Bruls et al. [18] used the global modal parameterization technique to reduce model-order of flexible multi-body dynamics. The procedure in both aforementioned methods is to divide the motion into two parts of rigid and elastic.

On account of the elasticity of the arms, there are degrees of freedom on which no actuation acts (Passive degrees of freedom). Consequently, flexible-arm robots fall into the category of underactuated mechanical systems. That is, systems with a lower number of control inputs compared to the number of degrees of freedom. Designing controllers for underactuated systems is an open problem. Control designs for such systems are dynamics-dependent while dynamical characteristics of the system play an essential role in the development of a control strategy. As an example of the existing control methods for underactuated systems, Mahmut Reyhanoglu et al. presented a theoretical scheme for modeling and control of underactuated systems. As one of the requirements of their approach, the non-integrable acceleration relations are required to be satisfied [19]. As another example, Zhang and Tarn designed a hybrid switching control strategy for nonlinear and underactuated mechanical systems [20].

Feedback linearization methods are wildly used for nonlinear systems. Nevertheless, As a result of the underactuated nature of flexible arms, the exact feedback linearization method [21] cannot be employed for such mechanical systems. Hence, the partial feedback linearization method [22] is considered as one the most suitable way to cope with underactuated systems. That is, the entire system can be either linearized with respect to the active degrees of freedom (collocated problem) or with respect to the passive degrees of freedom (non-collocated problem). Be that as it may, such a control strategy fails to handle uncertainties as control inputs to feedback-linearized systems depend on the governing equations.

In this paper, a novel adaptive robust nonlinear controller is designed for a class of multi-link flexible arms subject to uncertainties and unmodeled dynamics of the actuators. The contribution of the current research work is to utilize the system's real-time responses to improve the accuracy of the available mathematical model. Additionally, despite uncertainties and underactuated nature of the system, the proposed controller is able to track desired trajectories asymptotically. That is, despite the presence of uncertainties and disturbance sources, the tracking error converges to zero. To do such, first, an adaptive robust controller is designed for the active degrees of freedom. Meanwhile, adaptive laws are proposed to estimate the system uncertainties as well as approximating actuators uncertain dynamics. Furthermore, another adaptive robust controller is designed to control the passive degrees of freedom where the stabilities of the overall closed-loop system, in both cases, in the presence of uncertainties are established. In addition, due to the flexibility of the arms, the extraction of the system dynamics is demanding and, for the most part, should be done analytically. In the present approach, two adaptive signals are introduced and synthesized such that the cumbersome analytical part of the model extraction can be done numerically.

DYNAMIC EQUATIONS

In the current study, a multi-link flexible-arm robot is considered for modeling and control purposes. The investigated system is assumed to include m links containing l elastic modes. Therefore, the total degrees of freedom come to n=m+l.

Equations governing a flexible arm using the Euler-Bernoulli beam can be written as follows

$$\begin{pmatrix} M_{11} & M_{12} \\ M_{21} & M_{22} \end{pmatrix}\begin{pmatrix} \ddot{q}_1 \\ \ddot{q}_2 \end{pmatrix} + C(q,\dot{q}) + G(q) = \begin{pmatrix} 0 \\ T \end{pmatrix} \tag{1}$$

Where $M(q) \in \mathbb{R}^{n \times n} = \begin{pmatrix} M_{11} & M_{12} \\ M_{21} & M_{22} \end{pmatrix}$ is the positive definite inertial matrix; $C(q,\dot{q}) \in \mathbb{R}^n$ includes Coriolis and centrifugal terms; $G(q) \in \mathbb{R}^n$ contains gravitational terms; $q^{\mathrm{T}} = [q_1 \; q_2]^{\mathrm{T}}$ is the generalized coordinates vector where $q_1 \in \mathbb{R}^l$ denotes the passive degrees of freedom, and $q_2 \in \mathbb{R}^m$ represents the active degrees of freedom. $T \in \mathbb{R}^m$ is the torque generated by the actuators acting on the active degrees of freedom. $C(q,\dot{q})$ and G(q) can be calculated as [23]

$$C(q,\dot{q}) = \dot{M}(q)\dot{q} - \frac{1}{2}\frac{\partial\left(\dot{q}^{\mathrm{T}} M(q)\dot{q}\right)}{\partial q} \tag{2}$$

$$G(q) = \frac{\partial U}{\partial q} \tag{3}$$

where U denotes potential energy of the system. Extracting the dynamic equations of flexible robot arms is time consuming and can be analytically complicated. Due to its analytical nature, there exists a high chance of computational errors as the equations are derived. The extraction of the dynamic equations can be divided into two phases: 1) Extracting the portions of the dynamics that are straightforward and demand less calculations such as kinetic and potential energy. It is to be noted that the inertia matrix M(q) can be arrived at directly from the rearrangement of the kinetic energy of the system. 2) Obtaining the portions of the dynamics which require lengthy analytical calculations such as G(q) and $C(q,\dot{q})$. It is worth noting that calculating G(q) and $C(q,\dot{q})$ require calculus of variations which can be cumbersome for systems with a large degrees of freedom. The idea of this paper is to propose a way to procure G(q) and $C(q,\dot{q})$ numerically using the system responses. For this purpose, (1) is rewritten as

$$\begin{pmatrix} M_{11} & M_{12} \\ M_{21} & M_{22} \end{pmatrix}\begin{pmatrix} \ddot{q}_1 \\ \ddot{q}_2 \end{pmatrix} + \begin{pmatrix} h_1(q,\dot{q}) \\ h_2(q,\dot{q}) \end{pmatrix} = \begin{pmatrix} 0 \\ T + d(t) \end{pmatrix} \tag{4}$$

where

$$h_1(q,\dot{q}) = \dot{M}_{11}(q)\dot{q}_1 + \dot{M}_{12}(q)\dot{q}_2 - \frac{1}{2}\frac{\partial\left(\dot{q}^{\mathrm{T}} M(q)\dot{q}\right)}{\partial q_1} + \frac{\partial U}{\partial q_1} \tag{5}$$

$$h_2(q,\dot{q}) = \dot{M}_{21}(q)\dot{q}_1 + \dot{M}_{22}(q)\dot{q}_2 - \frac{1}{2}\frac{\partial\left(\dot{q}^{\mathrm{T}} M(q)\dot{q}\right)}{\partial q_2} + \frac{\partial U}{\partial q_2} \tag{6}$$

and d(t) is an unknown vector added to the equations to account for uncertain dynamics of the actuators. It is also assumed that $\|d(t)\| \leq \gamma$ where γ is an unknown constant.

In order to complete the dynamic equations given in (4), $h_1(q,\dot{q})$ and $h_2(q,\dot{q})$ should be computed. As it follows from (5) and (6), $h_1(q,\dot{q})$ and $h_2(q,\dot{q})$ are expressed in terms of the matrix M, system states, and its potential energy. Therefore, one can use the measurement of the states along with the kinetic and potential energy of the system and use numerical computations to obtain h_1 and h_2. Hence, the system dynamics is computed as it runs without going through analytical calculations. However, such dynamic equations are prone to numerical and measurement errors. Thus, a special measure should be taken to improve the accuracy of such computations. To compensate for such errors, two unknown auxiliary signals P and S are employed as follows

$$M_{11}\ddot{q}_1 + M_{12}\ddot{q}_2 + H_1 P = 0 \tag{7}$$

$$M_{21}\ddot{q}_1 + M_{22}\ddot{q}_2 + H_2 S = T \tag{8}$$

where $P \in \mathbb{R}^l$ and $S \in \mathbb{R}^m$ are two unknown auxiliary signals introduced to account for numerical and measurement errors; $H_1(q,\dot{q}) \in \mathbb{R}^{l\times l}, H_2(q,\dot{q}) \in \mathbb{R}^{m\times m}$ are two diagonal matrices constructed as $H_1(q,\dot{q}) = diag\left[h_1^2, h_1^2, \cdots, h_1^n\right]$ and $H_2(q,\dot{q}) = diag\left[h_2^2, h_2^2, \cdots, h_2^n\right]$; h_i^j denotes the jth element of the vector h_i where i=1,2 . Given the physical and geometric properties of flexible arm robots, it is not restrictive to assume that both H_1 and H_2, as defined above, have full ranks. Therefore, their column space can be utilized as a basis for the vectors h_1 and h_2 at any given time. Thus, for any exact value of h_1 and h_2 at any given time, there exist two unique vectors P and S such that $h_1=H_1P$ and $h_2=H_2S$. It is clear that the vectors P and S change in values over time. Hence, their values need to be updated according to the system response. Therefore, an adaptive procedure is developed for each of these vectors along with designing a controller in Section 3. Consequently, as the system runs, its response is used to update the values of P and S.

CONTROL DESIGN

In this section, an adaptive controller is designed for the considered system in the presence of uncertainties. Alongside the adaptive controller, two adaptive laws are derived for updating the values of P and S which are used to complete the dynamic equations. As it was shown in the previous section, (7) and (8) exhibit the dynamic equations governing an n-degree of freedom flexible-arm, with m active and l passive degrees of freedom. The first objective is to design a controller for the active degrees of freedom.

Control Design for Active Degrees of Freedom

Employing (7), (8) can be rewritten as follows

$$M_{21}\left(M_{11}^{-1}\left(-M_{12}\ddot{q}_2 - H_1 P\right)\right) + M_{22}\ddot{q}_2 + H_2 S = T + d \tag{9}$$

Two matrices N and R are defined as $R = M_{21}M_{11}^{-1}$ and $N = M_{22} - RM_{12}$. Consequently, (9) comes to

$$N\ddot{q}_2 + H_2 S - RH_1 P = T + d \tag{10}$$

As N is positive definite, (10) can be expressed as follows

$$\ddot{q}_2 = -\bar{H}_2 S + \bar{H}_1 P + \bar{U} + N^{-1}d \tag{11}$$

where $\bar{H}_2 = N^{-1}H_2$; $\bar{H}_1 = N^{-1}RH_1$, and $\bar{U} = N^{-1}T$. Tracking error and its dynamics can be defined as $e = q_2 - q_{2d}$ and $\dot{e} = \dot{q}_2 - \dot{q}_{2d}$, respectively; q_2d is the desired trajectory for the active degrees of freedom. A measure of the tracking error is also defined as

$$Q = \lambda_1 e + \dot{e} \tag{12}$$

where $\lambda_1 \in \mathbb{R}^{m\times m}$ is a positive definite matrix. The estimation errors of the uncertainty vectors are shown by $\tilde{P}$ and $\tilde{S}$ which are defined as $\tilde{S} = \bar{S} - S$ and $\tilde{P} = \bar{P} - P$; $\bar{S} \in \mathbb{R}^m$ and $\bar{P} \in \mathbb{R}^l$ are the estimations of the uncertainty vectors. Using the definitions of $\tilde{S}$ and $\tilde{P}$ (11) is rewritten as follows

$$\ddot{q}_2 = -\bar{H}_2\left(\bar{S} - \tilde{S}\right) + \bar{H}_1\left(\bar{P} - \tilde{P}\right) + \bar{U} + N^{-1}d \tag{13}$$

Considering (12), $\dot{Q}$ is calculated as follows

$$\dot{Q} = \lambda_1\left(\dot{q}_2 - \dot{q}_{2d}\right) - \bar{H}_2\left(\bar{S} - \tilde{S}\right) + \bar{H}_1\left(\bar{P} - \tilde{P}\right) + \bar{U} + N^{-1}d - \ddot{q}_{2d} \tag{14}$$

To proceed with the control design, the following Lyapunov function is proposed

$$V = \frac{1}{2}\left(Q^{\mathrm{T}}Q + \tilde{P}^{\mathrm{T}}\tilde{P} + \tilde{S}^{\mathrm{T}}\tilde{S} + \tilde{\gamma}^2\right) \tag{15}$$

where $\tilde{\gamma} = \hat{\gamma} - \gamma$ is the Euclidean norm of the estimation error of the uncertain dynamics of the actuators. Taking time-derivative of (15) gives

$$\begin{aligned} \dot{V} = {} & Q^{\mathrm{T}}\left(\lambda_1\dot{q}_2 - \lambda_1\dot{q}_{2d} - \bar{H}_2\left(\bar{S} - \tilde{S}\right) + \bar{H}_1\left(\bar{P} - \tilde{P}\right) + \bar{T} + u + N^{-1}d - \ddot{q}_{2d}\right) \\ & + \dot{\tilde{P}}^{\mathrm{T}}\tilde{P} + \dot{\tilde{S}}^{\mathrm{T}}\tilde{S} + \tilde{\gamma}\dot{\tilde{\gamma}} \end{aligned} \tag{16}$$

where $\bar{U}$ is decomposed into $\bar{U} = \bar{T} + u$. Given the structure of (16), the following control and adaptive laws are proposed

$$\bar{T} = -\psi Q - \lambda_1\left(\dot{q}_2 - \dot{q}_{2d}\right) + \bar{H}_2\bar{S} - \bar{H}_1\bar{P} + \ddot{q}_{2d} \tag{17}$$

$$\dot{\tilde{P}}^{\mathrm{T}} - Q^{\mathrm{T}}\bar{H}_1 = 0 \tag{18}$$

$$\dot{\tilde{S}}^{\mathrm{T}} + Q^{\mathrm{T}}\bar{H}_2 = 0 \tag{19}$$

where $\psi \in \mathbb{R}^{m\times m}$ is a positive definite matrix; (17) is the proposed control law, and (18)-(19) are the adaptive laws to update the two auxiliary signals introduced in (7)-(8). Substituting (17)-(19) into (16) yields

$$\dot{V} = -Q^{\mathrm{T}}\psi Q + Q^{\mathrm{T}}u + Q^{\mathrm{T}}N^{-1}d + \tilde{\gamma}\dot{\tilde{\gamma}} \tag{20}$$

(20) can be rewritten as follows

$$\dot{V} \le -\lambda_{\min}(\psi)\|Q\|^2 + \|Q\|\left\|N^{-1}\right\|(\hat{\gamma} - \tilde{\gamma}) + Q^{\mathrm{T}}u + \tilde{\gamma}\dot{\tilde{\gamma}} \tag{21}$$

where ||.|| stands for the Euclidean norm. Since N is a bounded positive definite matrix, the adaptive law to update $\tilde{\gamma}$ is proposed as

$$\dot{\tilde{\gamma}} = \|Q\|\left\|N^{-1}\right\| \tag{22}$$

Employing (22), (21) becomes

$$\dot{V} \le -\lambda_{\min}(\psi)\|Q\|^2 + \|Q\|\left\|N^{-1}\right\|\hat{\gamma} + Q^{\mathrm{T}}u \tag{23}$$

To establish the asymptotic stability of the closed-loop system, the second part of the control law is introduced as follows in (24). It is to be noted that the second part of the control law compensates for the actuators uncertain dynamics.

$$u = -\frac{\hat{\gamma}^2 \left\| N^{-1} \right\| Q}{\hat{\gamma} \left\| Q \right\| \left\| N^{-1} \right\| + \sigma(t)} \tag{24}$$

where $\sigma(t) \in L_1$ is an arbitrary signal where $\int_0^\infty \sigma(t)\mathrm{d}t < \infty$. By substituting (24) into (23), the following is obtained

$$\dot{V} \leq -\lambda_{\min}(\psi) \left\| Q \right\|^2 + \frac{\hat{\gamma} \left\| Q \right\| \left\| N^{-1} \right\| \sigma(t)}{\hat{\gamma} \left\| Q \right\| \left\| N^{-1} \right\| + \sigma(t)} \tag{25}$$

Since $\left\| Q \right\|, \left\| N^{-1} \right\|, \sigma(t)$ are positive, the following inequality is concluded

$$\frac{\hat{\gamma} \left\| Q \right\| \left\| N^{-1} \right\| \sigma(t)}{\hat{\gamma} \left\| Q \right\| \left\| N^{-1} \right\| + \sigma(t)} \leq \sigma(t) \tag{26}$$

Hence, (25) is simplified as follows

$$\dot{V} \leq -\lambda_{\min}(\psi) \left\| Q \right\|^2 + \sigma(t) \tag{27}$$

Taking integral from both sides of (27) gives

$$\lambda_{\min}(\psi) \int_0^t \left\| Q \right\|^2 \mathrm{d}\tau \leq V(0) - V(t) + \int_0^t \sigma(\tau)\mathrm{d}\tau \tag{28}$$

Considering the fact that $\sigma(t) \in L_1$, the following is achieved

$$\lambda_{\min}(\lambda_2) \int_0^\infty \left\| Q \right\|^2 \mathrm{d}t \leq V(0) - V(\infty) + a \tag{29}$$

where a is a positive constant. It is observed that the left side of (29) is non-negative, thus, it follows that (V V (0) − ∞( )) is bounded. As a result, it is deduced that the right side of (29) remains bounded as well. Consequently, QL∈$_2$. It is also easy to check that $\dot{Q} \in L_\infty$. Therefore, it is concluded from Barbalat's lemma that $\lim_{t\to\infty} Q(t) = 0$ which implies $\lim_{t\to\infty} e(t) = 0$. Thus, it was proven that adopting the extracted control laws (17), (24), and adaptive laws (18), (19)) and (22) guarantees the asymptotic stability of the closed-loop system. That is, the system is able to track desired trajectories asymptotically in the presence of model and actuators uncertainties.

Control Design for Passive Degrees of Freedom

In this section, a robust nonlinear controller is designed for the passive degrees of freedom alongside extracting adaptive laws to update uncertainties vectors. To do such, (8) is rewritten as

$$\ddot{q}_2 = M_{22}^{-1}\left(-M_{21}\ddot{q}_1 - H_2 S + T + d\right) \tag{30}$$

Substituting (30) into (7) yields

$$F\ddot{q}_1 - GH_2 S + h_1 P + GT + Gd = 0 \tag{31}$$

where $G = M_{12}M_{22}^{-1}$; $F = M_{11} - M_{12}M_{22}^{-1}M_{21}$. It follows from the structure of M that M_{22} is nonsingular. In addition, it can be easily shown that F is a positive definite matrix as well. Therefore (31) can be expressed as follows

$$\ddot{q}_1 = KS - DP - V - F^{-1}Gd \tag{32}$$

where $K = F^{-1}GH_2$, $D = F^{-1}H_1$, $J = F^{-1}GT$, and J is a control signal yet to be defined. The actual control input can be retrieved from J as follows

$$T = G^{-1}FJ \tag{33}$$

It is concluded from (33) that G should be nonsingular. Thus, according to the definition of G, M_{12} needs to be invertible as well. Hence, in order to extract a non-singular control law for the passive degrees of freedom, the following assumption is made:

Assumption 1 It is assumed that the number of passive degrees of freedom is less or equal than the number of active degrees of freedom.

The tracking error for the passive degrees of freedom and its dynamics are defined as $\bar{e} = q_1 - q_{1d}$ and $\dot{\bar{e}} = \dot{q}_1 - \dot{q}_{1d}$ where q_{1d} is the desired trajectory for the passive degrees of freedom. Moreover, a measure of the tracking error is defined as $\bar{Q} = \lambda_2\bar{e} + \dot{\bar{e}}$ where $\lambda_2 \in \mathbb{R}^{l\times l}$ is a positive definite matrix. Using the definitions of $\tilde{S}, \tilde{P}$ which are defined in the previous section, (32) is expressed as follows

$$\ddot{q}_1 = K\left(\bar{S} - \tilde{S}\right) - D\left(\bar{P} - \tilde{P}\right) - J - F^{-1}Gd \tag{34}$$

Based on the structure of (34), the following Lyapunov function is introduced

$$\bar{V} = \frac{1}{2}\left(\bar{Q}^{\mathrm{T}}\bar{Q} + \tilde{P}^{\mathrm{T}}\tilde{P} + \tilde{S}^{\mathrm{T}}\tilde{S} + \tilde{\gamma}^2\right) \tag{35}$$

The time-derivative of (35) is calculated as

$$\dot{V} = \bar{Q}^{\mathrm{T}}\left(\lambda_2\left(\dot{q}_1 - \dot{q}_{1d}\right) + K\bar{S} - D\bar{P} - J_1 - J_2\right) + \left(\dot{\tilde{P}}^{\mathrm{T}} + \bar{Q}^{\mathrm{T}}D\right)\tilde{P}$$
$$+\left(\dot{\tilde{S}}^{\mathrm{T}} - \bar{Q}^{\mathrm{T}}K\right)\tilde{S} + \tilde{\gamma}\dot{\tilde{\gamma}} - \bar{Q}^{\mathrm{T}}\left(F^{-1}Gd - \ddot{q}_{1d}\right) \tag{36}$$

where $J = J_1 + J_2$. Given (36), the following control and adaptive laws are proposed

$$J_1 = \bar{\psi}\bar{Q} + \lambda_2\left(\dot{q}_1 - \dot{q}_{1d}\right) + K\bar{S} - D\bar{P} - \ddot{q}_{1d} \tag{37}$$

$$\dot{\tilde{P}}^{\mathrm{T}} + \bar{Q}^{\mathrm{T}}D = 0 \tag{38}$$

$$\dot{\tilde{S}}^{\mathrm{T}} - \bar{Q}^{\mathrm{T}}K = 0 \tag{39}$$

Where $\bar{\psi} \in \mathbb{R}^{l \times l}$ is a positive definite matrix? Substituting (37)-(39) into (36) and taking the Euclidean norm form both sides lead to the following

$$\dot{V} \leq -\lambda_{\min}\left(\bar{\psi}\right)\left\|\bar{Q}\right\|^2 + \left\|\bar{Q}\right\|\left\|F^{-1}\right\|\left\|G\right\|(\hat{\gamma} - \tilde{\gamma}) - \bar{Q}^{\mathrm{T}}J_2 + \tilde{\gamma}\dot{\tilde{\gamma}} \tag{40}$$

To make (40) negative semi-definite, the following control and adaptive laws are proposed

$$J_2 = \frac{\hat{\gamma}^2\bar{Q}\left\|F^{-1}\right\|\left\|G\right\|}{\hat{\gamma}\left\|\bar{Q}\right\|\left\|F^{-1}\right\|\left\|G\right\| + \sigma(t)} \tag{41}$$

$$\dot{\hat{\gamma}} = \left\|\bar{Q}\right\|\left\|F^{-1}\right\|\left\|G\right\| \tag{42}$$

where σ(t) is a positive arbitrary signal belonging to L_1 . Substituting (40)-(42) into (40) gives

$$\dot{V} \leq -\lambda_{\min}\left(\psi\right)\left\|Q\right\|^2 + \sigma(t) \tag{43}$$

Similar to the stability analysis presented in the previous section, it is concluded from (43) that $\lim_{t\to\infty}\bar{Q}(t) = 0$, which implies that $\lim_{t\to\infty}\bar{e} = 0$. Hence, it is guaranteed that the tracking error converges to zero asymptotically in the presence of uncertainties.

SIMULATION RESULTS

In this section, a simulation is conducted to validate the analytical results. The simulated case study is a one-link flexible-arm robot with a payload at the tip of the arm. It is assumed that the system is subject to uncertainties and unmodeled dynamics in the actuators. The strategy is, first, to control the active degrees of freedom to their desired values. Then, the control law and adaptive laws switch to the non-collocated control to dampen the vibrations of the flexible arm. The physical properties of the investigated arm are as follows: Arm s Length = 1 m; Density = 7850 kg/m^3; Beams section area = 2e−4 m^2; Elasticity = 2.07e11; Load mass = 0.1 kg. The actual values uncertainty signals are S=8.2 and P=5.5.

It is shown in Figure 1 that the flexible arms hub angle is successfully controlled to the desired angle. It is also observed from Figure 1 that the hubs angle remains in its desired angle after the controller switches.

In Figure 2, the vibrations of the flexible arm are displayed, and its zoomed-in version is depicted in Figure 3. It follows from Figure 3 that the arm residual vibrations are dampened as the controller switches.

Figure 4 and Figure 5 correspond to the estimation of the introduced auxiliary signals to compensate for uncertainties. As noticed from the figures, the estimation of the uncertain parameters have converged to their limits after about two seconds through the simulation.

Finally, the required actuator torque to control the arm is shown in Figure 6. It follows from the simulations results that the arm tip can be controlled to its desired position more accurately. The reason stems from the fact that the non-collocated control helps decrease the response time. It is because that it would dramatically take a longer time for the residual vibrations to dampen naturally.It was shown both analytically, in Section 3, and through simulation that the proposed controller can achieve asymptotic tracking in the presence of unknown and bounded disturbance sources.

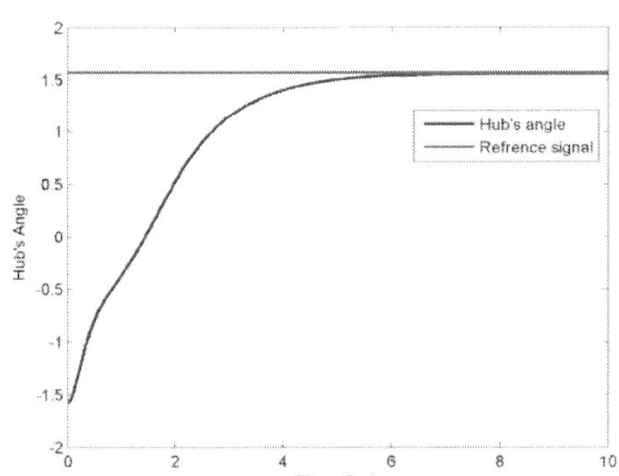

Figure 1. Arm's hub angle (rad).

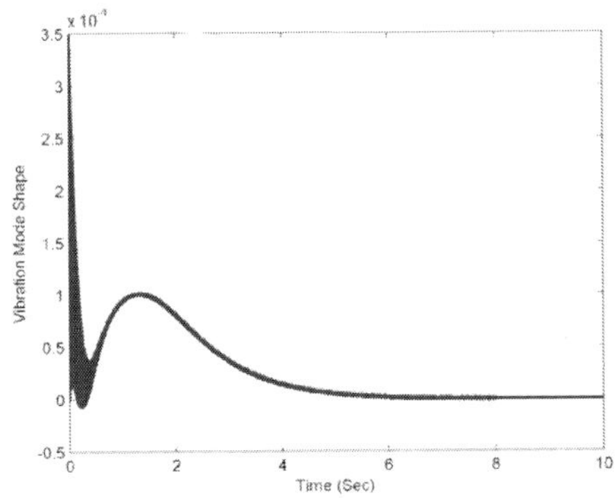

Figure 2. Flexible arm's vibrations (m).

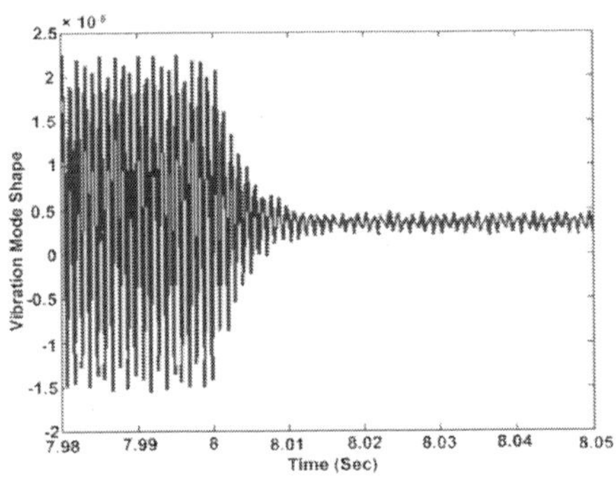

Figure 3. Zoomed-in version of the arm's vibrations.

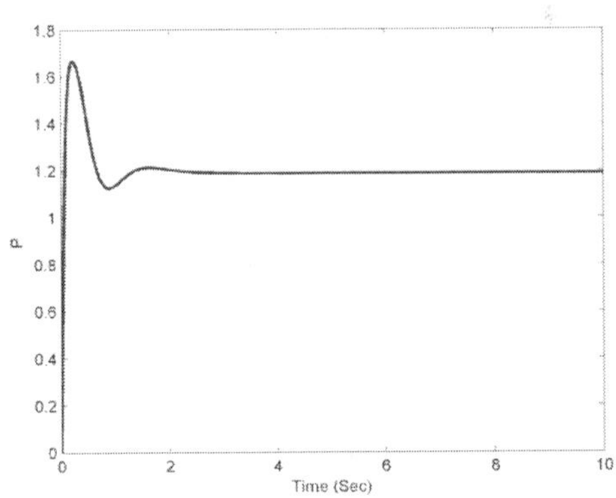

Figure 4. Auxiliary signal estimation (P).

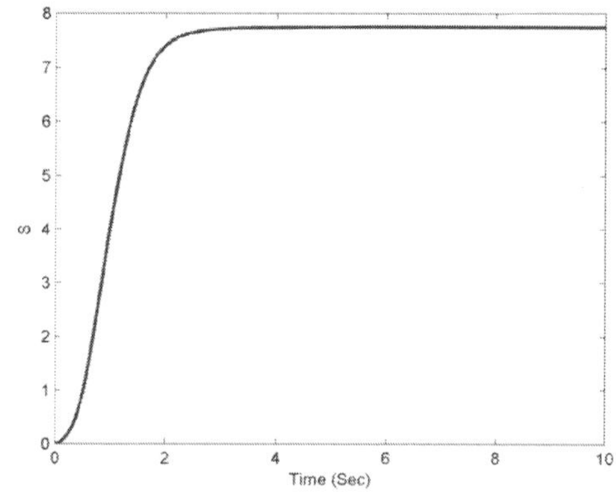

Figure 5. Auxiliary signal estimation (S).

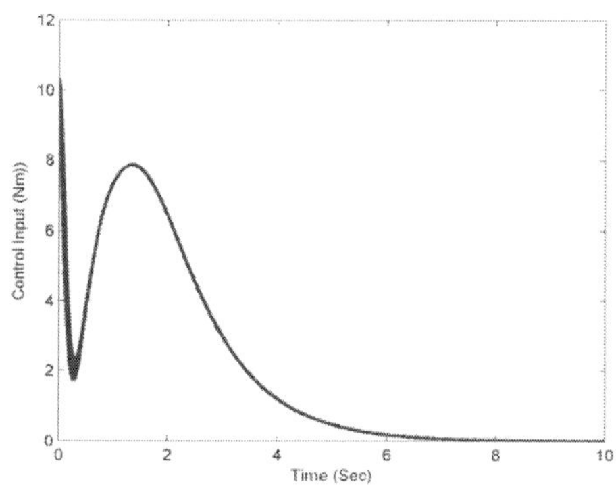

Figure 6. Control signal (Nm).

As a comparison to other recent robust approaches to the control problem of such flexible manipulators, authors in [24] linearize the dynamics to simplify the complexity of their model. However, the nonlinear dynamics of the system is considered in the present study, and the controller is designed accordingly. It is to be noted the asymptotic tracking in this study is established on the postulate that no prior knowledge of the bound of the disturbance is available. Be that as it may, existing methods in the literature merely guarantee the roundedness of control and state signals in case of unknown disturbance sources. That is, they cannot prove the output can converge to desired trajectories asymptotically. As an example, authors in [25] take a robust adaptive approach to control a flexible manipulator using the Lyapunov stability theory. Their analytical analysis suggests the roundedness of all signals in the presence of disturbance. Thus, given a desired trajectory, the output can stay in the vicinity of the trajectory. However, the convergence of the output to the desired trajectory is not guaranteed.

CONCLUSION

As presented in detail, an adaptive robust nonlinear controller was designed for a class of flexible-arm robots based on the Lyapunov stability theory. A new numerical technique was presented to facilitate the extraction of the governing equations of flexible-arm robots. It was shown how computational errors caused by numerical operations could be compensated and approximated as uncertainty signals. In the control design process, first, an adaptive controller was designed for the active degrees of freedom to enable the system to follow desired trajectories in the presence of uncertainties. Further, it was shown that under some feasible conditions, another adaptive robust controller could be designed for the passive degrees of freedom. Therefore, as the system was being controlled, its responses were utilized to improve the accuracy of the mathematical model. Thus, the presented method leads to improvements in the controller performance and increases tracking accuracy. Furthermore, due to a numerical approach to extracting the dynamics, the amount of analytical computations is reduced dramatically.

REFERENCES

1. Giorgio, I., Della Corte, A. and Del Vescovo, D. (2017) Modelling Exible Multi-Link Robots for Vibration Control: Numerical Simulations and Real-Time Experiments. Mathematics and Mechanics of Solids, 21, 1234-1246.
2. Kiang, C.T., Spowage, A. and Yoong, C.K. (2015) Review of Control and Sensor System of Exible Manipulator. Journal of Intelligent & Robotic Systems, 77, 187-213. https://doi.org/10.1007/s10846-014-0071-4
3. Dwivedy, S.K. and Eberhard, P. (2006) Dynamic Analysis of Exible Manipulators, a Literature Review. Mechanism and Machine Theory, 41, 749-777. https://doi.org/10.1016/j.mechmachtheory.2006.01.014
4. He, W., Ouyang, Y. and Hong, J. (2017) Vibration Control of a Exible Robotic Manipulator in the Presence of Input Deadzone. IEEE Transactions on Industrial Informatics, 13, 48-59. https://doi.org/10.1109/TII.2016.2608739
5. Kumar, A., Pathak, P.M. and Sukavanam, N. (2011) Reduced Model Based Control of Two Link Exible Space Robot. Intelligent Control and Automation, 2, 112-120. https://doi.org/10.4236/ica.2011.22013
6. Nani, V.-M. and Cires, I. (2014) Vibration Control for Two Air Compressor Units Used in Tandem. World Journal of Engineering and Technology, 2, 314-321. https://doi.org/10.4236/wjet.2014.24033
7. Sales, T., Rade, D. and De Souza, L. (2013) Passive Vibration Control of Exible Spacecraft Using Shunted Piezoelectric Transducers. Aerospace Science and Technology, 29, 403-412. https://doi.org/10.1016/j.ast.2013.05.001
8. Bai, K., Luo, M., Li, T., Wu, J., Yang, L., Liu, M. and Jiang, G. (2016) Active Vibration Adaptive Fuzzy Backstepping Control of a 7-Dof Dual-Arm of Humanoid Robot with Input Saturation. Journal of Intelligent & Fuzzy Systems, 31, 2949-2957. https://doi.org/10.3233/JIFS-169179
9. Zhang, Q., Li, C., Zhang, J. and Zhang, J. (2017) Smooth Adaptive Sliding Mode Vibration Control of a Exible Parallel Manipulator with Multiple Smart Linkages in Modal Space. Journal of Sound and Vibration, 411, 1-19. https://doi.org/10.1016/j.jsv.2017.08.052
10. Mottershead, J.E. and Ram, Y.M. (2006) Inverse Eigenvalue Problems in Vibration Absorption: Passive Modification and Active Control.

Mechanical Systems and Signal Processing, 20, 5-44. https://doi.org/10.1016/j.ymssp.2005.05.006

11. Svensson, J.L., Andersson, P.B. and Kropp, W. (2010) On the Design of Structural Junctions for the Purpose of Hybrid Passive Active Vibration Control. Journal of Sound and Vibration, 329, 1274-1288. https://doi.org/10.1016/j.jsv.2009.11.002
12. Coleman, M. and McSweeney, L. (2004) Analysis and Computation of the Vibration Spectrum of the Cartesian Exible Manipulator. Journal of Sound and Vibration, 274, 445-454. https://doi.org/10.1016/j.jsv.2003.09.036
13. Luo, Z.-H., Kitamura, N. and Guo, B.-Z. (1995) Shear Force Feedback Control of Exible Robot Arms. IEEE Transactions on Robotics and Automation, 11, 760-765. https://doi.org/10.1109/70.466607
14. Herman, P. (2009) Strict Lyapunov Function for Sliding Mode Control of Manipulator Using Quasi-Velocities. Mechanics Research Communications, 36, 169-174. https://doi.org/10.1016/j.mechrescom.2008.09.010
15. Dadfarnia, M., Jalili, N., Xian, B. and Dawson, D.M. (2004) A Lyapunovbased Piezoelectric Controller for Exible Cartesian Robot Manipulators. Journal of Dynamic Systems, Measurement, and Control, 126, 347-358. https://doi.org/10.1115/1.1767854
16. Liu, X., Schauer, T., Goldack, A. and Schlaich, M. (2017) Multi-Modal Active Vibration Control of a Lightweight Stress-Ribbon Footbridge Based on Subspace Identication. IFAC-Papers Online, 50, 7058-7063. https://doi.org/10.1016/j.ifacol.2017.08.1352
17. Aarts, R.G. and Jonker, J.B. (2002) Dynamic Simulation of Planar Exible Link Manipulators Using Adaptive Modal Integration. Multibody System Dynamics, 7, 31-50. https://doi.org/10.1023/A:1015271000518
18. Bruls, O., Duysinx, P. and Golinval, J.-C. (2007) The Global Modal Parameterization for Non-Linear Model-Order Reduction in Exible Multibody Dynamics. International Journal for Numerical Methods in Engineering, 69, 948-977. https://doi.org/10.1002/nme.1795
19. Reyhanoglu, M., van der Schaft, A., McClamroch, N.H. and Kolmanovsky, I. (1999) Dynamics and Control of a Class of Underactuated Mechanical Systems. IEEE Transactions on Automatic Control, 44, 1663-1671. https://doi.org/10.1109/9.788533
20. Zhang, M. and Tarn, T.-J. (2003) A Hybrid Switching Control

Strategy for Nonlinear and Underactuated Mechanical Systems. IEEE Transactions on Automatic Control, 48, 1777-1782. https://doi.org/10.1109/TAC.2003.817911

21. Chen, J.L. and Chang, W.-D. (2009) Feedback Linearization Control of a Twolink Robot Using a Multi-Crossover Genetic Algorithm. Expert Systems with Applications, 36, 4154-4159. https://doi.org/10.1016/j.eswa.2008.01.048
22. Arisoy, A., Gokasan, M. and Bogosyan, O. (2006) Tip Position Control of a Exible-Link Arm. 9th IEEE International Workshop on Advanced Motion Control, 445-450.
23. Olfati-Saber, R. (2001) Nonlinear Control of Underactuated Mechanical Systems with Application to Robotics and Aerospace Vehicles. PhD Dissertation, Massachusetts Institute of Technology.
24. Mohamed, Z., Khairudin, M., Husain, A. and Subudhi, B. (2016) Linear Matrix Inequality-Based Robust Proportional Derivative Control of a Twolink Exible Manipulator. Journal of Vibration and Control, 22, 1244-1256. https://doi.org/10.1177/1077546314536427
25. He, X., He, W. and Sun, C. (2017) Robust Adaptive Vibration Control for An uncertain Exible Timoshenko Robotic Manipulator with Input and Output Constraints. International Journal of Systems Science, 48, 2860-2870. https://doi.org/10.1080/00207721.2017.1360963

CHAPTER 7

Interactive Heuristic D* Path Planning Solution Based on PSO for Two-Link Robotic Arm in Dynamic Environment

Firas A. Raheem, Umniah I. Hameed

Control and Systems Department, University of Technology, Baghdad, Iraq

ABSTRACT

This paper is devoted to find an intelligent and safe path for two-link robotic arm in dynamic environment. This paper focuses on computational part of motion planning in completely changing dynamic environment at every motion sample domains, since the local minima and sharp edges are the most common problems in all path planning algorithms. In addition, finding a path solution in a dynamic environment represents a challenge for the robotics

Citation:- Raheem, F. and Hameed, U. (2019), "Interactive Heuristic D* Path Planning Solution Based on PSO for Two-Link Robotic Arm in Dynamic Environment". World Journal of Engineering and Technology, 7, 80-99. doi: 10.4236/wjet.2019.71005.

researchers, so in this paper, a proposed mixing approach was suggested to overcome all these obstructions. The proposed approach methodology for obtaining robot interactive path planning solution in known dynamic environment utilizes the use of modified heuristic D-star (D*) algorithm based on the full free Cartesian space analysis at each motion sample with the Particle Swarm Optimization (PSO) technique. Also, a modification on the D* algorithm has been done to match the dynamic environment requirements by adding stop and return backward cases which is not included in the original D* algorithm theory. The resultant interactive path solution was computed by taking into consideration the time and position changes of the moving obstacles. Furthermore, to insure the enhancement of the final path length optimality, the PSO technique was used. The simulation results are given to show the effectiveness of the proposed method.

Keywords: D* Algorithm, Particle Swarm Optimization (PSO), Path Planning, Two-Link Arm, Known Dynamic Environment

INTRODUCTION

The words "path planning" refer to find a path solution in robotics where the process is totally or sub automated. It is used to indicate the type of computational process for moving a robot from one place to another with respect to obstacles. In other words, it represents the search for initial feasible path which is considered as the first step to solve the path planning problem and includes finding a path for a robot that must move from initial given start point to the goal point which is also given as the destination position. The environment usually contains a fully de-fined set of obstacles. Therefore, the robot does not collide with any of the obstacles [1]. Generally, obstacles cannot always be static. When the obstacle changes its location over time, it called dynamic. Dynamic environment represents a revolution and important side in modern automation. It redounds in many important diverse applications like air and sea traffic control negotiating, freeway traffic, intelligent vehicles, automated assembly, and automated wheelchairs. Because of existence the moving objects, the robot needs to make decision quickly for avoiding possible collisions [2] [3]. There are three types of dynamic environment. The first type is called a known dynamic environment when all the information regarding the obstacles motions, sizes, shapes, and locations is completely known. The second type is called a partially known dynamic environment when not all information about the obstacles exists at the planning time. In this state, it needs to calculate the robot motion according to insufficient information about the environment. It is important to know when the robot's mission is to complete a map of the environment by scanning the area to make sure if there is a new object found. Finally, when there is nothing to know about the obstacles completely, it's called the totally unknown and dynamic environment and this needs some intelligent methods for sensing and making a proper decision [3].

Due to the rapid increase in the use of robots in different life and industry applications, path planning in dynamic environment is still considered as an actual challenge and needed to be solved interactively for both mobile and manipulator robots. The proposed method in this paper ensures finding an intelligent, proper, safe and relatively optimal path depending on the dynamic environment complexity. The structure of this paper contains seven sections in addition to the introduction, Related Research work, two-link robot arm kinematics, free Cartesian space analysis, proposed method, simulation results, and conclusions.

RELATED RESEARCH WORK

To overcome the dynamic environment challenges, many path planning algorithms have been used in the literature. Here we started with Anthony Stentz [4] whom used The D* algorithm which is represented the dynamic A*, it aims to create a short path in real-time by incrementally reforming paths for the robot's state as a new information is found. The D* extension which concentrate the repairs of significantly reduction the whole time required to subsequent and calculate the initial path re-planning operations which this paper aims to, followed by Yueshi Shen and Knut Huper [5] who suggested using a method to calculate the optimal joint trajectories for multi-DOF manipulator robot performing constrained motion task. The path planning is completed in two stages, firstly converting the variational calculus problem into a finite dimensional optimization problem, using Newton's method to compute the joint trajectory's inter-mediate points. Secondly, interpolating these points to form the joint curve and adjust it to ensure that the motion constraint will be satisfied throughout the entire trajectory. These calculations are easier to apply for multi-DOF manipulators than the Tow-Link robot arm because multi-DOF manipulators are more flexible.

Many other path planning algorithms were applied for solving dynamic environment problems to find an optimal path for robots in an environment that is only partially known and continuously changing. A method was suggested for generating a collision-free near-optimal path for an autonomous mobile robot in a dynamic environment containing moving and static obstacles using the neural network and fuzzy logic with a genetic algorithm. The neural network has a limited data size therefore, it cannot solve path in bigger environment furthermore, the simulation results show that the method is efficient and gives near-optimal and not optimal path reaching the target position of the mobile robot [6]. An improved virtual force field (VFF) algorithm approach was proposed for the real-time robot path planning, where the environment is unknown and changing. An area ratio parameter with the consideration of the robot size and obstacles was used. Furthermore, a fuzzy control module was added to deal with the problem of obstacle avoidance in dynamic environments. Due to the local minimum problem and the higher computational complexity, the path could not reach optimality [7].

From the same point of view, an algorithm for specifying Cartesian constraints with a single and dual arm operation is proposed based on Rapidly-Exploring Random Tree (RRT). One parameter is used to be the common in all RRT based on planers is the range. The range represents a threshold for the samples drawn newly and their small distance values after that those will add to the tree. Range value will be affective on the planning time which means that it should be selected carefully since the higher values causing short time planning which is advantage but as a result, it also cause a tensive motions and the smaller values give a long time planning and it composed more points in the tree. Also, more checking will be needed for the collision that may occur in the space; therefore it needs extra collision checking [8].

Most of the existing literature in dynamic environment considers only the traditional path planning algorithms and there are only a few studies that consider the heuristic algorithms like Anirudh Vemula [9] whom uses a heuristic based graph search planning algorithm that uses adaptive dimensionality consider time dimension corresponding to region where potential dynamic obstacle collisions can occur. Their results show the approach can only model simple interactions and fail to generalize for complex crowded settings.

Inspired by the above discussion, the contribution of the proposed method in this paper can be summarized in the following points:

1) Multi free Cartesian space analysis has been computed at each motion sample to identify the feasible workspace area.
2) Modifying the heuristic D* algorithm by adding stop case and return backward case for taking into consideration the obstacles

position and time which does not used before. These modification cases not included in the original D* theory.

3) Finding a path solution utilizing the proposed method in this paper that use the modified heuristic D* algorithm based on the PSO optimization technique in known dynamic environment.

4) Applying the proposed method to simulate the path solution for the two-link robot arm as case study, which is more difficult than the mass-point case study usually used in path planning researches.

5) The resulting path solution from this proposed method can solve the problem of finding an optimal path in dangerous, harsh, toxic environments and inaccessible areas.

TWO-LINK ROBOT ARM KINEMATICS

Kinematics is the science of motion that treats the subject without regard to the forces that cause it. Two types of kinematics problems can be classified; the first one is the forward kinematics problem which is concerned with the relationship between the individual joints of the robot manipulator and the position and orientation of the tool or end-effector [10].

The second is the inverse kinematics problem of the serial manipulators has been studied for many decades. It is needed in the control of manipulators. Solving the inverse kinematics is computationally expensive and generally takes a very long time in the real-time control of manipulators. Tasks to be performed by a manipulator are in the Cartesian space, whereas actuators work in joint space. Cartesian space includes orientation matrix and a position vector. However, joint space is represented by joint angles. The conversion of the position and orientation of a manipulator end-effector from Cartesian space to joint space is called an inverse kinematics problem [11]. The planar manipulator is shown in Figure 1(a) to illustrate the 2-DOF. The planar manipulators have l1l1 and l2l2 as their link lengths and θ1θ1 and θ2θ2 represent joint angles with x and y as task coordinates.

The forward kinematics equations are shown below:

$$x = l_1 \cos\theta_1 \cos\theta_1 + l_2 \cos(\theta_1 + \theta_2) \quad (1)$$

$$y = l_1 \sin\theta_1 \sin\theta_1 + l_2 \sin(\theta_1 + \theta_2) \quad (2)$$

In 2-DOF planner manipulator, the Geometric Solution Approach of inverse kinematics is used for obtaining the joint variables for all the points in Cartesian space regardless their orientation. The calculations of θ1θ1 and θ2θ2 for the 2-DOF manipulator could be found by applying the following equations.

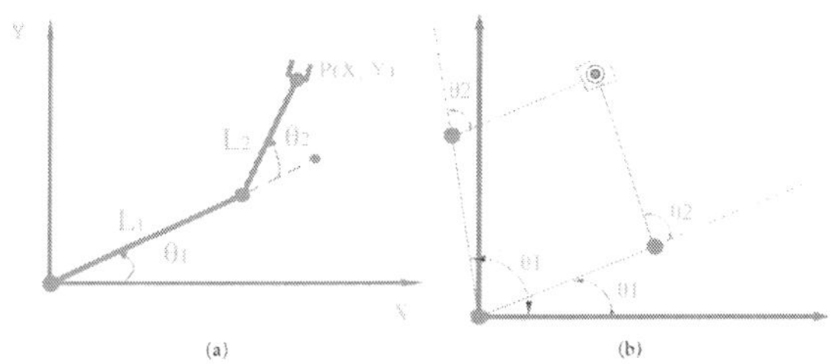

Figure 1. (a) Two degrees of freedom manipulator; (b) The configurations of two-link arm elbow up and elbow down.

Utilizing the cosines law, where the angle is given by:

$$\cos\theta_2 = \frac{p_x^2 + p_y^2 - l_1^2 - l_2^2}{2l_1l_2} = D \tag{3}$$

Since sin θ_2 is

$$\sin\theta_2 = \mp\sqrt{1-D^2} \tag{4}$$

To solve θ_2, it has to use the $\cos^{-1}$ function. Although it is more benefical to use tan−1tan−1 for numerical solution, in the software implementation, we will use the function $\tan^{-1}$ (b/c) which is ATAN2 (b, c). This function has a uniform accuracy over the range of its arguments, returns a unique value for the angle depending on the signs of b and c, and gives the correct solution if b and/or c is zero [11].

Subsequently, the two conceivable solutions for θ_2 can be written as:

$$\theta_2 = \tan^{-1}\frac{\mp\sqrt{1-D^2}}{D} \tag{5}$$

At last, θ1θ1 can be solved by:

$$\theta_1 = \tan^{-1}(y/x) - \tan^{-1}\frac{l_2\sin\theta_2}{l_1 + l_2\cos\theta_2} \tag{6}$$

where l_1 and l_2 are lengths of the first and the second links, respectively.

FREE CARTESIAN SPACE ANALYSIS

The result of inverse kinematics has been used to build and analyze the workspace in this thesis. The specific points that the end-effector can reach, can be defined as free Cartesian space (FCS). The angles (θ1 and θ_2) which were found by inverse kinematics are related to these specific points directly. The shape and volume of the FCS in an environment, which contains many obstacles, are changed depending on the number of obstacles, position and size in addition to the joints limits. In this paper, Table 1 shows the modeling and simulation. These restrictions impact and limit movement of manipulator and additionally split up the workspace into a reachable region and an inaccessible one.

Table 1. Theoritical range of the two-link robotic arm case study.

Link No.	Joint 1	Joint 2
Ranges	$0 \leq \theta 1\theta 1 \leq 360$	$- \quad 9 \quad 0$ $\leq \theta 2\theta 2 \leq 90$

To find all possible solutions, we should compute the FCS and that done was by analyzing the environment's points. The center point and the diameter of the obstacles are the basic issues upon which the checking function depends.

Moreover, there are three possible states for each point in the whole environment. Firstly and as shown in Figure 1(b), there are two solutions to each point, the elbow up solution and the elbow down solution.

But in the second state, there is just one solution, either elbow up or elbow down, and in the third one, there is no solution because the point lies out of the reachable area of the manipulator or on an obstacle, or the one or two links collide with an obstacle [12].

The resulted areas are separated into three parts, the first one contain the points which can be reached by the elbow up solution, and the second area contains the points which can be reached by the elbow down solution, and the third one represents the unreachable points that contain points outside the accessible range of manipulator workspace, points in obstacle's body and points that make collision with the obstacles.

The biggest factor that affects the free Cartesian is the length of the arm link [12].

PROPOSED METHOD: D* ALGORITHM BASED ON PSO

After applying the D* algorithm to find the non-interactive path solution in known dynamic environment and studying the produced path, it has been proposed to make the environment interactive considering the position and time of the obstacles to enhance the path planning. Employing the D* algorithm in the known obstacle position and time dynamic environment, to find the robot path needs to build a map with its limitation, start and target locations and the obstacles locations and dimensions. The proposed planning for the two-link robot arm in dynamic environment is shown in Figure 2. In the interactive path solution of known dynamic environment, in each move of the robot from node to node in each motion sample the free Cartesian space (FCS) will be calculated. As a result, the movement range of the robot will be defined and described accurately. It is a more intelligent behavior having multiple FCS with each motion sample the obstacles move. The new FCS, which was calculated, gives an indication to find the feasible area for the robot to move on. This process takes time but gives an accurate decision for the environment and high accuracy to the robot movement in the environment.

Since the process is off-line and after constructing the environment, the algorithm will start its work when the robot firstly stands on the target node and starts the initial expand to its 8-connected neighbor. Each node will have an initial cost function where the robot will choose the least cost function node to move on and put it in the closed list.

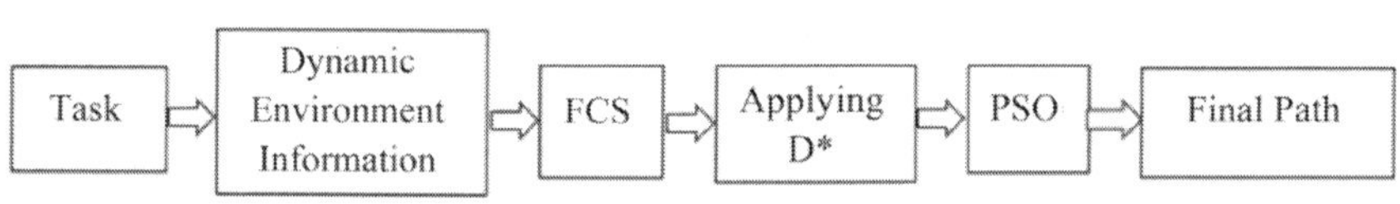

Figure 2. Proposed planning for the two-Link robotic arm in dynamic environment.

The other nodes will go to the open list with taking into consideration the obstacles having a huge cost function and sorting the initial locations in the closed list. With each motion samples, the closed list will be updated to catch the new obstacle location. This means, the closed list is changing dynamically over motion samples to save the obstacle nodes in the current motion sample only, and the process complete like that until reaching the start node. When the search is completed the robot will move from the start node with respect to the closed list instantaneously to have information about free node for moving on towards the target.

$$f = g + h \quad (7)$$

where f is objective function and g is the estimated cost from start point while h is the cost to the goal.

The velocity of the moving obstacle is calculated by:

$$v_x = \frac{\Delta x}{MS \times T_s}, \quad v_y = \frac{\Delta y}{MS \times T_s}$$

where Δx represents the change of distance in X-axis, Δy represents the change of distance in Y-axis and Ts represents the sample time which assumed to be 0.01 sec.

After finding the D* path, the PSO optimization technique is used for enhancing the path, removing sharp edges and making it the shortest because the D* path has stairs shape sometimes. The principle of PSO technique depends on generated specified number of particles in random positions in the specified workspace, also the velocity of those particles is nominated randomly, each particle has a memory that stores all the best position have been visited before, in addition to the fitness in that position which has been improved over time [13].

In each iteration of PSO technique, the pos_{ij} and $veli,_{jt}$ vectors of particle i is modified in each dimension j in order to lead the particle i toward either the personal best vector ($pbesti_{ij}$) or the swarm's best vector ($gbest_{ij}$)

$$vel_{i,j}^{t+1} = w \times vel_{i,j}^{t} + c_1 \times r_{1,j} \times \left(pbest_{ij} - pos_{ij}\right) + c_2 \times r_{2,j} \times \left(gbest_{ij} - pos_{ij}\right)$$

(8)

The position of each particle (the point that is chosen from D* path) is updated by using the new velocity for that particle, according to equation below:

$$pos_{ij}^{t+1} = pos_{ij}^{t} + vel_{ij}^{t+1}$$

(9)

where c_1 and c_2 are the cognitive coefficients ($c_1+c_2 \leq 4$), and $r_{1,j}$ and $r_{2,j}$ are random real numbers between [0, 1], the inertia weight w controls the particle momentum.

The value of vel_{ij} is clamped to the range $[-vel_{max}, vel_{max}]$ to reduce the probability of leaving the search space by the particle. If the search space is defined by the bounds $[-pos_{max}, pos_{max}]$, then the value of vel_{max} is typically set so that $vel_{max} = k * pos_{max}$, where $0.1 \leq k \leq 1.0$.

A large inertia weight (w) eases the local when its value is large and eases the global search when its value is small [14].

The first step to operate the PSO is to set the parameters of PSO ($c_1 = 1$, $c_2 = 3$, $w = 1$) which are found by a trial and error technique to reach the best results. Now, the D* path enters the PSO work, which will represent the D* path as the search space and choose the optimal points from it taking into account the general shape for the D* path and without distorting the original form. After determining the PSO search space which is the D* path matrix, the search process begins with defining the number of via points (the points that is chosen from D* path) that search for the new points according to the search space dimension and the PSO parameters. These via points are chosen randomly from the search area (D* path) and will represent the position parameters in the PSO equation.

The cubic spline interpolation equation has been used to connect the best chosen points from the PSO optimization technique to generate the corresponding smoothed path. The cost function of PSO, which is the path length, is computed. Then the elementary velocity is set, and updating the local and global parameters are the next step. This procedure is reiterated until the given points that are looked over the path are finished. The next step is updating the velocity and position equations, for converging the via points into the best cost, and also updating the local and global parameters. The procedure is repeated for the number of specified populations and the entire process is repeated for a number of specified iterations. The global values that construct the shortest path are then allocated. The entire process flowchart is represented in Figure 3. After getting the smoothed path, the robot will move also interactively with position and time on the new PSO path.

SIMULATION RESULT

First Environment

Here we have applied the two-Link robot arm as a case study. The arm length is equal to the half of the environment with 50 cm for each link and the arm has 360 degrees. The obstacle motion behavior is shown in Figure 4. Since the multi FCS is applied as clarified in Figure 5, where the column (5(a)) represents the elbow down FCS analysis at some selected motion samples; 1, 4, 5, 7, and 8, while in the column (5(b)) and (5(c)) represents the elbow up FCS analysis and the gathering between elbow up and elbow down FCS analysis respectively at the same motion samples. In each FCS figure the colored area (blue, red or green) is represented the free and safe area for moving and the white area is represented the unreachable area. Therefore, it has calculated the elbow up and elbow down FCS to enable the robot from making decision which free space is the suitable for moving.

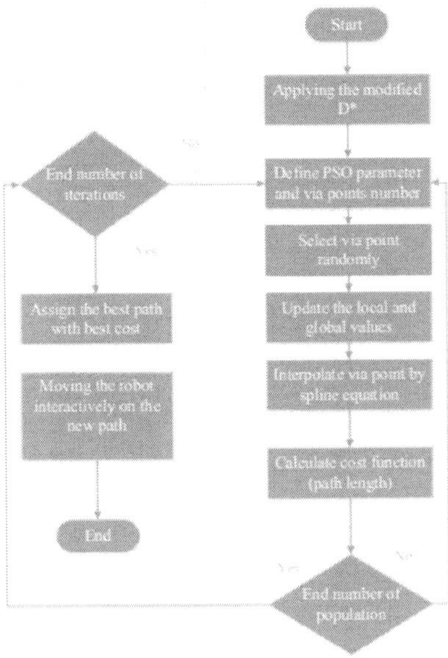

Figure 3. Proposed D* via Point flowchart for position and time consideration dynamic environment.

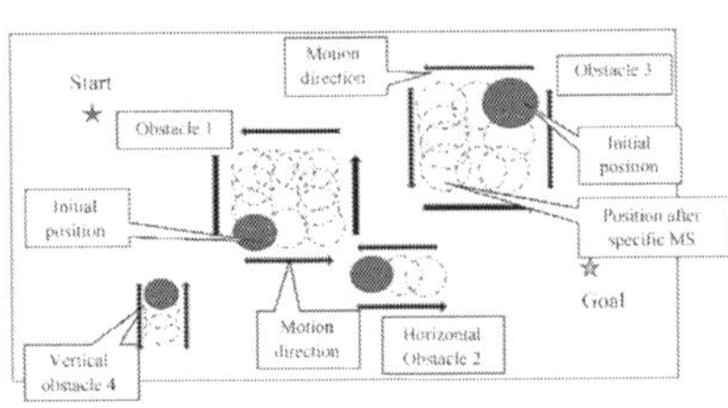

Figure 4. Obstacle motion behavior.

After computing these calculations the robot arm could move more freely because it could detect with each FCS the free area to move on. Figure 6shows the two-link robot arm D* path environment where part (a) represents the beginning of the planning while the end-effector is on the start point, part (b) is the arm location after 8 motion sample from planning process. At part (c) the arm is return backward at motion sample 15 to avoid collision, part (d) and

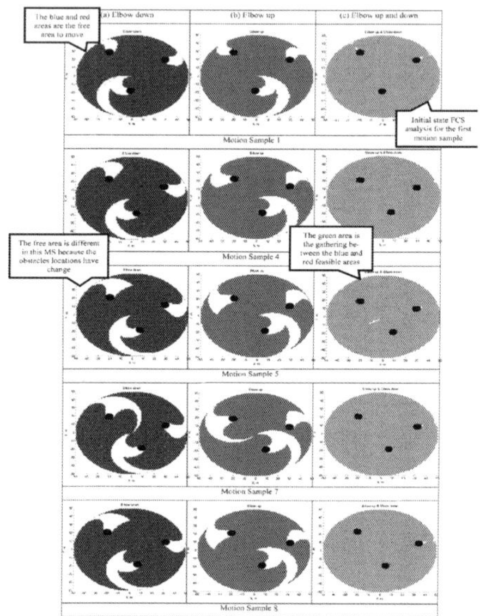

Figure 5. FCS for the D* path and PSO path where (a) represents elbow down, (b) represents elbow up and (c) represents gathering between elbow up and elbow down.

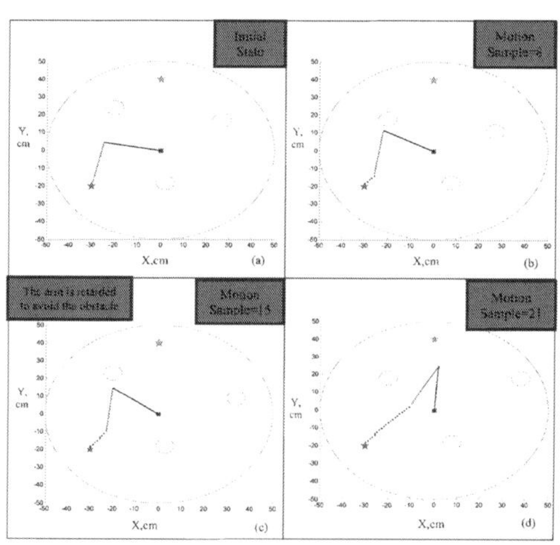

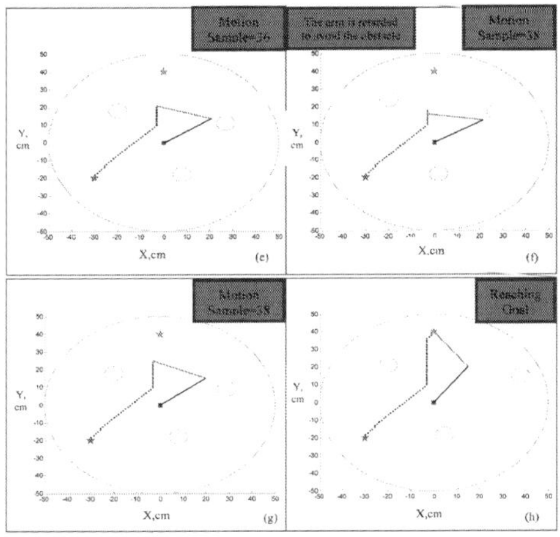

Figure 6. The Two-link robot arm result for the D* path.

(e) showed the planning after 21 and 36 motion sample respectively and after 38 motion sample the arm is returned backward to evade the obstacle then it reached the goal point safely as shown in part (g) and (h). Where the process is off-line, the PSO optimization technique is used to get rid of sharp edge and shorten the path as shown in Figure 7(a), where the blue line represents the D* path and the black line represents the PSO path. Also, Figure 7(b) clarified that the D* path based on PSO decreased by 3.2846 cm long from the original path. Figure 8 shows the D* based on PSO path planning process from the initial state in part (a) where the end-effector is on the start location to the part (f) where the process has completed by reaching goal location. At part (b) where the motion sample equal 7 the arm has restarted backward to safe itself from collision. In addition, Figure 9 and Figure 10 represent the two link robot arm joint angles and change in these angles for the optimized path respectively. The same procedure will be done for the second environment.

These environments featured the interactive style motion where the robot should make a decision and must be stopped in some situations if it is needed, in this case, the stop state and the backward motion guarantee a collision free path. Table 2 clarifies the environment information.

Second Environment

For more verification, a second more complicated environment with four moving obstacles in such different way in comparing with the first environment and more close to the robot links. Two from these obstacles move in a square motion behavior as in Figure 4. Since the multi FCS is applied as clarified in Figure 11, where the column (11(a)) represents the elbow down FCS analysis at some selected motion samples; 1, 5, 7, 8, and 11, while in the columns (11(b)) and (11(c)) represents the elbow up FCS analysis and the gathering between elbow up and elbow down FCS analysis respectively at the same motion samples.

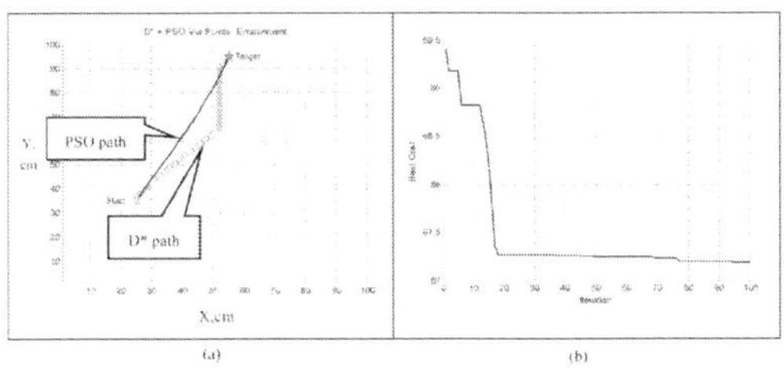

Figure 7. (a) Final D* and PSO path and (b) the path cost function.

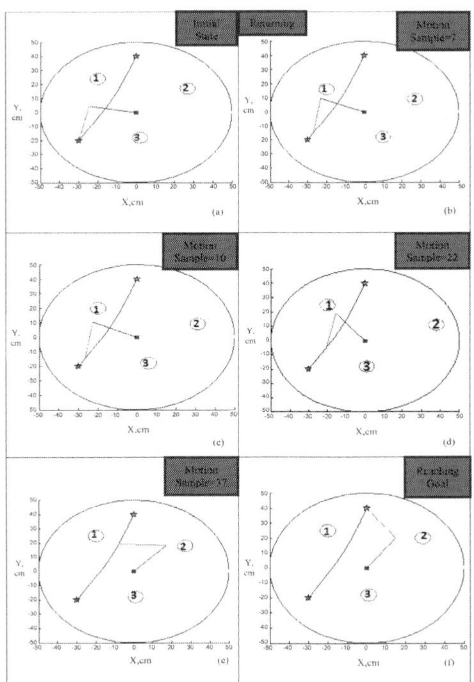

Figure 8. Two-link robot arm result for the D* based on PSO path.

In each FCS figures the colored area (blue, red or green) indicates the free and safe area for moving robot links while the white area indicates the unreachable area.

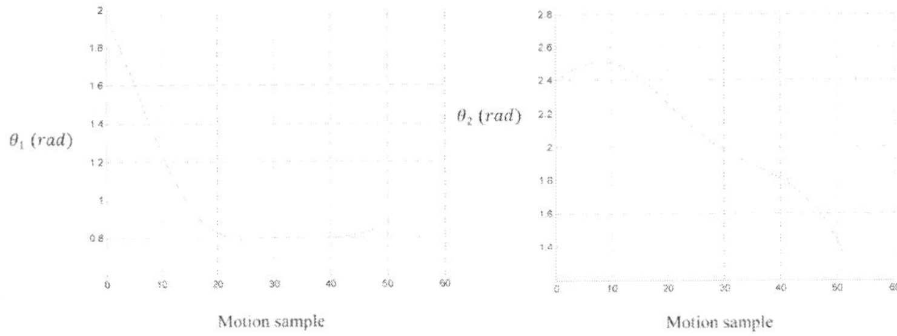

Figure 9. The two-link robot arm joint angles for the optimized path.

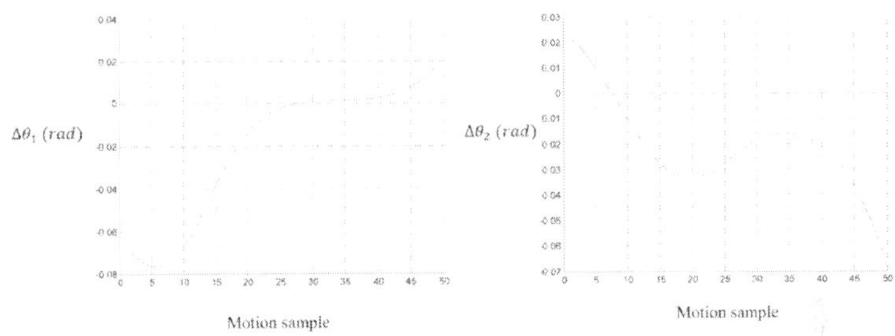

Figure 10. The change in joint angles for the optimized path.

Table 2. Environments information.

Environment	1	2
Obstacles number	3	4
Population	50	50
iteration	100	100
D* path	72.4264 cm	78.2132 cm
PSO path	69.6865 cm	73.8408 cm

Therefore, the robot can take a proper decision for selecting the next points in a free and intelligent path inside the permitted motion area. Figure 12 shows the two-link robot arm D* path environment where part (a) represents the beginning of the planning while the end-effector is on the start point, part (b) shows the arm location after 11 motion samples, part (c) indicates that the arm is returning backward case at the motion sample 21 to avoid collision, part (d) and

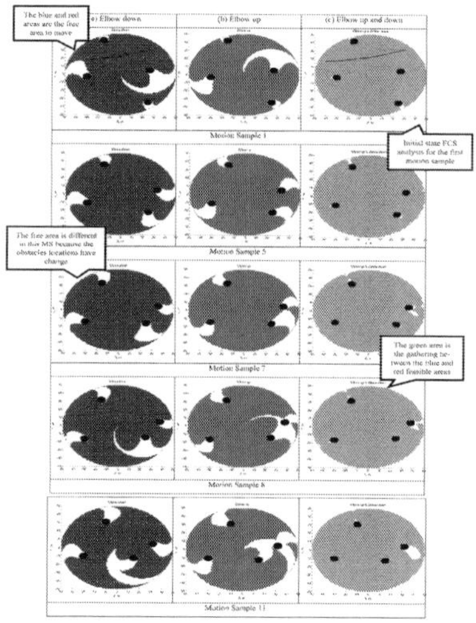

Figure 11. FCS for the D* path and PSO path for the second environment where (a) represents elbow down, (b) represents elbow up and (c) represents gathering between elbow up and elbow down.

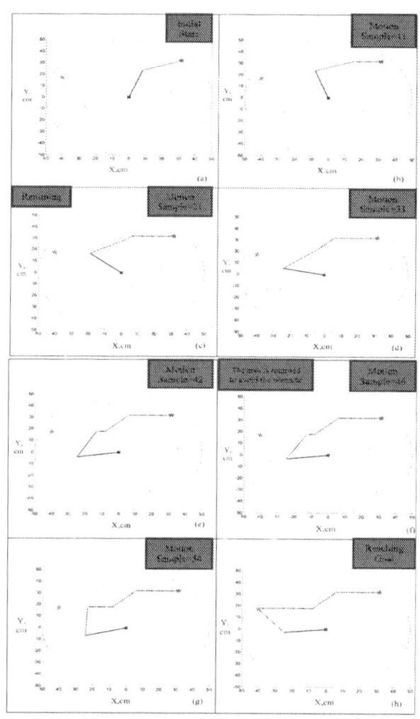

Figure 12. The two-link robot arm result for the D* path for the second environment.

part (e) show the planning after 33 and 42 motion samples respectively, while after 46 motion samples, the arm is returned backward again to evade the closest obstacle before reaching the goal point safely as shown in part (g) and (h).Because of the known position and time motion behavior of the obstacles, the process of the path planning can be made off-line, therefore the PSO optimization technique can plays an important role for finalizing this process and finding the best possible shortest and optimal path to get rid of the heuristic sharp path edges. Both of the D* and PSO paths are shown in Figure 13(a), where the blue line represents the D* path and the black line represents the PSO path. Also, Figure 13(b) represents the optimization cost function progress and proves that the final D* based on PSO path length was shorter by 4.3724 cm than the original path. Figure 14 confirms the path safety validation by showing clearly all the robot configurations during motion according to the optimized D* based on PSO path from the initial point as in part (a) to the goal point as in part (f). In addition, Figure 15 and Figure 16 represent the two link robot arm joint angles and the changes in these angles during the optimized path respectively. Also, Table 2 clarifies the environment information.In comparison with the nearest research literature [4] that try to solve the same problem for mobile robot using the D* heuristic method, our proposed method has different results because of our original modification that applied on D* algorithm which guarantees the interactivity and safety of the path and because of finalizing the path using PSO optimization technique.

CONCLUSION

This paper proposed an optimized heuristic approach for solving the problem of two-link robot arm path planning in known position and time changes dynamic environment. This approach utilizes the use of heuristic D* algorithm for finding the best path and the use of PSO optimization technique for getting the final optimal path.

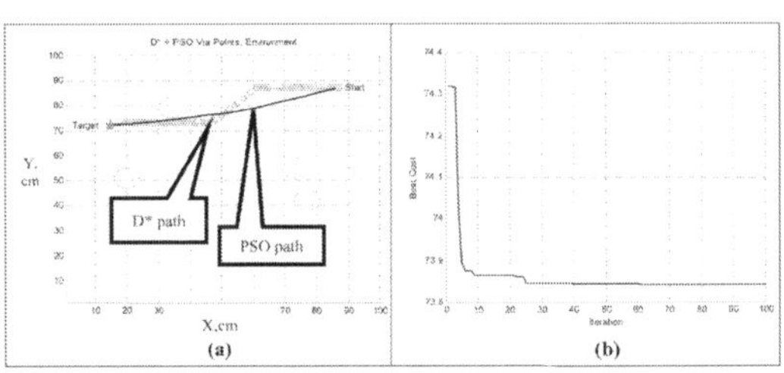

Figure 13. (a) Final D* and PSO path and (b) the path cost function for the second environment.

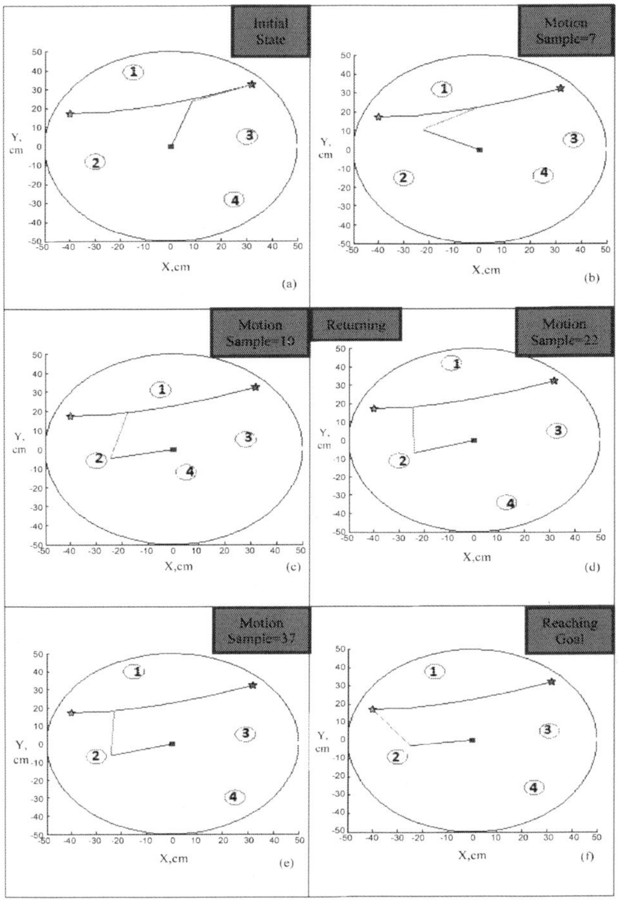

Figure 14. The two-link robot arm result for the D* based on PSO path for the second environment.

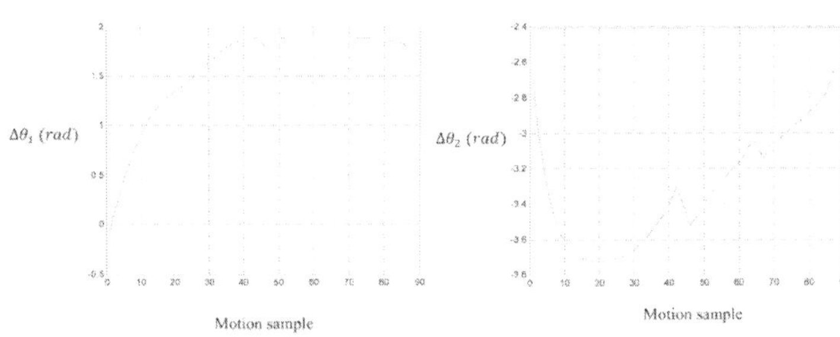

Figure 15. The two-link robot arm joint angles for the optimized path of the second environment.

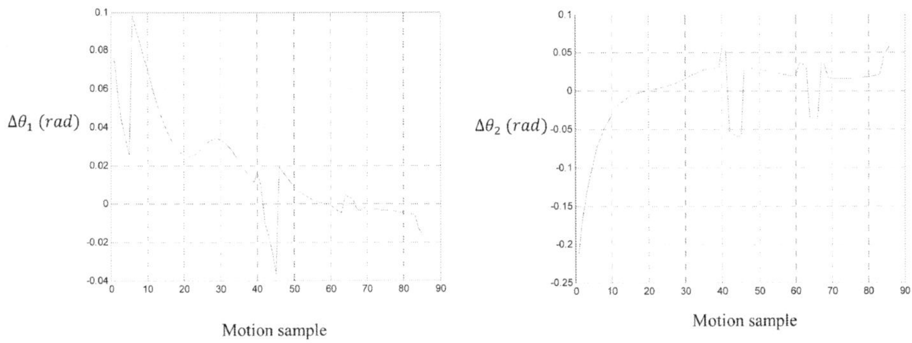

Figure 16. The change in joint angles for the optimized path of the second environment

Also, a modification on the D* algorithm has been done by adding a stop case and return backward case to ensure the path solution interactivity response to the obstacles position and time continuous changes during robot motion. From this approach, it was possible to get an intelligent path in spite of the environmental difficulties where a completely collision free path could be found at the expense of optimality. Therefore, the PSO optimization technique has been used to enhance the final path by removing the sharp edges until reaching path optimality. The result clearly proves the advantage of using this proposed methodology since the robotic arm has preserved the intelligence interactive behavior even after applying PSO Technique. The length of the final path was smart, smooth, safe, and optimal, and traditional D* algorithm could not find such a path. Our proposed method has a new approach of solution that leads to different results since it keeps the motion behavior of the robot to be intelligent and interactive with dynamic environment changes. The future directions for developing this proposed method include utilizing other intelligent methods and theories.]

Conflicts of Interest

The authors declare no conflicts of interest regarding the publication of this paper.

REFERENCES

1. Miao, H., Tian, Y.-C. and Feng ,Y.M. (2013) Robot Path Planning in Dynamic Environments using a Simulated Annealing Based Approach. Faculty of Science and Technology, Queensland University of Technology, Brisbane.
2. Raheem, F.A. and Bader, M.M. (2017) Development of Modified Path Planning Algorithm Using Artificial Potential Field (APF) Based on PSO for Factors Optimization. American Scientific Research Journal for Engineering, Technology, and Science, 37, 316-328.
3. Fujimura, K. (1991) Motion Planning in Dynamic Environments. Springer-Verlag, Tokyo. https://doi.org/10.1007/978-4-431-68165-6
4. Stentz, A. (1995) The Focused D* Algorithm for Real-Time Replanning. Proceedings of the International Joint Conference on Artificial Intelligence, Montreal, 20-25 August 1995, 1652-1659.
5. Shen, Y.S. and Huper, K. (2004) Optimal Joint Trajectory Planning for Manipulator Robot Performing Constrained Motion Tasks. Department of Information Engineering, Australian National University, National ICT Australia Limited, Canberra.
6. Khelchandra, T. and Huang, J. (2015) Robot Motion Planning with Neuro-Genetic- Fuzzy Approach in Dynamic Environment. International Journal of Information Technology, 21, 1-23.
7. Ni, J.J., Wu, W.B., Shen, J.R. and Fan, X.N. (2014) An Improved VFF Approach for Robot Path Planning in Unknown and Dynamic Environments. Mathematical Problems in Engineering, 2014, Article ID: 461237. https://doi.org/10.1155/2014/461237
8. Asadi, B. (2015) Single and Dual Arm Manipulator Motion Planning Library. IEEE/RSJ International Conference on Intelligent Robots and Systems (IROS), Hamburg, 28 September-2 October 2015.
9. Vemula, A. (2017) Safe and Efficient Navigation in Dynamic Environments. Carnegie Mellon University, Pittsburgh.
10. Spong, M.W., Hutchinson, S. and Vidyasagar, M. (2005) Robot Modeling and Control. 1st Edition, Wiley, Hoboken.
11. Craig, J.J. (2005) Introduction to Robotics Mechanics and Control. 3rd Edition, Pearson Prentice Hall Pearson Education, Inc., Upper Saddle River, NJ.
12. Sadiq, A.T., Raheem, F.A. and Abbas, N.A.F. (2017) Optimal Trajectory Planning of 2-DOF Robot Arm Using the Integration of PSO Based on

D* Algorithm and Cubic Polynomial Equation. The First International Conference for Engineering Researches, Baghdad, March 2017, 458-467.

13. Boonyaritdachochai, P., et al. (2010) Optimal Congestion Management in an Electricity Market Using Particle Swarm Optimization with Time-Varying Acceleration Coefficients. Computers & Mathematics with Applications, in press.
14. Poli, R., Kennedy, J. and Blackwell, T. (2007) Particle Swarm Optimization: An Overview. Swarm Intelligence, 1, 33-57.

CHAPTER

8

Optimal Task Placement of a Serial Robot Manipulator for Manipulability and Mechanical Power Optimization

Rogério Rodrigues dos Santos[1], Valder Steffen, Jr.[1], Sezimária de Fátima Pereira Saramago[2]

[1]School of Mechanical Engineering, Federal University of Uberlândia, Uberlândia, Brazil

[2]Faculty of Mathematics, Federal University of Uberlândia, Uberlândia, Brazil

ABSTRACT

Power consumption and accuracy are main aspects to be taken into account in the movement executed by high performance robots. The first aspect is

Citation:- R. Santos, V. Steffen and S. Saramago, "Optimal Task Placement of a Serial Robot Manipulator for Manipulability and Mechanical Power Optimization," Intelligent Information Management, Vol. 2 No. 9, 2010, pp. 512-525. doi: 10.4236/iim.2010.29061.

important from the economical point of view, while the second is requested to satisfy technical specifications. Aiming at increasing the robot performance, a strategy that maximizes the manipulator accuracy and minimizes the mechanical power consumption is considered in this work. The end-effector is constrained to follow a predefined path during the optimal task positioning. The proposed strategy defines a relation between mechanical power and manipulability as a key element of the manipulator analysis, establishing a performance index for a rigid body transformation. This transformation is used to compute the optimal task positioning through the optimization of a multicriteria objective function. Numerical simulations regarding a serial robot manipulator demonstrate the viability of the proposed methodology.

Keywords: Optimal task placement, Optimal robot path planning, Multicriteria optimization

INTRODUCTION

Minimization of production costs and maximization of productivity are some of the major objectives of industrial automation. In this scenario, serial robot manipulators have been proved to be very useful tools.

Due to the augmenting use of serial robot manipulators to perform a number of tasks in industry, requirements concerning higher precision, improved productivity, reduced costs, and better manufacturing quality become very important. To effectively explore all the potential of the robot, its path planning is a subject of major concern.

There are several works in the literature dedicated to different approaches concerning this subject. Regarding the off-line path planning approach, in Saramago [1] a solution of moving a robot manipulator with minimum cost along a specified geometric path in the presence of moving obstacles is presented. The optimal traveling time and the minimum mechanical energy of the actuators are considered together to form a multiobjective function to be minimized along the process.

In Santos [2] a strategy to determine the trajectory for a defined movement taking into account the requirements of torque, velocity, operation time and robot positioning along the movement is proposed. The analysis is performed as an off-line path planning through spline interpolation techniques in the manipulator joint space. In the paper of Chiddarwar [3] an off-line path planning for coordinated manipulation is proposed. The swept sphere volume technique is used to model multiple robots and static obstacles. In

Santos [4] the trajectory of two manipulators while manipulating a single object in a collaborative task and the object placement are written as an optimization problem. End-effector positioning and torque requirements are considered together in an optimal control formulation. Accuracy and energy consumption are improved during the path planning. Also, the flexibility effects of manipulators working in a vertical workspace are taken into account, and joint limits are considered in a box-constrained objective function to ensure the movement feasibility at the optimal configuration.

In von Stryk [5] an optimal control path-planning strategy for which the dynamics of the robot and the total traveling time are considered as objectives to be optimized is proposed. In Santos [6] an optimal control strategy that considers the presence of moving obstacles in the workspace is presented.

The problem of time-optimal control along a specified path has been investigated for several authors (Bobrow [7], Slotine [8]). In Constantinescu [9] a smooth time optimal strategy for serial robot manipulators is presented.

The paper of Seraji [10] addressed a geometric approach to determine the appropriate base location from which a robot can reach a target point. The work of Zeghloul [11] presented some kinematic performance criteria for the optimal placement of robots, and proposed a general optimization method to determine the placement of manipulators automatically.

In Feddema [12], an algorithm for determining the optimal placement of a robotic manipulator for minimum time coordinated motion is proposed. In Park [13] a study that characterizes a set of desired goal positions and a left-invariant distance metric parametrized by length scale is studied.

The paper of Chirikjian [14] proposed several metrics to be used to generate interpolated sequences of motions of a solid body. In Martinez [15] an analysis regarding the metric of a rigid displacement obtained from the kinematic mapping is presented and other metrics for the set of spatial and planar displacements are proposed. The work of Tabarah [16] proposed an optimization strategy for determining the optimal base for a given manipulator, and two cooperating manipulators, following a continuous path with constant velocity. They considered the constant velocity as a constraint. A strategy for velocity minimization is proposed by Zhang [17], based on a neural network solver. A kinematic planning scheme is reformulated as a quadratic problem that is solved by using a real-time algorithm, and applied to a Puma 560 robotic arm.

In Abdel-Malek [18] a strategy regarding the placement of a robot manipulator aiming at reaching a specified set of target points is described. The placement of a serial manipulator in a given workspace is achieved by defining the position and orientation of the manipulator's base with respect to a fixed reference frame. The strategy is based on characterizing the robot placement by adjusting a constrained cost function that represents the workspace to a set of target points.

With the aim of increasing productivity in the path following, industrial robotic applications have been addressed in the literature by determining path-constrained time-optimal motions, by taking into account the torque limits of the actuators. In these formulations, the joint actuator torques are considered as controlled inputs and the open loop control schemes result in bang-bang or bang-singular-bang controls (Bobrow [7], Chen [19], Shiller [20]). The paper of Xidias [21] proposed the generalization of the task scheduling problem for articulated robots in a constrained 2D environment. An algorithm for the optimum collision-free movement is proposed.

However, the increase of the speed may result (in some cases) in increasing the robot end-effector vibration or decreasing its accuracy. This is observed due to several factors, such as the bang-bang nature of the control, joint flexibility, joint friction, gear arrangement, or even a combination of them.

Vibration is not desired because it can both degrade the system performance and decrease the actuator lifetime. A related and negative effect of vibration is the poor accuracy of the system.

The present contribution discusses an alternative way to improve the overall performance of serial robot manipulators. The use of manipulability and mechanical power related indexes to achieve the best task placement is proposed.

As the basic idea of manipulability consists in describing directions in the task or joint space that maximizes the ratio between some measure of effort in joint space and a measure of performance in task space, the present contribution consists in the optimal positioning of the task to be performed, which results in the maximization of the robot manipulability and the minimization of the required mechanical power.

This will lead to accuracy maximization and power consumption minimization, since manipulability can be understood as an ability measurement (Fu [22]) and mechanical power is related to power consumption (Saramago [1]).

In the present paper both kinematics and dynamics features are considered in the proposed task placement strategy. From the best of the authors' knowledge, this problem was not previously addressed in the literature, despite its importance and applicability.

The position of the end-effector is fixed according to a Cartesian reference system, and described as a sequence of sucessive positions. The optimization process applies a translational and rotational transformation matrix with respect to a given reference frame. As a result, an optimal sequence of kinematical robot positions is determined.

The outline of the paper is the following. In Section 2, a review about manipulability is presented. In Section 3, the mechanical power concept is discussed.

A strategy to describe the task and its positioning in the design space is proposed in Section 4. Additionally in this section the design variables are defined. In Section 5 the multicriteria programming formulation is presented. In Section 6, the optimization strategy is established and the numerical experiments are performed in Section 7. Finally, the conclusions are drawn in Section 8.

MANIPULABILITY

Some factors should be taken into account when the posture of the manipulator in the workspace is determined for performing a given task during operation. An important factor is the ease of arbitrarily changing the position and orientation of the end-effector at the tip of the manipulator.

As an approach for evaluating quantitatively the ability of manipulators from the viewpoint of kinematics, the concepts of manipulability ellipsoid and manipulability measure (Yoshikawa [23]) are presented.

Consider a manipulator with n degrees of freedom. The joint variables are denoted by an n-dimensional vector, $q = [q_1, q_2, ..., q_n]^T$. An m-dimensional vector $r = [r_1, r_2, ..., r_m]^T$, $(m \leq n)$ describes the position and orientation of the end-effector.

The relation between the velocity vector v corresponding to r and the joint velocity $\dot{q}$ is given by:

$$\mathbf{v} = \mathbf{J(q)\dot{q}} \qquad (1)$$

where J(q) is the Jacobian matrix, computed from the kinematics description of the robot.

The set of all end-effector velocities v that are realizable by joint velocities such that the Euclidean norm of $\dot{\mathbf{q}}$ satisfies $\|\dot{\mathbf{q}}\| \leq 1$ defines an ellipsoid in the m-dimensional Euclidean space. In the direction of the major axis of the ellipsoid, the end-effector can move at high speed, and in the direction of the minor axis, the end-effector can move only at low speed. Also, the larger the ellipsoid is, the faster the end-effector can move.

Since this ellipsoid represents an ability of manipulation, it is called the manipulability ellipsoid.

The principal axes of the ellipsoid can be found by making use of the singular-value decomposition of J(q).

One of the representative measures for the ability of manipulation derived from the manipulability ellipsoid is the volume of the ellipsoid. Since this volume is proportional to the eigenvalues of the Jacobian, it can be seen as a representative measure.

Therefore, the manipulability measure, $f_{1,}$ for the manipulator configuration, q, is defined as

$$f_1 = \sqrt{\det(\mathbf{J}(\mathbf{q})\mathbf{J}(\mathbf{q})^T)} \quad (2)$$

Generally $f_1 \geq 0$ holds, and $f_1 = 0$ if and only if rank $J(q) < m$.

There is a direct relation between singular configuration and manipulability (through the Jacobian concept). According to this relationship, it can be shown that the larger the manipulability measure, the larger the ability of avoiding singular configurations.

MECHANICAL POWER

The consideration of the dynamics behavior of a serial robot manipulator is of great importance for its path planning. This information allows a detailed analysis and consequently the development of a precise control specification.

Many efficient schemes have been proposed to model the dynamics of rigid multibody mechanical systems (Richard [24], Mata [25], Mata [26]).

The dynamics model can be obtained explicitly through algebraically computation, or numerically through iterative computation.

The techniques based on the Newton-Euler method starts from the dynamics of all individual parts of the system. They look at the instantaneous or infinitesimal aspects of the motion, using vector quantities such as Cartesian velocities and forces.

Alternatively, the Euler-Lagrange based methods starts from the kinetic and potential energy of the entire system, by considering the states of the system during a finite time interval. This approach works with scalar quantities, namely the energies.

Independently of the approach used, at the end of the process the generalized forces are determined:

$$\mathbf{M(q)\ddot{q}} + \mathbf{c(\dot{q}, q)} + \mathbf{g(q)} = \tau \tag{3}$$

where $\mathbf{q}, \mathbf{\dot{q}}$ and $\mathbf{\ddot{q}}$ are the joint position, velocity and acceleration, respectively. M(q) is the joint space mass matrix and $\mathbf{c(\dot{q}, q)}$ is the vector of Coriolis and centrifugal forces. Vector g(q) is the vector of gravitational forces and τ is the generalized force vector.

The energy that is necessary to move the robot is an important design issue, because in real applications energy supply is limited and any energy reduction leads to smaller operational costs. Due to the relationship that exists between energy and force, the minimal energy can be estimated from the generalized force, $\tau_i(t)$, that is associated to each joint i at the time instant $t_0 \geq t \geq t_f$.

In the present contribution, for a manipulator with n degrees of freedom, the interval between the initial time t_0 and the final time t_f is represented by a set of N points and the computation is numerically carried out through a recursive Newton-Euler formulation (Craig [27]). Then, the mechanical power is used for design purposes as defined by:

$$f_2 = \sum_{j=1}^{n} \int_{t_0}^{t_f} \left[\tau_j(t)\ \dot{q}_j(t)\right]^2 dt \tag{4}$$

This expression is representative of the phenomenon under study because it considers both the kinematics and the dynamic aspects of the trajectory, simultaneously (Saramago [1]).

TASK SPECIFICATION

The task is specified as a set S of N Cartesian points (x, y, z) and the respective time step. Hence,

$$S = \left\{(t, x, y, z) \middle| t_0 \leq t \leq t_f, (x, y, z) \in \Re^3\right\}. \tag{5}$$

In the case of non-redundant manipulators there exist a finite number of configurations that satisfies the end-effector positioning specification. This path-following problem is a highly constrained task, and the inverse kinematics problem will have four solutions or less.

Considering the need of movement continuity together with a time-fixed specification at each step and a relative small step size, the point-to-point path planning strategy may fail due to the complexity of the constraints.

Therefore, a task positioning optimization is proposed as an alternative to increase the robot performance without changing the position of the task reference points with respect to each other. In other words, all the points will be moved simultaneously through a rigid body transformation to find the best position from the point of view of the ability of the manipulator.

To achieve this goal, the use of a homogeneous transformation A is proposed, which in this case is defined as

$$\mathbf{A} = \begin{bmatrix} \cos\beta & -\sin\beta & 0 & \delta_x \\ \sin\beta & \cos\beta & 0 & \delta_y \\ 0 & 0 & 1 & \delta_z \\ 0 & 0 & 0 & 1 \end{bmatrix} \quad (6)$$

where β is a rotation angle in the z axis (given in rad) and δ_x, δ_y and δ_z are translations along the reference axes (given in meters), respectively. If desired, it is possible to extend this formulation to consider the rotation with respect to the reference axes.

The physical meaning of considering a rotation angle with respect to x or y is that the robot's base plane will not be parallel to the ground's reference plane. In other words, the robot's z axis (its height) is not normal to the ground's reference plane (xy plane). As positioning two planes according to a precise angle may not be possible in a number of practical situations (even when it is possible, it is not an easy procedure to be done, in general terms), the rotation along the z axis only is considered, maintaining the robot's base plane parallel to the task's base plane.

After defining the set S that describes the task, the inverse kinematics with respect to each Cartesian point is computed.

The inverse kinematics computation can be achieved algebraically, by considering the analytical model of the manipulator and its geometry, or numerically. With the aim of presenting a general procedure, in the present contribution the inverse kinematics problem is solved numerically through an inverse problem formulation. This approach results an efficient and general procedure that works satisfactorily for manipulators exhibiting arbitrary complexity (Santos [28]).

The obtained path planning presents an end-effector positioning error, while the robot executes the path. The interest is focused on errors due to task positioning, as the inverse kinematics can generally be solved with a sufficient precision.

Once the objective is specified, it is necessary to define the optimization domain, i.e., the design space. As the design variables are defined in a theoretical space (defining a rigid body transformation) and the positioning constraints are defined in the Cartesian space (the workspace), a constraint parameter is included in the formulation.

Defining the theoretical positioning error at each point as

$$error_i = \begin{cases} 0 \text{ if } \mid P_{end} - T_0^n P_{base} \mid < \varepsilon \\ \mid P_{end} - T_0^n P_{base} \mid \text{ otherwise} \end{cases} \quad (7)$$

where P_{base}, P_{end} and T_0^n are the robot base point, the end-effector point, and the kinematics representation of the manipulator, respectively. The penalty parameter used as a constraint in the following optimization formulation is given by

$$s_E = \sum_{i=1}^{N} error_i \quad (8)$$

and represents the sum of all end-effector positioning errors during the movement.

Equation (7) means that the end-effector error is considered null when the accuracy is better than the one specified by ε (given in meters). Otherwise, this error corresponds to the distance from the end-effector specified position to the position obtained from the direct kinematics calculation.

In some robotic implementations it is necessary to plan the layout of the workspace, i.e., it is required to locate the robot base in such a way that dexterity is maximized at (or around) given targets. It is worth mentioning that such cases are also covered by the present formulation. Since the optimal design ((β, δ_x, δ_y, δ_z) in Equation (6)) means a coordinate transformation of the end-effector path regarding the fixed robot base, it is also possible to adjust the location of the robot base regarding the fixed end-effector reference through the relation ($-\beta$, $-\delta_x$, $-\delta_y$, $-\delta_z$).

MULTICRITERIA PROGRAMMING

In problems with multiple criteria, one deals with a design variable vector x, which satisfies all constraints and makes the scalar performance index that is calculated from the m components of an objective function vector f(x) as small as possible. This goal can be achieved by the vector optimization problem:

$$\min_{\mathbf{x}\in\Omega}\{\mathbf{f}(\mathbf{x}) \mid \mathbf{h}(\mathbf{x}) = \mathbf{0}, \mathbf{g}(\mathbf{x}) \leq \mathbf{0}\} \tag{9}$$

where $\Omega \subset \Re^n$ is the function domain (the design space).

An important feature of such multiple criteria optimization problem is that the optimizer has to deal with objective conflicts (Eschenauer [29]). Other authors discuss the same topic by defining compromise programming, because there is no unique solution to the problem (Vanderplaats [30]).

The Proposed Formulation

To consider together the manipulability and the total mechanical power, the following scalar objective function is proposed

$$f_3 = k\frac{f_1^0}{f_1} + (1-k)\frac{f_2}{f_2^0}, \quad 0 \leq k \leq 1 \tag{10}$$

where f_1 and f_2 are given by Equations (2) and (4), respectively. The reference values f_1^0 and f_2^0 are obtained from the computation of the manipulability and mechanical power sum at the initial task positioning. Therefore, in the ideal case the objective function value is lower than 2, where $f_1^0/f_1 \leq 1$ (the manipulability was increased) and $f_2/f_2^0 \leq 1$ (the mechanical power was decreased).

The specification of a reference value to both the manipulability measure and the mechanical power are of paramount importance in the construction of the multicriteria objective function.

The inverse kinematics error is included in the problem as an equality constraint in the optimization formulation, through the expression $s_E = 0$ (Equation (8)).

It should be pointed out that the proposed formulation is able to handle the case in which the task is outside the robot workspace. This configuration reflects in the inverse kinematics error, which is taken into account as a constraint (Equation (8)) of the optimization formulation.

Additionally, initial and final velocity specifications are considered as

$$\dot{q}_j(t_0) = 0, \quad j = 1,..n. \tag{11}$$

$$\dot{q}_j(t_f) = 0, \quad j = 1,...n. \tag{12}$$

for each joint j in the initial and final times t_0 and t_f, respectively. To represent a rest-to-rest motion, zero initial and final velocities are considered above.

OPTIMIZATION STRATEGY

An important feature of the proposed analysis is the computation of highly nonlinear equations. Despite the good performance of classical nonlinear programming techniques, the global optimal design is hardly reached due to the existence of local minima in the design space.

With the aim of improving the robustness of the optimization process, the use of a global methodology is considered. The tunneling algorithm (Levy [31,32]) is a heuristic methodology designed to find the global minimum of a function. It is composed of a sequence of cycles, each cycle consists of two phases: a minimization phase having the purpose of lowering the current function value, and a tunneling phase that is devoted to finding a new initial point (other than the last minimum found) for the next minimization phase. This algorithm was first introduced by Levy [31], and the name derives from its graphic interpretation.

In summary, the computation evolves according to the following phases:

a) Minimization phase: Given an initial point q^0, the optimization procedure computes a local minimum q^* of f(q). At the end, it is considered that a local minimum is found.

b) Tunneling phase: Given q^* found above, since it is a local minimum, there exists at least one $\mathbf{q}^0 \in \Omega$, such that

$$f\left(\mathbf{q}^0\right) \leq f\left(\mathbf{q}^*\right), \mathbf{q}^0 \neq \mathbf{q}^* \tag{13}$$

In other words, there exists $q^0 \in Z = \{q \in \Omega - \{q^*\} \mid f(q) \leq f(q^*)\}$. To move from q^* to q^0 along the tunneling phase, it is defined a new initial point $q = q^* + \delta$, $q \in \Omega$ that is used in the auxiliary function

$$F(\mathbf{q}) = \frac{f(\mathbf{q}) - f(\mathbf{q}^*)}{[(\mathbf{q} - \mathbf{q}^*)^T(\mathbf{q} - \mathbf{q}^*)]^\eta} \tag{14}$$

which has a pole in q* , for a sufficient large value of η By computing both phases iteratively, the sequence of local minima may lead to the desired global minimum..

Different values for η are suggested for being used in Equation (14) to avoid undesirable points and prevent the search algorithm to fail (Levy [31,32]). In the present case the stop criterion is five consecutive iterations without further improvement in the minimal objective value.

The minimization phase is performed by using a constrained formulation, where the objective function is given by Equation (10) and the constraint function is defined by Equation (8). The velocity specification (Equations (11) and (12)) is achieved through the determination of a suitable interpolation polynomial function using the joint position coordinates.

In Equation (14), during the tunneling phase, only the unconstrained formulation given by Equation (10) is considered. This unconstrained nonlinear programming problem is solved by using the BFGS variable metric method (Luenberger [33], Vanderplaats [30]). The constrained formulation of the minimization phase (Equations (10), (11) and (12)) is solved by using the modified method of the feasible directions (Vanderplaats [34]).

The algorithms are implemented in FORTRAN by using the optimization library DOT (Vanderplaats [35]).

NUMERICAL SIMULATIONS

The numerical simulations performed to illustrate the presented methodology use two robot models, namely the Scara and the Puma 560 robot manipulators. These two types of serial robot manipulators are classical configurations in the field. The case studies have been proposed due to their complexity as compared to those normally found in the literature. The following results considers only optimal profiles whose sum of the end-effector error is zero (Equation (8)), given $\varepsilon = 0.001$ m (Equation (7)). This means that the inverse kinematics solution is found for all positions with the required precision. Different tolerance values may result from different optimal profiles.

As different values for the weighting factor k lead to different optimization problems (Equation (10)), the value of the global minimum is normalized according to the following equation

$$f_4 = \frac{f_1^0}{f_1} + \frac{f_2}{f_2^0} \qquad (15)$$

This performance index enables comparing the improvement of each objective without the weighting effect, and is used in the subsequent analysis.

In summary, the global optimization for each weighting factor ($k = 0, 0.1, 0.2, 0.3, ..., 1$) is performed according to Equation (10) and the comparison among the optimal designs obtained is performed by using Equation (15). Thus, eleven global optimizations for each task have been carried out. The maximum inverse kinematics error was defined as being lower than $\varepsilon = 0.001$ m in Equation (7).

A Scara manipulator was used in the first and second numerical experiments. The third numerical experiment refers to a Puma 560 manipulator. By considering the ability of describing a complex path, the data for a set of Cartesian specification points (Equation (5)) is determined through a cubic spline interpolation of the parametric coordinates of the desired path. Therefore, all the values are obtained from a suited set of Cartesian points.

The path for each experiment is arbitrarily defined with the aim of exploring the efficiency of the methodology in performing a complex task. All the obtained results consider only configurations for which inverse kinematics errors are lower than ε, i.e., $s_E = 0$ (Equation (8)).

Each value k in Equation (10) defines a new optimization problem. To compare the values given by different optimization formulations, the normalization of the results, despite the weighting factor, is taken into account as established by f_4 (Equation (15)).

First Experiment

In this experiment, a subset of 7 reference points is defined: (0, 0.4, 0, –0.3), (1, 0.4, –0.1, –0.3), (2, 0.2, –0.1, –0.3), (3, 0.2, –0.2, –0.3), (4, 0.4, –0.25, –0.3), (5, 0.4, –0.2, –0.3) and (6, 0.2, –0.25, –0.3). Comparing these values with Equation (5), the initial task is inside a rectangle defined by $0.2 \leq x \leq 0.4$, $-0.3 \leq y \leq -0.1$ and $z = -0.3$. The corresponding traveling time is defined by $0 \leq t \leq 6$. These data are computed by interpolating cubic splines for t, x, y and z, respectively. The interpolated data encompass a set of $N = 27$ points (Equation (5)) that describes the path-following problem, as presented in **Figure 1**.

At the original position, the values of the design variables are: $\beta = 0$ rad, $\delta_x = 0$ m, $\delta_y = 0$ m, and $\delta_z = 0$ m. These values are used as an initial guess for the optimization process (Equation (10)).

The global minimum values obtained are presented in **Figure 2**.

The optimal values of manipulability and mechanical power indexes demonstrate that there is no clear dependence between the variables. Good results are provided for k = 0.2, 0.5, 0.6, and 0.9.

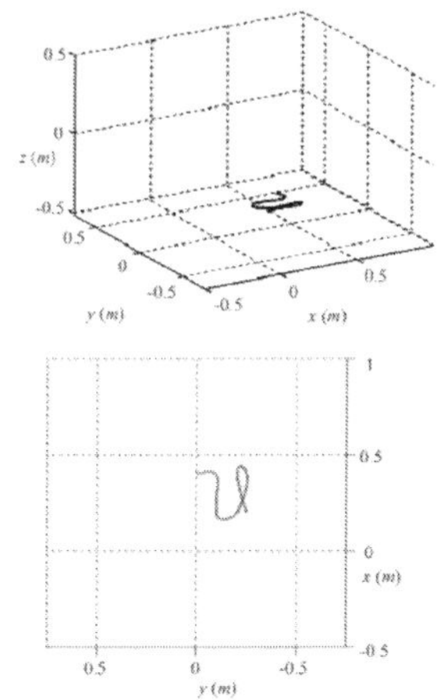

Figure 1. The proposed path-following problem from two different perspectives.

Details about the minimal performance index, k = 0.9, are presented in **Table 1**. The overall performance was improved 28%, while the minimal manipulability and the mechanical power were improved 62% and 18%, respectively.

Some of the values provided by other weighting factors may result in further improvement of the manipulability or mechanical power, separately. However, the present configuration is a global minimum of the performance index f_4.

The improvement obtained during the tunneling process is shown in Figure 3.

After the 6th iteration of the tunneling process, no further improvement was achieved during the next five consecutive iterations. This behavior defines the stop criteria for the global optimization search. The above result highlights the good performance of the method, as satisfactory results are obtained after a small number of iterations.

Figure 4 shows the robot performing the path following task and Figure 5 presents the robot optimal positioning, according to the optimal design variable values given in Table 1.

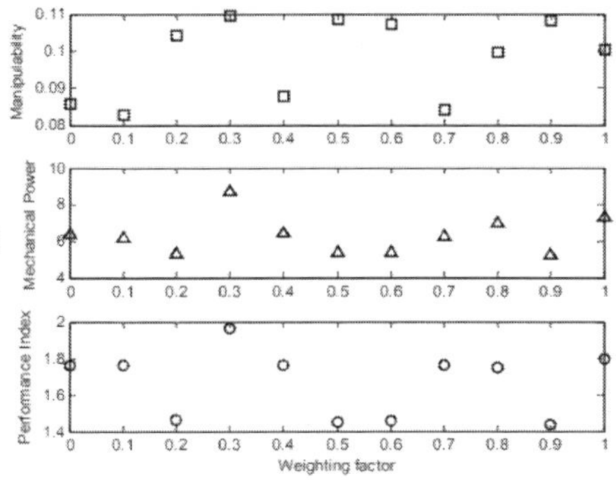

Figure 2. Optimal values obtained by using different weighting factors.

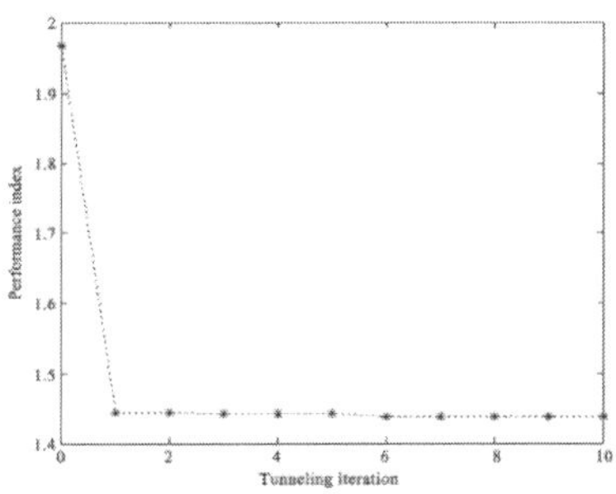

Figure 3. Performance index along the tunneling iterations (for k = 0.9).

Table 1. Initial and optimal parameter values (for k = 0.9).

Parameter	*Initial value*	*Optimal value*
Minimal manipulability f_1	0.067	0.108
Mechanical power f_2 (*Watts*)	6.451	5.261
Performance index f_3	2.000	1.436
Design variables (β, δ_x, δ_y, δ_z)	(0, 0, 0, 0)	(0.107, -0.013, -0.227, 0.055)

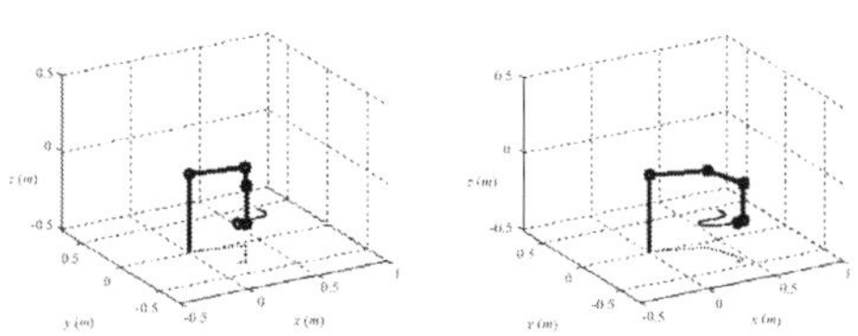

Figure 4. Robot performing the path following task.

The joint coordinates at each time step are obtained from inverse kinematics computation. Once the task is considered as a set of consecutive Cartesian points with continuous transition between them, the corresponding joint coordinates presents a smooth behavior. **Figure 6** shows the initial and optimal joint coordinates of the first three links and the corresponding manipulability index.

Once the inverse kinematics is iteratively computed for each end-effector positioning, it is possible to include the obstacle avoidance feature in the above strategy. The corresponding procedure for the inverse kinematics computation is addressed in Santos [28], for example.

Second Experiment

In this experiment a subset of 8 reference points is defined: (0, 0.2, 0, –0.3), (1, 0.4, –0.1, –0.3), (2, 0.4, –0.05, –0.3), (3, 0.2, –0.1, –0.3), (4, 0.25, –0.15, –0.3), (5, 0.3, –0.17, –0.3), (6, 0.27, –0.13, –0.3) and (7, 0.26, –0.2, –0.3). By comparing these values with Equation (10), the initial task is inside a rectangle defined by $0.2 \leq x \leq 0.4$, $-0.2 \leq y \leq 0$ and $z = -0.3$. The corresponding traveling time is defined by $0 \leq t \leq 7$. These data are computed by interpolating cubic splines for t, x, y and z, respectively. The interpolated data encompass a set of N = 54 points

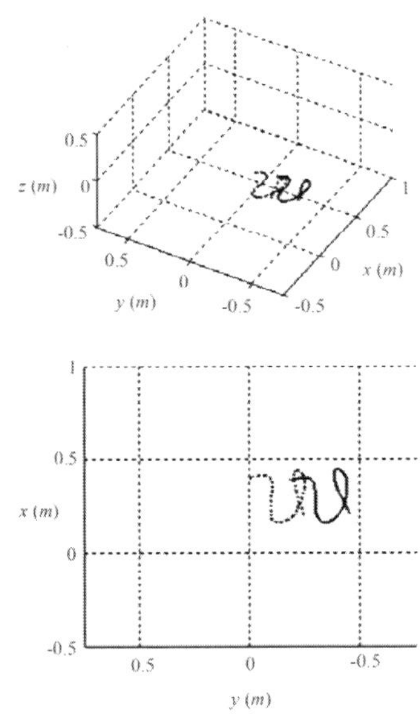

Figure 5. Proposed (---) and optimal (—) paths from two different perspectives.

(Equation (5)) that describes the path-following problem, as presented in Figure 7” target=”_self”> Figure 7.

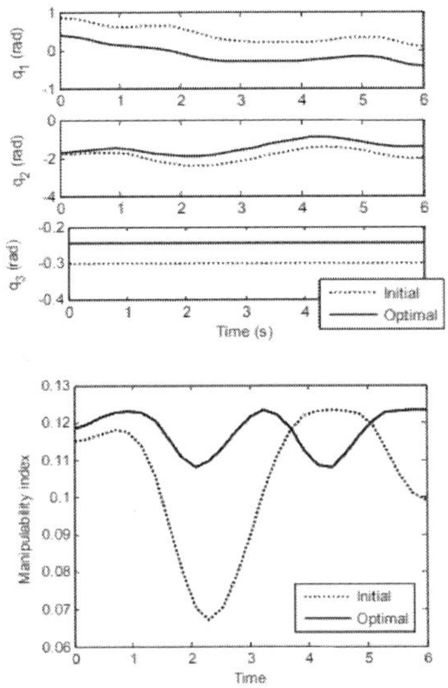

Figure 6. Joint coordinates and the corresponding manipulability index (Scara manipulator).

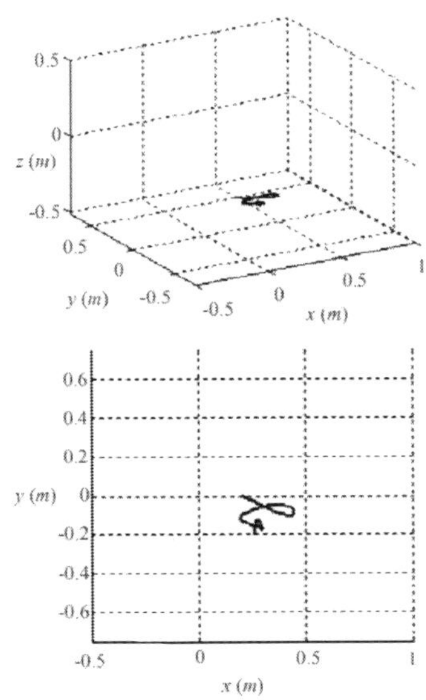

Figure 7. The proposed path-following problem from two different perspectives.

At the original position, the values of the design variables are: $\beta = 0$ rad, $\delta_x = 0$ m, $\delta_y = 0$ m, and $\delta_z = 0$ m. These values are used as an initial guess for the optimization process (Equation (10)). Global optimizations were performed for eleven different values of the weighting factor k, (k = 0, 0.1, 0.2, ..., 1) (see Equation (10)).

The minimum values obtained by the optimization process are presented in Figure 8" target="_self"> Figure 8.

In this experiment good results are provided by k = 0.3, 0.4, and 0.7. Details about the minimal performance index, k = 0.7, are presented in Table 2. The overall performance was improved 17%, while the minimal manipulability and the mechanical power were improved 46% and 3%, respectively.

Eventually, the values provided by other weighting factors may result in further improvement either of the manipulability or the mechanical power, separately. However, the above configuration is a global minimum of the performance index f_4.

The design improvement achieved by the tunneling process is shown in Figure 9" target="_self"> Figure 9.

After iteration 18 of the tunneling process, no further improvement was achieved for five consecutive iterations, defining the stop criterion of the global optimization search.

Figure 10 shows the robot performing the path-following task and Figure 11 presents the optimal positioning, according to the optimal design values presented in Table 2.

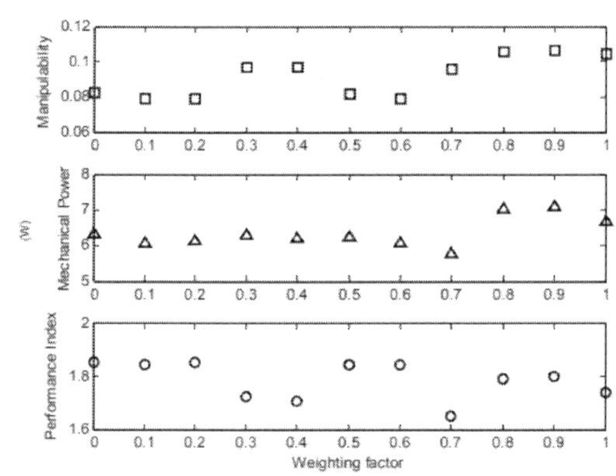

Figure 8. Optimal values obtained by using different weighting factors.

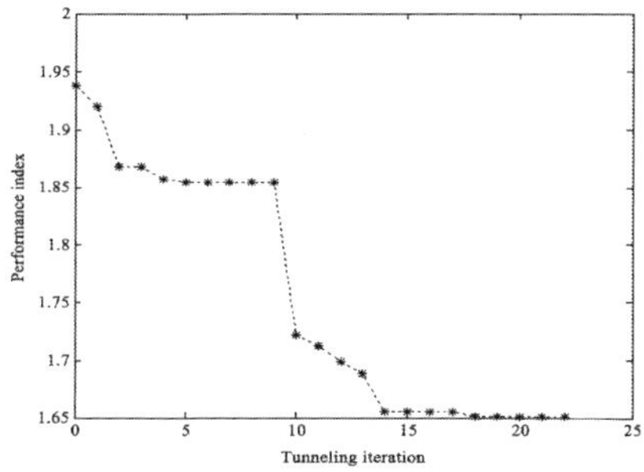

Figure 9. Performance index along the tunneling iterations (for k = 0.7).

Table 2. Initial and optimal parameter values (for k = 0.7).

Parameter	*Initial value*	*Optimal value*
Minimal manipulability f_1	0.065	0.095
Mechanical power f_2 (*Watts*)	5.989	5.773
Performance index f_3	2.000	1.650
Design variables (β, δ_x, δ_y, δ_z)	(0, 0, 0, 0)	(0.044, 0.084, -0.116, 0.066)

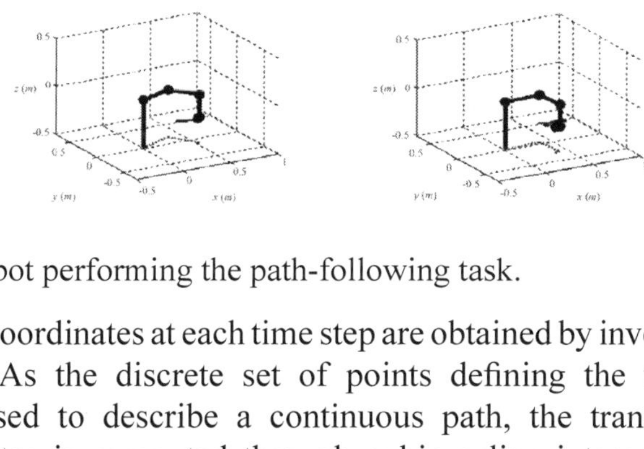

Figure 10. Robot performing the path-following task.

The joint coordinates at each time step are obtained by inverse kinematics computation. As the discrete set of points defining the task (Equation (5)) is supposed to describe a continuous path, the transition between joint coordinates is computed through cubic spline interpolation. **Figure 12** shows the initial and optimal joint coordinates for the first three links and the corresponding manipulability index.

For the examples presented above, the task was defined as being parallel to the xy plane, i.e., the path is parallel to the ground.

Third Experiment

In this experiment, a subset of 7 reference points is defined: (0, 0.4, 0.1, 0), (1, 0.4, 0, 0), (2, 0.4, 0, –0.4), (3, 0.4, –0.3, –0.4), (4, 0.4, –0.4, 0), (5, 0.4, –0.3, 0), and (6, 0.4, –0.4, –0.4). By comparing these values with Equation (5), the initial task is inside a rectangle defined by x = 0.4, –0.4 ≤ y ≤ 0.1, and –0.4

$\leq z \leq 0$. The corresponding traveling time is defined by $0 \leq t \leq 6$. These data are computed by interpolating cubic splines for t, x, y and z, respectively. The interpolated data encompass a set of N = 54 points (Equation (5)) that describe the path-following problem, as presented in **Figure 13**.

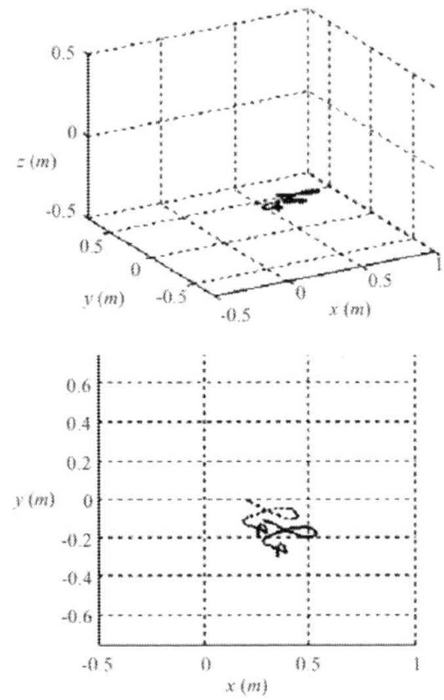

Figure 11. Proposed (---) and optimal (—) paths from two different perspectives.

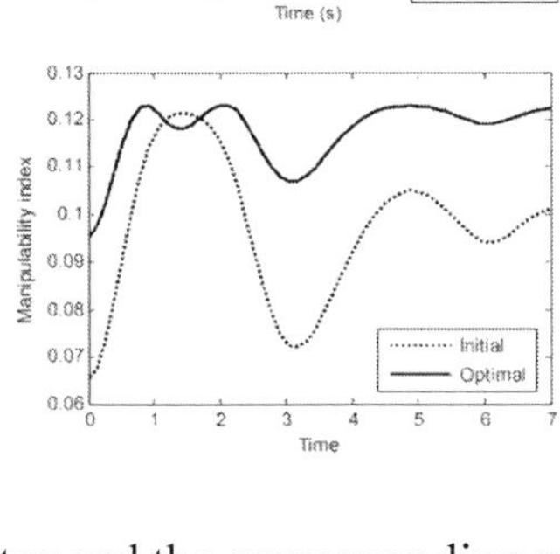

Figure 12. Joint coordinates and the corresponding manipulability index (Scara manipulator).

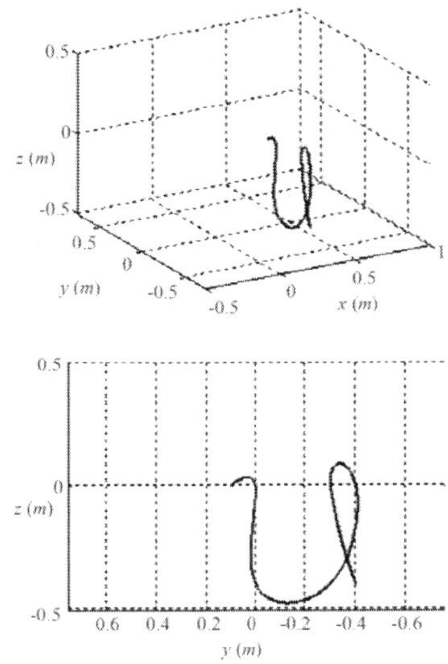

Figure 13. The proposed path-following problem from two different perspectives.

At the original position, the values of the design variables are: $\beta = 0$ rad, $\delta_x = 0$ m, $\delta_y = 0$ m, and $\delta_z = 0$ m. These values are used as an initial guess for the optimization process (Equation (10)).

The optimal values of both the manipulability and the mechanical power indexes demonstrate that there is no significant difference among most of the results presented. The best values are provided by k = 0.4 and 0.5.

Details about the minimal performance index, k = 0.5, are presented in Table 3” target=”_self”> Table 3. The overall performance was improved 40%, while the minimal manipulability and the mechanical power were improved 64% and 41%, respectively.

Some of the values obtained by using other weighting factors may result in further improvement of either the manipulability or the mechanical power, separately. However, the above configuration is a global minimum of the performance index f_4.

The performance provided by the tunneling process is shown in Figure 14.

After the 5th iteration of the Tunneling process, no further improvement was achieved for the next five conescutive iterations. This behavior defines the stop criteria for the global optimization search. In this case, it is possible to see that good results are obtained in a few iterations.

Figure 15 shows the robot performing the path following task and Figure 16 presents the optimal positioning, according to the optimal design variable values given in Table 3.

Table 3. Initial and optimal parameter values (for k = 0.5).

Parameter	*Initial value*	*Optimal value*
Minimal manipulability f_1	0.053	0.087
Mechanical power f_2 (*Watts*)	1461.114	862.289
Performance index f_4	2.000	1.202
Design variables (β, δ_x, δ_y, δ_z)	(0, 0, 0, 0)	(0.391, 0.101, 0.169, 0.418)

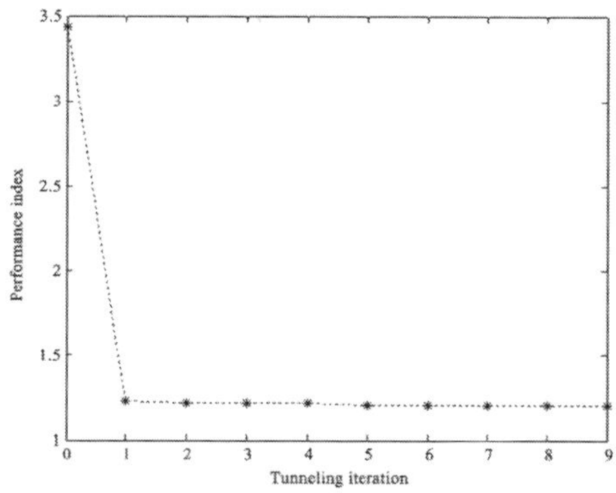

Figure 14. Performance index along the tunneling iterations (for k = 0.5).

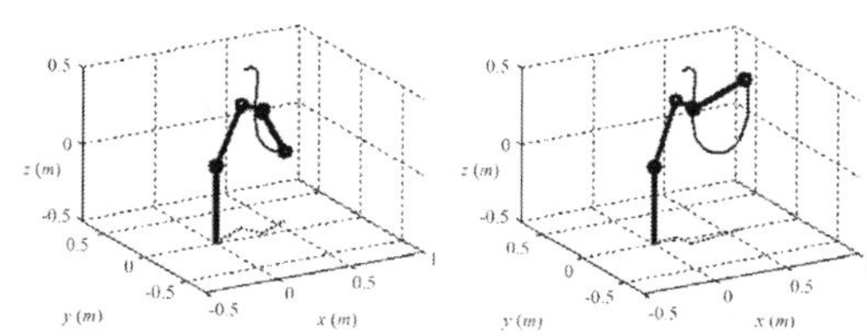

Figure 15. Robot performing the path-following task.

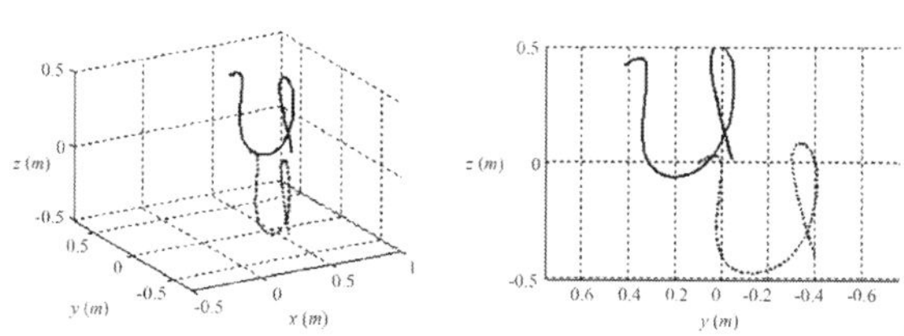

Figure 16. Proposed (---) and optimal (—) paths from two different perspectives.

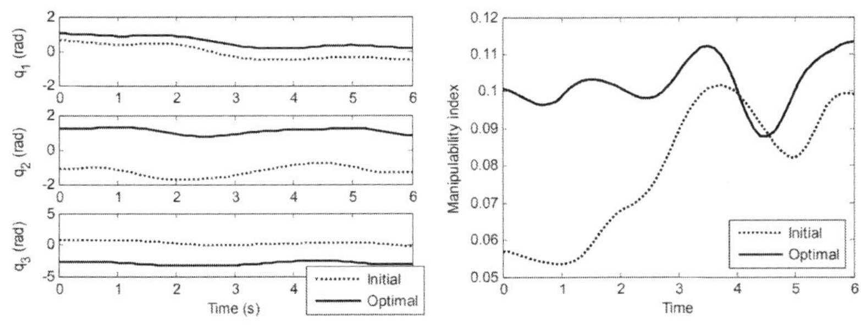

Figure 17. Joint coordinates and the corresponding manipulability index (Puma 560 manipulator).

The joint coordinates at each time step are obtained by inverse kinematics computation. As the task is considered as a set of consecutive Cartesian points with continuous transition, the corresponding joint coordinates presents a smooth behavior. Figure 17 shows the initial and optimal joint coordinates for the first three links and the corresponding manipulability index.

The previous example considered a task initially defined as being parallel to the xz plane, i.e., the path is parallel to a vertical plane. It can be noticed that the optimal design corresponds to rotating the task about 0.391 rad anticlockwise with respect to the reference frame.

CONCLUSIONS

In this work the path-following problem of robot manipulators was addressed. An approach to increase the manipulability while decreasing dynamics requirements was proposed by using optimization techniques.

The concept of manipulability was revisited. The basic idea behind manipulability analysis consists in describing directions either in the task space or the joint space that optimize the ratio between a measure of effort in the joint space and a measure of performance in the task space. In this work a classical formulation to describe the relation between the two spaces (task and joint spaces) was considered.

Since the end-effector presents various performances due to a number of factors (control specification, joint flexibility, joint friction, etc.), a general methodology to reduce high torques and to increase manipulability was proposed. The corresponding equation considers the overall performance of the entire manipulator, as the determinant of the Jacobian is proportional to the product of its singular values, and the mechanical power index is obtained by a sum involving all the joints. The methodology also includes the manipulator positioning error analysis, through the definition of constraint functions in the optimization model.

The serial manipulator usually performs differently in different zones of the workspace. Considering also the cases for which different joint velocity and torque profiles may result different performance levels, a better position into the workspace (in the sense that the performance indexes are optimized) for performing a specified task was determined.The optimization problem was based in a global search heuristic, improving the robustness of the process with respect to the initial guess.Numerical experiments showed that the task repositioning successfully increases the manipulability and reduces the total mechanical power, simultaneously. As changes in the task positioning may result different associated manipulability indexes and different joint velocity and torque profiles, the methodology searches for an optimal way of using the robot possibilities (as described by the performance indexes) by finding an optimal position into the workspace to perform the task.

In some practical cases, the manipulated objects cannot be displaced because of physical constraints such as weight or geometry. Even in such cases the proposed strategy is suitable to be applied by changing the position of the manipulator itself. The corresponding changes in the robot's base are given by the optimal profile by using the opposite sign of the optimal values, i.e., $(-\beta, -\delta_x, -\delta_y, -\delta_z)$. It is also possible to adjust the analysis to consider different objectives, as the total torque instead of the mechanical power, for example. The steps to be followed remain the same.

In summary, the main contributions of this paper are the simultaneous analysis of both the required mechanical power and the manipulability (dynamics and kinematics aspects), together with a global deterministic heuristic to achieve the optimal task placement.

As the proposed methodology is suitable for serial manipulators of arbitrary complexity, the authors believe that the present methodology represents a contribution concerning general optimal robot path planning and design.

ACKNOWLEDGEMENTS

The authors are thankful to the agencies FAPEMIG and CNPq (INCT-EIE) for partially funding this work.

REFERENCES

1. S. F. P. Saramago and V. Steffen Jr., "Optimization of the Trajectory Planning of Robot Manipulators Taking Into-Account the Dynamics of the System," Mechanism and Machine Theory, Vol. 33, 1998, pp. 883-894.
2. R. R. D. Santos, V. Steffen Jr. and S. F. P. Saramago, "Robot Path Planning: Avoiding Obstacles," In: 18th International Congress of Mechanical Engineering, Ouro Preto-MG, Brazil, 2005.
3. S. S. Chiddarwar and N. R. Babu, "Offline Decoupled Path Planning Approach for Effective Coordination of Multiple Robots," Robotica, Vol. 28, No. 4, 2010, pp. 477- 491.
4. R. R. D. Santos, V. Steffen Jr. and S. F. P. Saramago, "Optimal Path Planning and Task Adjustment for Cooperative Flexible Manipulators," ABCM Symposium Series in Mechatronics, Associação Brasileira de Engenharia e Ciências Mecânicas, ABCM, Vol. 3, 2008, pp. 236-245.
5. O. Von Stryk and M. Schlemmer, "Optimal Control of the Industrial Robot Manutec r3," Computational Optimal Control, International Series of Numerical Mathematics, Vol. 115, 1994, pp. 367-382.
6. R. R. D. Santos, V. Steffen Jr. and S. F. P. Saramago, "Robot Path Planning in a Constrained Workspace by Using Optimal Control Techniques," In: III European Conference on Computational Mechanics, Lisbon, Portugal, 2006, pp. 159-177.
7. J. E. Bobrow, S. Dubowsky and J. S. Gibson, "Time-Optimal Control of Robotic Manipulators along Specified Paths," The Intemational Journal of Robotics Research, Vol. 4, 1995, pp. 3-17.
8. J. E. Slotine and H. S. Yang, "Improving the Efficiency of Time-Optimal Path-Following," IEEE Transaction on Robotics and Automation, Vol. 5, No. 1, 1989, pp. 118- 124.
9. [9] D. Constantinescu and E. A. Croft, "Smooth and TimeOptimal Trajectory Planning for Industrial Manipulators along Specified Paths," Journal of Robotic Systems, Vol. 17, 2000, pp. 233-249.
10. H. Seraji, "Reachability Analysis for Base Placement in Mobile Manipulator," Journal of Robotic Systems, Vol. 12, 1995, pp. 29-43.
11. S. Zeghloul and J. A. Pamanes, "Multi-Criteria Optimal Placement of Robots in Constrained Environments," Robotica, Vol. 11, No. 2, 1993, pp. 105-110.
12. J. T. Feddema, "Kinematically Optimal Robot Placement for Minimum

Time Coordinated Motion," Proceedings of the IEEE International Conference of Robotics and Automation, Minneapolis, 1996, pp. 3395-3400.

13. F. C. Park, "Distance Metrics on the Rigid-Body Motions with Applications to Mechanicsm Design," ASME Journal of Mechanical Design, Vol. 117, 1995, pp. 48-54.
14. G. S. Chirikjian and S. Zhou, "Metrics on Motion and Deformation of Solid Models," ASME Journal of Mechanical Design, Vol. 120, 1998, pp. 252-261.
15. J. M. R. Martinez and J. Duffy, "On the Metrics of Rigid Body Displacement for Infinite and Finite Bodies," ASME Journal of Mechanical Design, Vol. 117, No. 1, 1995, pp. 41-47.
16. B. Tabarah, B. Benhabib, R. Fenton and R. Cohen, "Cycle-Time Optimization for Single-Arm and Two-Arm Robots Performing Continuous Path Operation," 21st Biennial Mechanisms Conference, Chicago IL, 1990, pp. 401-406.
17. Y. Zhang, and K. Li, "Bi-Criteria Velocity Minimization of Robot Manipulators Using LVI-Based Primal-Dual Neural Network and Illustrated Via PUMA560 Robot Arm," Robotica, Vol. 28, No. 4, 2010, pp. 525-537.
18. K. Abdel-Malek and W. Yu, "On the Placement of Serial Manipulator," Proceedings of DETC00 2000 ASME Design Engineering Technical Conferences, Baltimore, MD, 2000, pp. 1-8.
19. Y. Chen and A. A. Desrochers, "Structure of Minimum-Time Control Law for Robotic Manipulators with Constrained Paths," IEEE International Conference on Robot and Automat, Scottsdale, Az, USA, 1989, pp. 971- 976.
20. Z. Shiller and H. H. Lu, "Computation of Path Constrained Time Optimal Motions with Dynamic Singularities," ASME Journal of Dynamic Systems, Measurement, and Control, Vol. 114, 1992, pp. 34-40.
21. E. K. Xidias, P. T. Zacharia and N. A. Aspragathos, "Time-Optimal Task Scheduling for Articulated Manipulators in Environments Cluttered with Obstacles," Robotica, Vol. 28, No. 3, 2010, pp. 427-440.
22. K. S. Fu, R. C. Gonzales and C. S. G. Lee, "Robo-Tics: Control, Sensing, Vision and Intelligence," McGraw-Hill, New York, 1987.
23. T. Yoshikawa, "Manipulability of Robot Mechanisms," The

International Jouranl Robotics Research, Vol. 4, 1985, pp. 3-9.
24. M. J. Richard and C. M. Gosselin, "A Survey of Simulation Programs for the Analysis of Mechanical Systems," Mathematics and Computers in Simulation, Vol. 35, No. 2, 1993, pp. 103-121.
25. V. Mata, F. Benimeli, N. Farhat and A. Valera, "Dynamic Parameter Identification in Industrial Robots Considering Physical Feasibility," Journal of Advanced Robotics, Vol. 19, No. 1, pp. 101-120.
26. V. Mata, S. Provenzano, F. Valero and J. I. Cuadrado, "Serial Robot Dynamics Algorithms for Moderately Large Numbers of Joints," Mechanism and Machine Theory, Vol. 37, No. 8, pp. 739-755.
27. J. J. Craig, "Introduction to Robotics: Mechanics & Control," 2nd Edition, Reading, MA: Addison-Wesley, 1989.
28. R. R. D. Santos, V. Steffen Jr. and S. F. P. Saramago, "Solving the Inverse Kinematics Problem through Performance Index Optimization," In: XXVI Iberian Latin-American Congress on Computational Methods in Engineering, Guarapari-ES, 1985.
29. H. Eschenauer, J. Koski and A. Osyczka, "Multicriteria Design Optimization," Springer-Verlag, Berlin, 1990
30. G. N. Vanderplaats, "Numerical Optimization Techniques for Engineering Design," 3rd Edition, VR & D Inc., 2001.
31. A. V. Levy and S. Gomez, "The Tunneling Method Applied to Global Optimization. Numerical Optimization," (Ed. P. T. Boggs) R. H. Byrd and R. B. Schnabel, Eds., Society for Industrial and Applied Mathematics, 1985, pp. 213-244.
32. A. V. Levy and A. Montalvo, "The Tunneling Algorithm for the Global Minimization of Functions," The SIAM Journal on Scientific and Statistical Computing, Vol. 6, No. 1, 1985, pp. 15-29.
33. D. G. Luenberger, "Linear and Nonlinear Programming," 2nd Edition, Addison-Wesley, USA, 1984.
34. G. N. Vanderplaats, "An Efficient Feasible Direction Algorithm for Design Synthesis," American Institute of Aeronautics and Astronautics Journal, Vol. 22, No. 11, 1984, pp. 1633-1640.
35. G. N. Vanderplaats, "DOT-Design Optimization Tools Program," Vanderplaats Research & Development, Inc., Colorado Springs, 1995.

CHAPTER

9

Kinematical Analysis and Simulation of High-Speed Plate Carrying Manipulator Based on Matlab

Ke Wang, Jiping Zhou

School of Mechanical Engineering, University of Yangzhou, Yangzhou, China

ABSTRACT

In order to construct the more effective kinematics method for industry, by taking a high-speed plate handing robot as an example, the structure and parameters of the robot linkages are analyzed, and the standard Denavit-Hartenberg method is applied to establish the coordinates and the kinematic equation of the linkages. Depending on the graphics and matrix calculation

Citation:- K. Wang and J. Zhou, "Kinematical Analysis and Simulation of High-Speed Plate Carrying Manipulator Based on Matlab," Engineering, Vol. 4 No. 12, 2012, pp. 850-856. doi: 10.4236/eng.2012.412108.

ability of Matlab especially including the Robotics Toolbox, the handling robot has been modeled and its kinematics, inverse kinematics and the trajectory planning have been simulated. Therefore, the correctness of kinematic equation has been verified, meanwhile, the functions of displacement, velocity, acceleration and trajectory of all the joints are also obtained. In a further step, this has verified the validity of all the structure parameters and provided a reliable basis for the theoretical research on the design, dynamics analysis and trajectory planning of the manipulator control system.

Keywords: High-Speed Plate Carrying Manipulator; D-H Method; Robotics Toolbox; Trajectory Planning

INTRODUCTION

As a typical representative and a main technical means of information technology and advanced manufacturing technology, the complete set of automatic stamping processing line has become a high technology to which the developed countries have paid much attention. And its development level has become one of the most important standards to measure a nation's technical development. It has been extensively applied in the industry of mechanical manufacturing, nuclear, aerospace, energy and transportation, petroleum chemistry, building, electronic and etc. [1]. But, one key aspect of the automatic stamping processing line is to develop a mechanical manipulator with the characteristic of the high-speed and dynamic transmission.

At present, the sheet metal forming production line is operated at the high productive rate of 15 - 25 SPM. Removing the time spent by the press stamping operations, there is only about 2 s left to the robot manipulator. In such a short period of time, the manipulator has to operate at a high speed of 200 - 250 m/min, so that the operations such as loading, moving and unloading can be fulfilled. And this has put forward higher requirements to the mechanical structure, material friction characteristics, and structural dynamic characteristics of the robot manipulator.

Taking a high-speed plate carrying manipulator as the research object, this paper firstly analyzes the structure and connecting rod parameters; then adopts the standard D-H method [2] to establish the kinematics equation; and then discusses the positive and inverse kinematics algorithm, and the trajectory planning problems; finally in the environment of Matlab, the kinematics model is built to take kinematics simulation by using of Robotics Toolbox [3]. In simulation process, we can not only directly observe the robot motion, but also get the required data in the graphic form. Therefore, the virtual performance of the product can be tested in the conceptual design stage, so as to improve the design performance, reduce design cost and decrease product development time.

THE STRUCTURE DESIGN AND LINK PARAMETERS OF HIGH-SPEED PLATE CARRYING MANIPULATOR

The Design Requirements of High-Speed Plate Carrying Manipulator

The Workplace of High-Speed Plate Carrying Manipulator

The traditional stamping processing method relies on a stand-alone manual which is inefficient, inaccurate and insecurity. So it has already become increasingly unsuited to the requirements of modern mass production, especially when the requirements of the sheet metal processing annual output exceeds thousands tons, this contradiction is more prominent. The current stamping equipment is toward the trend of automation, sets and online development. With the rapid development of China's market economy, especially in the coastal areas where the shortage of skilled workers is severe, and labor costs rises sharply, many machinery manufacturers have the urgent demand for manufacturing automation, and require machine and equipment manufacturing industry to provide users with a complete set of on-line technical services to improve production efficiency and reduce labor costs. Therefore, the development of the technology of sheet metal production line is even more important. One of the core technologies of sheet metal stamping equipment line is the design of high-speed plate carrying manipulator. As shown in **Figure 1**, it is high-speed plate carrying manipulator in the application of stamping processing complete sets of equipment on-line system.

The Operation Process of High-Speed Plate Carrying Manipulator

According to the composition of the stamping process sets of on-line system and the role of the high-speed plate carrying manipulator, the working cycle of a manipulator contains nine action process, as shown in **Figure 2**:

1) From the point of origin, the left electromagnetic valve is energized after pressing the start button, then the manipulator moves to the left. It won't stop until it encounters the left limit switch.
2) Simultaneously the manipulator begins to drop after the decreased electromagnetic valve opens, then, it won't stop until it encounters the lower limit switch.
3) At the same time the clamp electromagnetic valve is energized, then the manipulator is clamped.
4) After the rise electromagnetic valve is energized, the manipulator begins to rise, then it won't stop until it encounters the rise limit switch.
5) Simultaneously the manipulator moves to the right after the right electromagnetic valve is energized, then, it won't stop until it encounters the right limit switch.
6) Simultaneously the manipulator begins to drop after the decreased electromagnetic valve is energized, then, it won't stop until it encounters the lower limit switch.
7) At the same time the clamp electromagnetic valve is opened, then, the manipulator is opened.
8) After the rise electromagnetic valve is energized, the manipulator begins to rise, then, it won't stop until it encounters the rise limit switch.

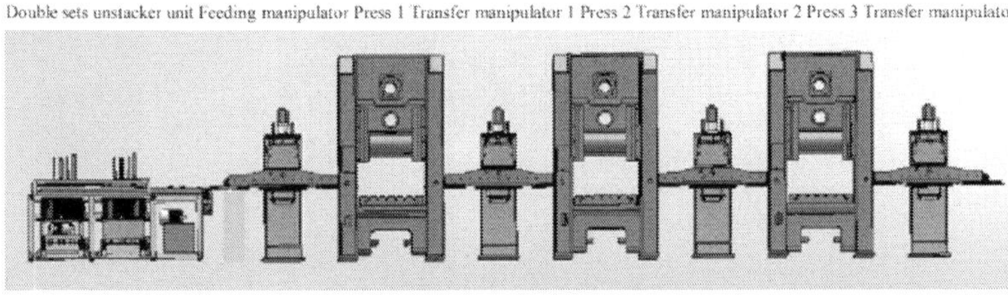

Figure 1. The composition of the stamping processing sets of on-line system.

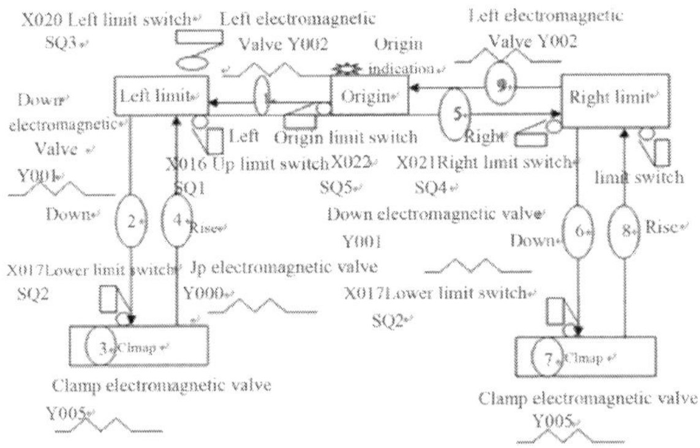

Figure 2. The operation process of the manipulator.

9) Simultaneously the manipulator moves to the origin after the left electromagnetic valve is energized, then, it won't stop until it encounters the left limit switch.

So far, the manipulator after 9 step action completes a cycle of its movement, and then, it continues the cycle work.

Based on the above analysis, we have developed a three-dimensional model of the manipulator as shown in Figure 3, which can be used to analyze the dynamic characteristics of the manipulator.

The Design Parameters of High-Speed Plate Carrying Manipulator

D-H Transformation

In order to describe the translational and rotational relationships between adjacent rods, Denavit and Hartenberg (1955) have proposed a matrix method to establish the possessed coordinate system for each rod in the linkage chain. This method is to establish a homogeneous transformation matrix for each of the join bar coordinate, which represents the relationship of the previous coordinate system of the bar, and the principle [4,5] is detailed as follows:

OXYZ: A fixed reference coordinate system of the fixed coordinates, is called the coordinate system.

$O_iX_iY_iZ_i$: Fixed connected with the member bar of number I of the robot, the origin of coordinate is at the center point of the joint of the I + 1th.

Identify and establish each coordinate system by following three rules.

1): Z_{i-1} axis along the motion shaft of the join of the i th.

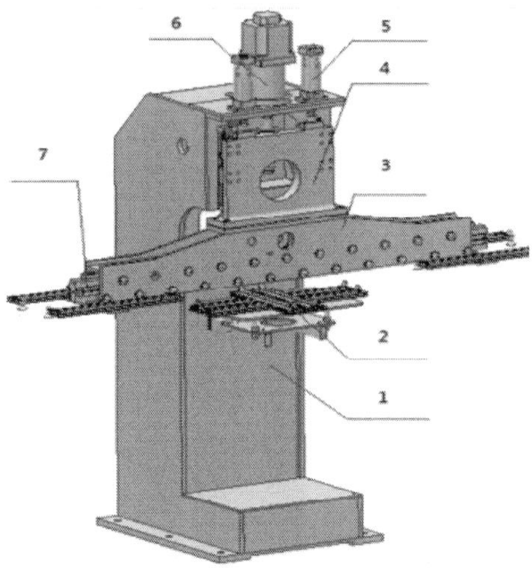

Figure 3. The assembly model of manipulator, 1-body, 2- feeder, 3-beam, 4-slipway, 5-balance cylinder, 6-servo motor, 7-X axis transmission system.

2): X_i axis vertical the axis of Z_i and Z_{i-1} and point to the direction of away from the Z_{i-1} axis.

3): Y_i axis according to the requirements of the right hand coordinate system to establish.

Meanwhile the notation of D-H of the rigid bar depends on the four parameters of the connecting rod.

The angle of two connecting rod θ_i : $X_{i-1} \rightarrow X_i$ around the corner of Z_{i-1}.

The distance of two connecting rod d_i: $X_{i-1} \rightarrow X_i$ along the distance of Z_{i-1}.

The length of the connecting rod a_i: $Z_{i-1} \rightarrow Z_i$ along the distance of X_{i-1}.

The torsion angle of the connecting rod: α_i: $Z_{i-1} \rightarrow Z_i$ around the corner of X_i.

For rotational joints, θ_i is the joint variables, the rest are joint parameters (remain unchanged); for mobile joints, d_i is the joint variables, the rest are joint parameters.

The Design Coordinate of Manipulator

High-speed plate carrying manipulator is mainly composed of vertical pillars (body), horizontal arm (beam), sliding table (Y axis transmission system) and base. Horizontal arm mounted on the machine body level can move around and can move up and down along the vertical pillars. By reference to the high-speed plate carrying manipulator in Figure 3, we establish the coordinate system of manipulator in Figure 4 based on the above analysis.

According to the D-H parameters method, four parameters are defined for each link: the connecting rod angle θ_i, the distance of two connecting rod d_i, the length of the connecting rod a_i, the connecting rod torsion Angle a_i, the D-H parameters of manipulator as is shown in Table 1.

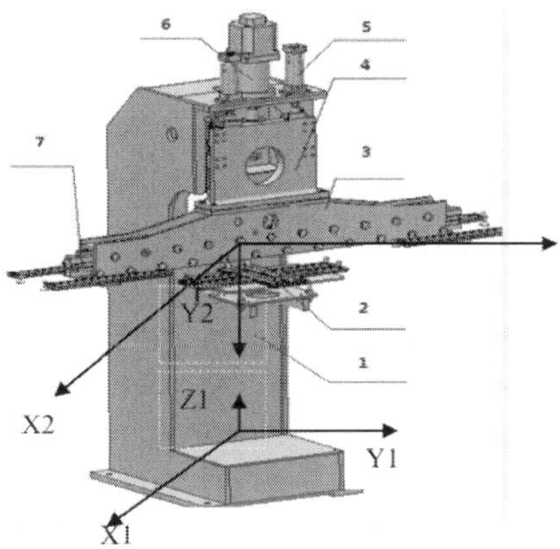

Figure 4. The D-H coordinate of high-speed plate carrying manipulator.

Table 1. The D-H parameters and joint variables of manipulator.

Joint	θ_i	d_i	a_i	d_i
1	0	d_1	0	–90
2	0	d_2	0	0

THE KINEMATICS SIMULATION ALGORITHM OF MANIPULATOR

The Kinematics of Manipulator

The so-called kinematics problem [4, 5] is given to the manipulator of each joint variable, and then obtains the position and posture of the end of the actuator, and its essence is to establish the kinematics equations. For the solution of kinematics equation, this paper uses homogeneous transformation matrix $^{i-1}A_i$ to describe the coordinate system of i relative to the position and pose of the coordinate system of the i – 1, this is the general formula for transformation matrix $^{i-1}A_i$.

$$^{i-1}A_i = \begin{bmatrix} \cos\theta_i & -\sin\theta_i\cos\alpha_i & \sin\theta_i\sin\alpha_i & a_i\cos\theta_i \\ \sin\theta_i & \cos\theta_i\cos\alpha_i & -\cos\theta_i\sin\alpha_i & a_i\sin\theta_i \\ 0 & \sin\alpha_i & \cos\alpha_i & d_i \\ 0 & 0 & 0 & 1 \end{bmatrix} \quad (1)$$

Now, putting the D-H parameters and joint variables of manipulator substitution of the **Table 1** in (1), we get the homogeneous transformation matrix ^0T_i of the i coordinate system relative to the position and posture of the base coordinate system, expressed as:

$$^0T_i = {}^0A_1\,{}^1A_2 \cdots {}^{i}\,{}^{1}A_i \quad (2)$$

Especially, when $i = 2$, $T = {}^0T_2$ can be obtained, it determines the position and posture of the end of the manipulator relative to the base coordinate system, the matrix of T can be expressed as:

$$T = {}^0T_2 = {}^0A_1{}^1A_2 = \begin{bmatrix} 1 & 0 & 0 & 0 \\ 0 & 0 & 1 & d_2 \\ 0 & -1 & 0 & d_1 \\ 0 & 0 & 0 & 1 \end{bmatrix} \quad (3)$$

Note:

$$^0A_1 = \begin{bmatrix} 1 & 0 & 0 & 0 \\ 0 & 0 & 1 & 0 \\ 0 & -1 & 0 & d_1 \\ 0 & 0 & 0 & 1 \end{bmatrix}, \; {}^1A_2 = \begin{bmatrix} 1 & 0 & 0 & 0 \\ 0 & 1 & 0 & 0 \\ 0 & 0 & 1 & d_2 \\ 0 & 0 & 0 & 1 \end{bmatrix}$$

The Inverse Kinematics Issue of Manipulator

The inverse kinematics issue of manipulator is defined as follows: with the known the position and posture of the end of the actuator, we need to solve the variables of each joint of the manipulator, here in our case the variables are d_1 and d_2.

The target point of the movement of the manipulator center is ${}^0P = [0 \;\; Y_0 \;\; Z_0 \;\; 1]^T$. By using the transitional joint, the manipulator's center point can coincide with the target point, and the target point can be expressed as: ${}^2P = [0 \;\; -d_1 \;\; d_2 \;\; 1]^T$, then we can establish the following equation:

$$ {}^0P = {}^0T_2 \, {}^2P \tag{4} $$

Combining (3) and (4) and solving the kinematics equation, we can get the following equation.

$$ d_2 = Y_0/2 \tag{5} $$

$$ d_1 = Z_0/2 \tag{6} $$

THE KINEMATICS SIMULATION OF MANIPULATOR

Use Link and Robot Functions to Establish the Manipulator's Object

1) Before the manipulators simulation, firstly input the manipulator's parameters by calling the function Link:

$L = ([\text{alpha}, A, \text{theta}, D, \text{sigma}], \text{standard})$

(note: alpha, A, theta, D represent the variables of α_i, α_i, θ_i and d_i respectively; "sigma" represents the joint types: 0 for the rotational joint, and 1 for the transitional joint; "standard" is for the standard D-H parameters. The function robot is used to create a manipulator object by using the Link function in the format of Robot (Link…). The commands for creating the high-speed plate carrying manipulators is expressed as follows:

```
L1 = ([-pi/2  0  0  0.08  1],'standard');
L2 = ([0  0  0  0.9  1],'standard');
r = robot({L1  L2},'Manipulator');
plot(r)
drivebot(r)
```

2) We can immediately see the three-dimensional view of the manipulator which can be used in the form manually for driving through the slider in the chart to drive the movement of the

manipulator, which is just like the actual control of the manipulator [3]. As shown in **Figure 5**, it has brought great convenience for manipulator's teaching and training.

The Test of Kinematic Model

In the Figure 5, by moving each slide in the bar to move the manipulator, the first joint movement can change the height of the manipulator in the vertical direction. The second joint can change the length of the manipulator in horizontal direction, So that we can obtain different position and posture. By adjusting the slides and the joint variables, we can get the approximate model corresponding to the actual structure in Figure 3, as shown in Figure 6.

In order to verify the correctness of the kinematics Equation (3), the geometric parameters and the joint variables of each rod of the manipulator are put into the kinematic equations to solve the setting position and posture of the coordinate system of the end link rod in relation to the base coordinate system. And then the corresponding coordinate values are input to the manipulator's trajectory planning, and the actual end position and posture information are compared with the results from solving the equation. Two groups of joint variable values are randomly chosen:

$$q_1 = [1.0053 \quad 0.50265]$$

$$q_2 = [-0.50265 \quad -0.50265]$$

The error is very small through the comparative analysis of the actual value and set point, which explains that the kinematics equation and the model are reliable and consistent.

The Simulation of Trajectory Planning

According to the requirements of the task, trajectory planning will designate in advance the robot's operating procedure and action process. The simulation method could describe in details the movement process of the industrial robot [6-10]. Planning can be made in the space of both joint and operation. We would introduce two main motions [4,5]: 1) PTP, Point-To-Point motion; 2) CP, Continuous-Path. With regard to continuous-path, not only the initial and final points of the manipulator need to be set, but also some other points between the two points (called path points) which must move along some specific path (path constraint) need to be indicated.

The design of the high-speed plate carrying manipulator adopts a point-to-point trajectory planning. Set A as the starting point, moving to B where a task is finished, and then set B as a starting point, moving to C where a certain task is completed, and then set C as a starting point, moving to point C to complete the preset tasks, and continue to move, and so on. Here, the move from point A to point B and from point B to point C does not exist any point between them, and there is not any requirement for the known path of the movement. Therefore, we can take this above planning as PTP planning.

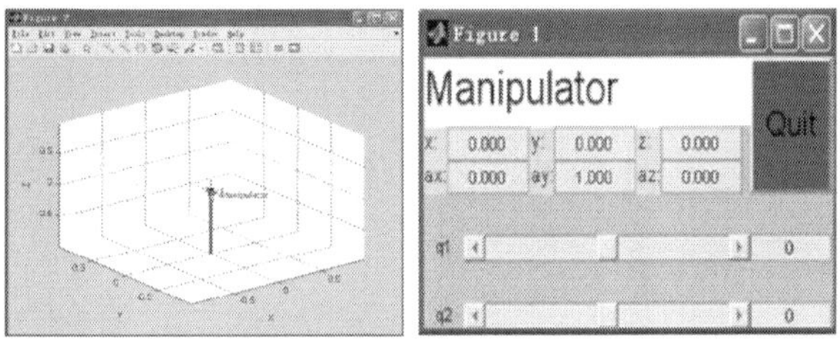

Figure 5. The three-dimensional diagram and slide block control chart of the high-speed plate carrying manipulator.

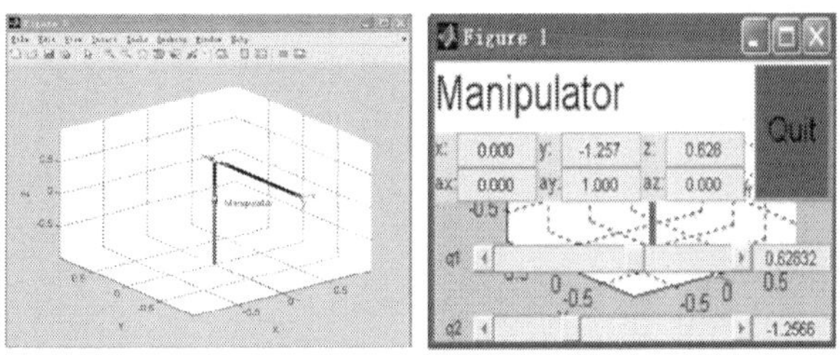

Figure 6. A state of high-speed plate carrying manipulator's three-dimensional diagram and slide block control chart.

The starting point is set to be $q_0 = [0 \quad 0]$, ending point $q_1 = [0.62832 \quad -1.2566]$, and between the two points the initial and final speeds are both zero, with the time of movement time: $t = 2$ s, the relative program as follows:

$q_0 = [0 \quad 0]; q_1 = [0.62832 \quad -1.2566]$

$t = [0: \quad 0.1: \quad 2]$

$q = \text{jtraj}(q_0, \quad q_1, \quad t)$

$\text{plot}(r, \quad q)$

By conducting the above program, we can observe the whole process of the high-speed plate carrying manipulator moving from Figures 5 and 6. We can also draw, by way of the function of

$$[q,\ qd,\ qdd] = \text{jtraj}(q_0,\ q_1,\ t), \text{plot}(t,\ q(:,\ i))$$

each joint's displacement curve, as shown in **Figure 7**(a) and (b). (Note: q represents the displacement, i represents the joint Numbers). Also we can draw speed curve as shown in Figures 7(c)-(d), acceleration curve as shown in Figures 7(e)-(f), through the corresponding functions:

$$\text{plot}(t,\ qd(:,\ i)), \text{plot}(t,\ qdd(:,\ i))$$

The Analysis Based on the Results of Simulation

From Figures 7(a) and (b), it can be seen that the displacement of the mobile joint 1 gradually changes from zero to 0.62832 m, movable joint 2 displacement from zero change to 1.2566 m; from Figures 7(c) and (d) we

(a) The displacement curve of joint 1

b) The displacement curve of joint 2

(c) The velocity curve of joint 1

(d) The velocity curve of joint 2

(e) The acceleration curve of joint 1

(f) The acceleration curve of joint 2

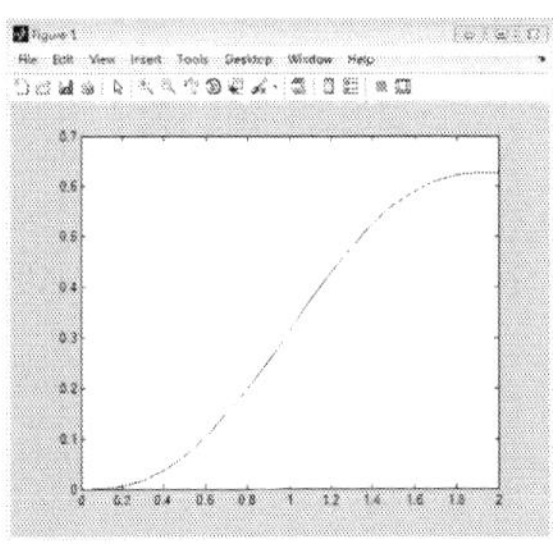

(a) The displacement curve of joint 1

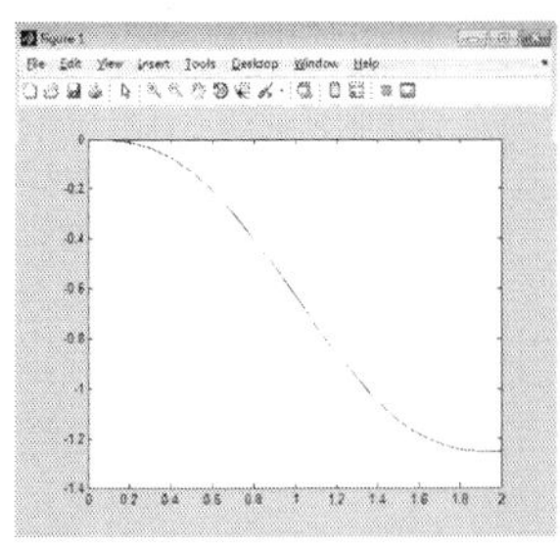

(b) The displacement curve of joint 2

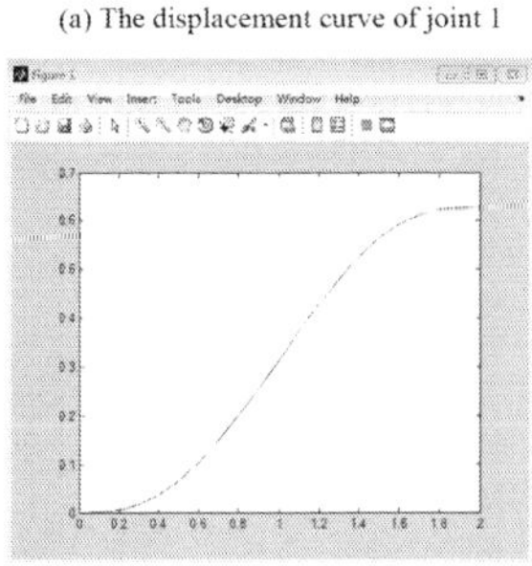

(c)The velocity curve of joint 1

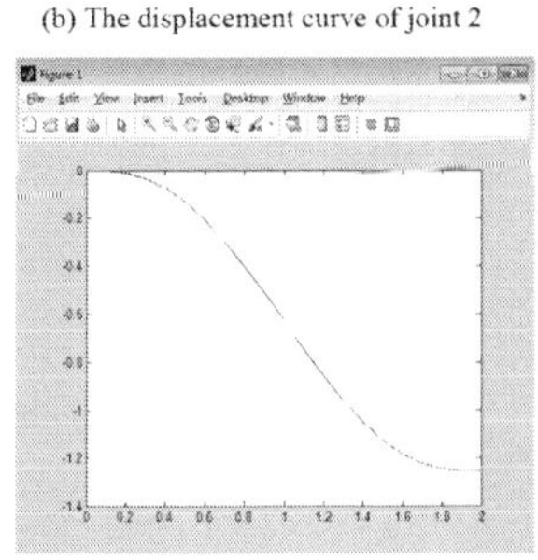

(d) The velocity curve of joint 2

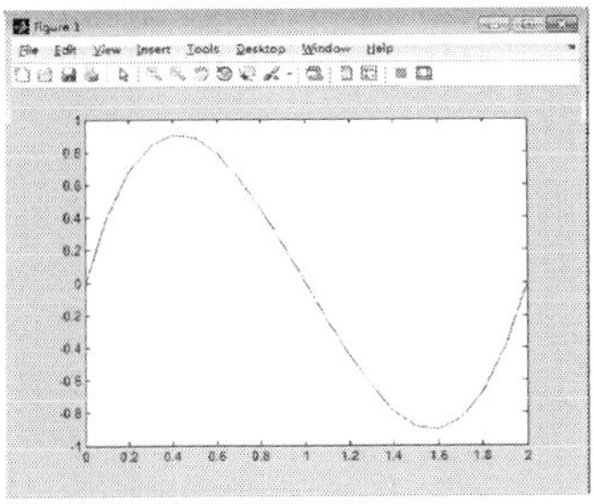

(e) The acceleration curve of joint 1

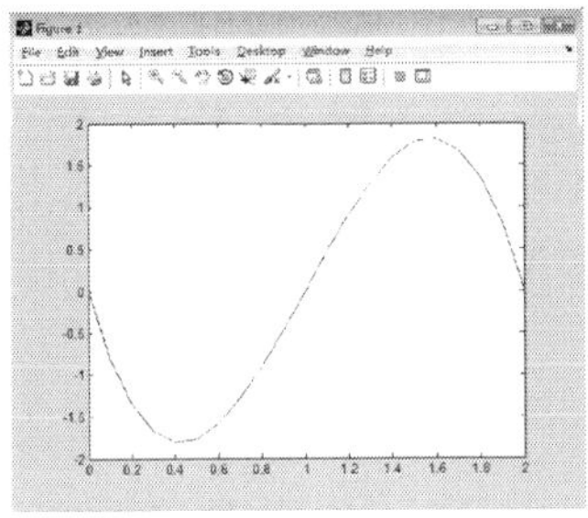

(f) The acceleration curve of joint 2

Figure 7. Displacement, velocity and acceleration curves of high-speed plate carrying manipulator.

can obtain that the initial and final speeds of the joints 1, 2 are zero, the maximum velocity appearing at the intermediate time t = 1 s; from Figures 7(e)-(f), it can be seen, the initial and final acceleration speeds are zeros, two maximum values in motion, with one positive and the other negative. The manipulator's displacement curve is smooth, and the curve of velocity and acceleration is continuous. Therefore, it is concluded that: in the process, the manipulator runs smoothly, and the whole body vibrates in a normal scope, which thus explains the validity of the design.

CONCLUSIONS

Ke Wang, Jiping Zhou The paper carries out a kinematical analysis of a highspeed plate carrying manipulator which is designed by using the module function of Robotics Toolbox in the Matlab. Based on this, we have achieved the following conclusions in three aspects: 1) the structure and parameters of the plate carrying manipulator have been designed; 2) the kinematics model has been established with the standard D-H method, and its positive and inverse kinematics have been analyzed; 3) by using Robotics Toolbox in Matlab, the kinematics model has been set up in order to make the study of the motions of manipulator easier and more apparent, and to conduct a kinematics simulation for the manipulator for obtaining the displacement, velocity and acceleration curve of each manipulator joint. Thus, this proves the validity of the design of manipulator structure parameters, as well as provides the reliable basis for the theoretical analysis in terms of the design, dynamics analysis, and trajectory planning of control system of the manipulator.

REFERENCES

1. Q. Wang, X. J. Hu and L. Z. Li, "The Kinematic Analysis and Simulation of the Humanoid Welding Manipulator Based on the MATLAB," Materials Science and Engineering College of Hefei University of Technology, 2011.
2. J. J. Craig, "Introduction to Robotics," Mechanical Industry Press, Beijing, 2006.
3. P. I. Corke, "A Robotics Toolbox for MATLAB," IEEE Robotics and Automation Magazine, Vol. 3, No. 1, 1996, pp. 24-32. doi:10.1109/100.486658
4. X. S. Jiang, "Introduction to Robotics," Liaoning Science and Technology Press, Shenyang, 1994.
5. Z. X. Cai, "The Robotics," Tsinghua University Press, Beijing, 2000.
6. J. X. Luo and G. Q. Hu, "The Kinematic Simulation Research of Robot Based on the MATLAB," Journal of Xiamen University, JCR Science Edition, Vol. 44, No. 5, 2005, pp. 640-644.
7. J. Han and L. Hao, "Trajectory Planning and Simulation of Robot in Joint Coordinate System," Journal of Nanjing University of Science and Technology, Vol. 24, No. 6, 2000, pp. 540-543.
8. Z. X. Wang and W. X. Fan, "The Kinematics Analysis and Simulation of Industrial Robot Based on the Matlab," Mechanical and Electrical Engineering, Vol. 29, No. 1, 2012, pp. 33-37.
9. The MathWorks Inc., "Matlab the Language of Technical Computing Version 6," The MathWorks Inc., 2002.
10. P. I. Corke, "Robotics Toolbox for Matlab (Release 7.1) [EB/OL]," 2002. http://www.cat.csiro.au/cmst/staff/pic/robot

CHAPTER 10

Flexible Impact Force Sensor

Chen Liu, Yukio Fujimoto, Yoshikazu Tanaka

Department of Transportation and Environmental Engineering, Hiroshima University, Higashi-Hiroshima, Japan

ABSTRACT

When human body is punched by boxing glove, both body surface and boxing glove deform in a complex shape. The purpose of this study is to develop a flexible sensor that can be used in such interface. Firstly, several mechanical phenomena, which are the cause of the error signal of the sensor, are discussed. These are the influences of out-of-plane bending deformation,

Citation:- Liu, C. , Fujimoto, Y. and Tanaka, Y. (2014), "Flexible Impact Force Sensor". Journal of Sensor Technology, 4, 66-80. doi: 10.4236/jst.2014.42008.

shear force caused by rubbing, shear force caused by the Poisson's effect of contact material, and the lateral compressive force caused by the overhanging deformation of flexible material. Then, a prototype sensor that can eliminate the error factors of these is developed. The sensor is a distribution type impact sensor in which sixteen sensor elements are arranged in a 4 × 4 matrix. Punching experiments using a boxing glove are carried out by installing the sensor on the load cell, on the concrete wall and on the sandbag. From the experiment, it is found that the impact force can be measured with good accuracy by using the sensor. Despite the fact that the sensor has inadequate distribution number of sensor elements, the sensor structure meets mechanical requirements for the flexible impact sensor.

Keywords: Flexible Sensor, Impact Force, Boxing, Sandbag, Distribution Type Sensor, Human Body

INTRODUCTION

When a boxer punches the human body, both body surface and boxing glove deform in a complex shape. At present, the accurate measurement of impact force in such interface is said to be difficult. The reason is because the complex deformation and the complex force distribution are induced in the flexible interface and they affect the sensor output. Up to now, many kinds of tactile sensors such as resistance type [1], piezo-resistance type [2] , capacitance type [3] and light intensity type [4] have been developed. Most of them are assuming to mount the sensor on the robot members. However, it has not been studied much about the influences of the deformation and the associated force distribution at the flexible interface on the accuracy of the sensor output. The reason of this is most of the tactile sensors are intended to use under a small force level and a small deformation level on the installation surface [5] [6].

By the way, it has been reported that peak impact force of about 4 kN is generated when a strong boxer punches the face of crash test dummy [7]. When the hitting target is sandbag, the interface of glove and sandbag will be deformed more greatly. A suitable sensor which can measure the impact force in such condition is not seen at present. The purpose of this study is to provide a sensor that can measure the impact force accurately at the interface of large deformation and large force.

In this study, firstly, difficulty of force measurement in flexible interface is discussed based on the experiments using a sheet type sensor (Chapter 2). Then, some mechanical factors necessary to achieve the flexible sensor are discussed (Chapter 3). These factors are the out-of-plane bending deformation of sensor, the shear force caused by rubbing on sensor surface, the shear force generated by contact friction, and the transverse compression caused by the overhanging deformation of the flexible material.

In Chapter 4, as a sensor that can overcome the mechanical problems in flexible interface, a flexible sensor of distribution type is developed. In Chapter 5, boxing punching tests are carried out using a prototype sensor by installing the sensor on the sponge rubber sheet which is placed on the load cell or the concrete wall. Further, the punching test is carried out by installing the sensor on a sandbag. In particular, the present study is focused to achieve a sensor structure that is hard to cause trouble under a big impact force like boxing punching.

MEASUREMENT OF FORCE ON SOFT MATERIAL SURFACE BY USING A SHEET-TYPE SENSOR

Spherical Surface Pressing on the Flexible Material Surface

As a simple case, spherical pressing experiment of a sensor installed on the flexible material surface was carried out. Figure 1(a) shows a sheet type sensor made from piezoelectric film (PVDF) of which both surfaces are covered by silicon rubber sheets [8]. When compressive force F is applied as shown in Figure 1(b), the rubber sheets shrink in the thickness direction and also be extended in the horizontal direction by the Poisson's effect. Since the elastic modulus of PVDF is much larger than that of rubber sheet, the PVDF is stretched by the rubber elongation in the horizontal direction. Then, the output signal proportional to the compressive force F is generated in the sensor.Figure 2(a) shows a sheet type flexible sensor with slit notches around the sensor edge. A single sheet type sensor is not possible to follow the out-of-plane deformation. In order to make out-of-plane deformation possible, one way is to process the slit notches around the sensor edge. The slit notches are expected to get a similar result with "preventing wrinkles of thin paper to wrap an apple".Figure 2(b) is the sensor covered by doubly stretch cloth on both surfaces. A polyethylene sheet is sandwiched in between the doubly stretch cloth. When the sensor with slit notches is directly placed on the flexible material and pressed by curved surface object, the output signal of the sensor is influenced by the contact friction between the sensor and the contact object. The stretch cloth and the polyethylene sheet play a role to reduce the influence of contact friction by slide effect.

Figure 3 shows the condition of compression test. Thick acryl plate was fixed on the top of a load cell. Sponge rubber plate with 40 mm thickness was placed on the acryl plate. The sensor was put on it and pressed by spherical body with 50, 100 or 150 mm diameter. The experiment was carried out by changing the compressive force.

Figure 4 compares the relationship between the peak force of the load cell and the sensor output. A good linear relationship is seen in the figure. Despite the change in the diameter of spherical body, the output gradient does not change much. The dashed line in the figure shows the output gradient under the condition that the sensor is placed directly on the flat acryl plate. The sensor detects the normal force on the curved surface, but the load cell detects the vertical component. The difference in gradient of the two straight lines is considered to be related with the detection characteristics.

Fist Pressing, Palm Pressing and Fist Punching on the Surface of Flexible Material

Next experiment was carried out under the condition that the contact interface deforms in a complex shape. Figure 5 shows the conditions of fist pressing, palm pressing and fist punching. The experiments were conducted by changing the compressive force. Figure 6 shows the relationship between the peak force of the load cell and the peak output of the sensor during compression. A quite large scatter of plots is seen in the figure.

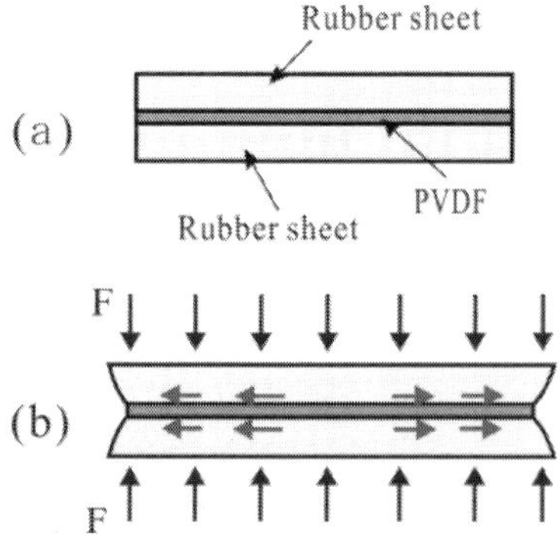

Figure 1. Sensor cross-section and its detection principle of compressive force.

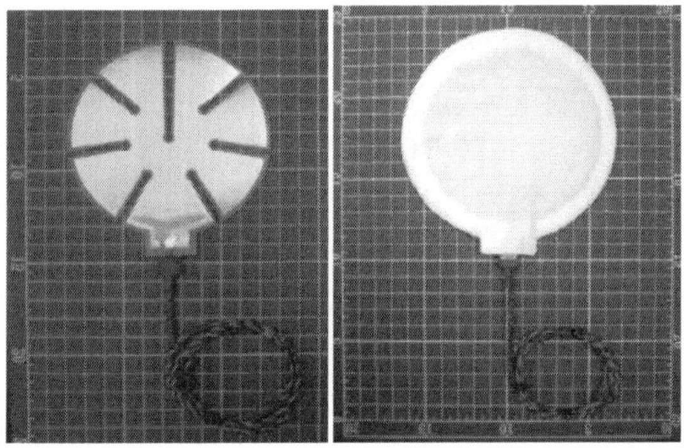

(a) (b)

Figure 2. Sheet type flexible sensor. (a) Main body of sensor with slits in outer edges; (b) Sensor covered by stretch cloth.

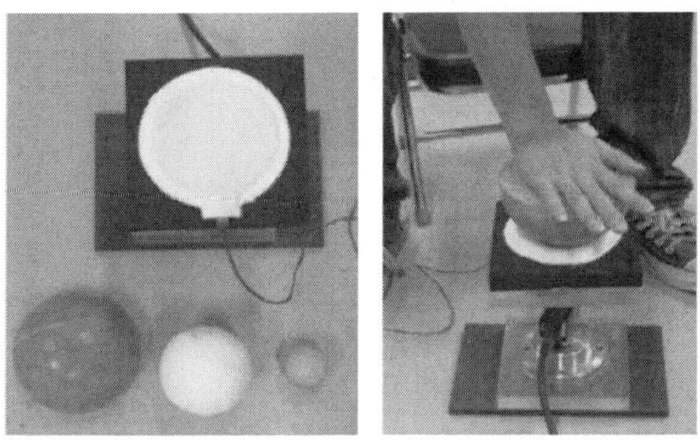

Figure 3. Spherical press of the sensor on the sponge rubber.

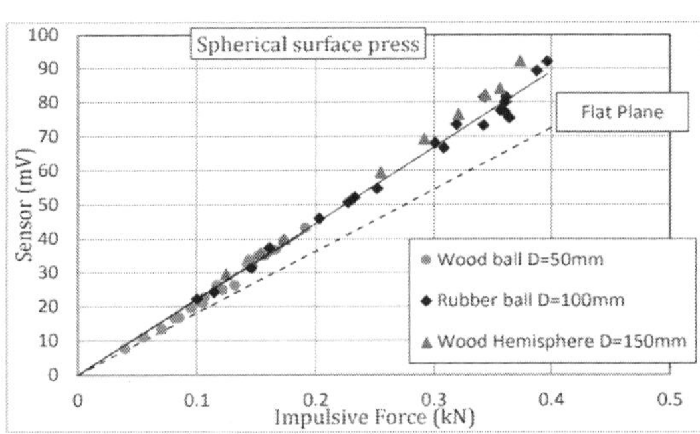

Figure 4. Sensor output vs. load cell output. Dashed line is the output gradient of the sensor on flat plane surface..

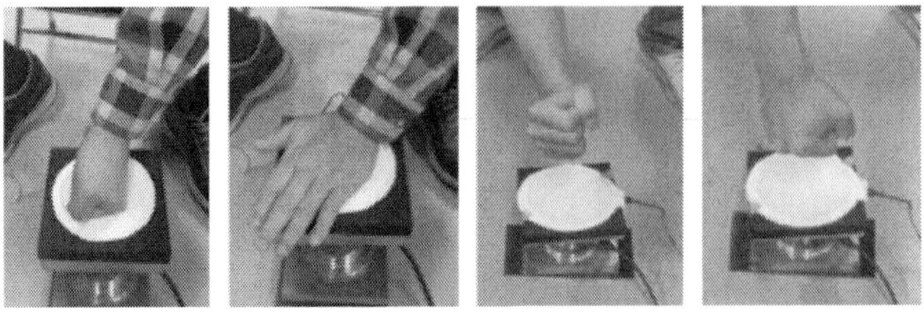

Figure 5. Fist press, palm press and fist punch of the sensor placedon the sponge rubber plate.

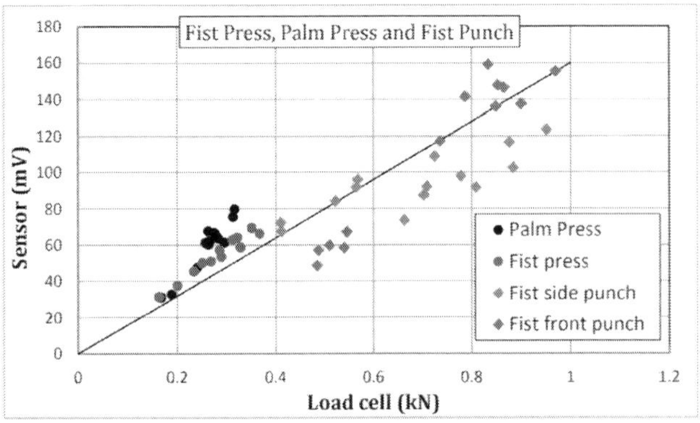

Figure 6. Relationship between the output peak of sensor and the peak force of load cell.

In the experiment of Section 2.1, the loading side was a hard material with simple contact surface and the receiving side was a flexible material. On the other hand, the contact surface of fist or palm is more complex and it changes in every trial. Further, rubbing force and inclined force are also working. The cause of the scatter of the output gradient can be considered as the influences of complex force distribution and the excessive out-of-plane deformation occurring in the sensor. This result means that a single sheet-type sensor is not suitable to achieve the measurement of impact force on the flexible material surface.

Figure 7 classifies the condition of impact force measurement into four levels, in view of the difficulties of measurement. Figure 7(a) is the condition that a hard object with simple contact geometry collides. Figure 7(b) is the condition that a soft object with simple contact geometry collides. In accordance with the deformation of soft object, in this condition, unexpected shear forces (rubbing force and contact friction) work at the interface. Figure 7(c) is the condition that a hard object with complex contact geometry collides. In this case, the out of-plane bending deformation of the sensor and the distributions of shear forces at the interface become more complex. Figure 7(d) is the condition that a soft object with complex contact geometry collides. In this case, stress distribution and deformation becomes most complicated. Boxing punching on sandbag surface is close to the condition of Figure 7(d).

MECHANICAL ISSUES IN IMPACT FORCE MEASUREMENT ON FLEXIBLE MATERIAL SURFACE

The objective of this research is to develop a sensor which can be used even when both objects of loading side and load receiving side are soft and the contact geometry is complicated. After repeating several experiments, the authors realized that the discussions about the error signals are necessary for the flexible sensor. They will be described in the following.

Influence of Out-of-Plane Bending Deformation of the Sensor Element

Figure 8 shows a sensor element used in the flexible sensor. A pair of silicon rubber sheet is adhered on the both surfaces of piezoelectric film (PVDF).

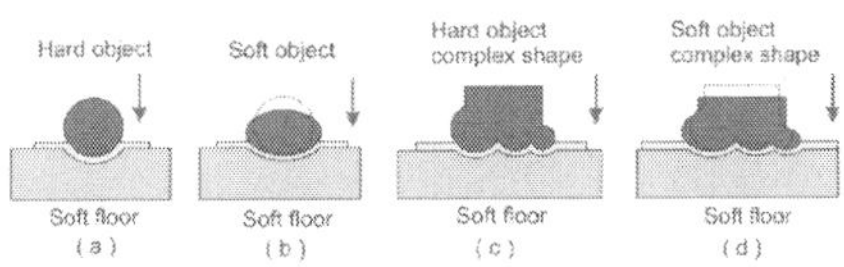

Figure 7. Difficulty of force measurement at flexible interface is classified into four levels depending on the shape of the impact object. (a) Hard object with simple contact geometry; (b) Soft object with simple contact geometry; (c) Hard object with complex contact geometry; (d) Soft object with complex contact geometry.

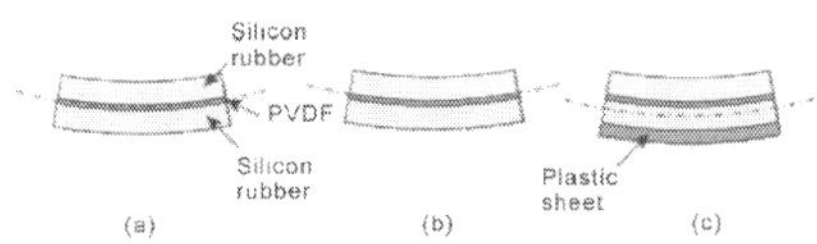

Figure 8. Error signal caused by the out-of-plane bending deformation of sensor element. (b) and (c) represent the shift of the neutral axis by bending deformation.

The bending neutral axis comes to the center of PVDF thickness, theoretically, due to the vertically symmetric structure. Therefore, the sensor output should not occur almost by bending deformation. However, in practice, by variation in the thickness of the adhesive layer and etc., the neutral axis is shifted slightly from the center of PVDF thickness [9]. Thus result in the occurrence of tensile or compressive strain in PVDF (Figure 8(b)). If a resin sheet like a flexible substrate is adhered on one side surface of the silicon rubber, the bending neutral axis is further shifted, and results in greater strain (Figure 8(c)). It is extremely difficult to control the strain of PVDF caused by bending deformation.

Influence of Rubbing Force on the Sensor Surface (Shear Force Caused by Rubbing)

Figure 9 shows a schematic view of a sensor element placed on the floor, on which a compressive force F on the left side and a rubbing force S on the right side are acting. The vertical reaction force R_F occurs below the F, and the shear force R_S against S occurs caused by the friction between floor and sensor element. As the result, tensile strain is produced in the region between the points F and S (shown by red arrow). The rubbing force S sometimes is generated as the horizontal component of inclined force which is applied to the sensor surface.

Influence of Contact Friction on the Sensor Surface (Shear Force Caused by the Poisson's Effect)

Figure 10(a) shows a sensor element sandwiched in between the metal plates. Since the elastic modulus of silicon rubber is smaller than those of PVDF and aluminum plate, the silicon rubber deforms into barrel-shape when compressive force F acts.

Figure 10(b) shows a state in which a pair of sponge rubber sheet is sandwiched between the silicon rubber and aluminum plate additionally and applied a compressive force F. Since the elastic modulus of sponge rubber is further smaller than that of silicon rubber, the sponge rubber deforms into bigger barrel-shape. Even if the same F is applied, the silicon rubber is stretched more in this condition. This means that the sensor output is affected by the Young's modulus of the contact material as well as its thickness.

As an example, Figure 11 shows the magnitude of sensor output in accordance with the change of contact material. The output is adjusted as 1.0 when the sensor element is sandwiched directly between the aluminum plates. A circular sensor element with 50 mm diameter is used. The silicon rubber on both sides of PVDF is

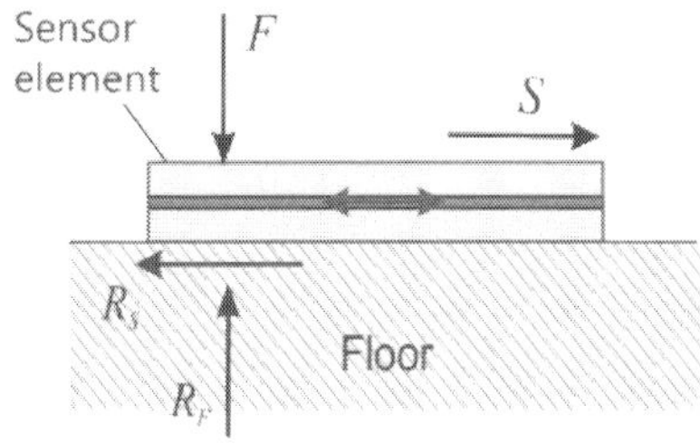

Figure 9. Example of the tensile strain (red arrow) generated in the sensor element caused by rubbing force. F is compressive force, S is rubbing force and R_F and R_S are the corresponding reaction forces.

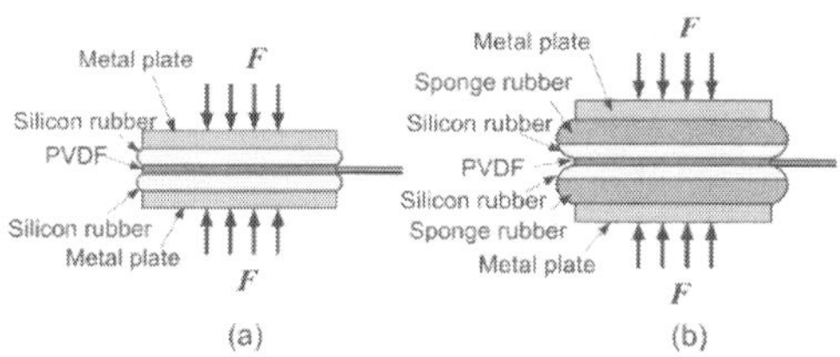

Figure 10. Error signal caused by the difference of contact friction.

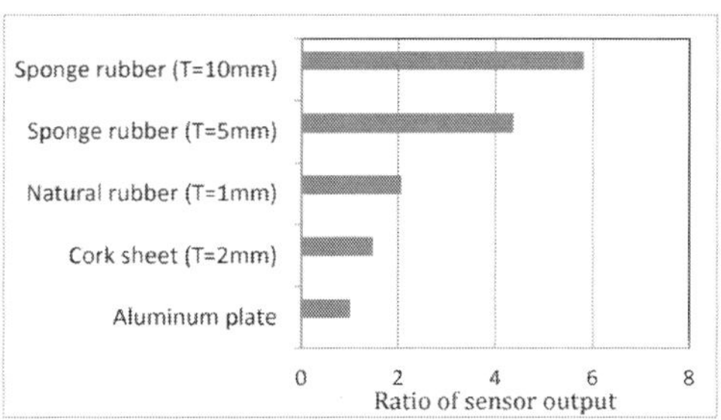

Figure 11. Variability of sensor signal caused by the difference of contact material.

1 mm thickness. Sponge rubber sheet (T = 5 mm, 10 mm thick) or natural rubber sheet (T = 1 mm) or cork plate (T = 2 mm) was sandwiched between the sensor element and the aluminum plate. As the contact material becomes softer, the sensor output increases as shown in the figure. In the case of thick sponge rubber (T = 10 mm), the magnitude becomes several times compared with the case of the aluminum plate.

Influence of Lateral Compression Caused by the Overhanging Deformation of Flexible Material

Figure 12 shows the state that a soft material in large size enough to cover the sensor element collides with the element. When the force F acts on the sensor surface, impact force (solid curve) in compressive direction is produced as shown in Figure 12(a) at first. However, by the succeeding deformation of soft material, a part exceeding the area of the sensor element may press the element from the lateral direction. By the transverse compressive force from the surrounding, the sensor generates a signal in opposite direction as shown by dashed curve of Figure 12(b). As a result, the sensor signal becomes the sum of solid and dashed curve as shown in Figure 12(c).

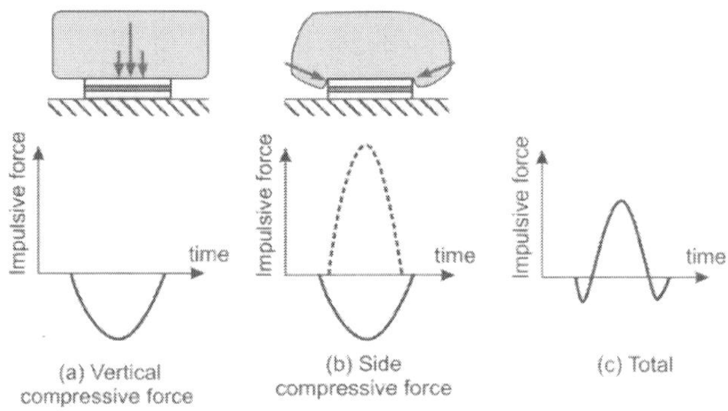

Figure 12. Lateral compression phenomenon caused bythe overhanging deformation of flexible material.

Figure 13 shows an example of such waveforms by the free drop test of soccer ball.

A circular sensor element with 50 mm diameter was placed on the load cell, and the ball was fallen onto the sensor. Figure 13 compares the sensor output (red curve) and the load cell output (black curve). When bottom of the ball collide the center of sensor, the sensor output is generated on the negative side.

After that, it turns into the opposite side in the halfway, since the overhanging part of deformed ball presses the sensor laterally.

Therefore, the sensor waveform becomes quite different from the load cell waveform. Such lateral compression occurs as well when the sensor bottom is in buried condition in soft material.

FLEXIBLE SENSOR MADE FROM PLURAL SENSOR ELEMENTS ARRANGED IN A MATRIX

In order to eliminate the above influences, in this study, we develop a sensor element of rigid plate type. The sensor elements are distributed in a matrix, and the seam portion of each sensor element is connected flexibly in order to ensure the flexibility. Further, a new manner to acquire the sensor signal without using electrical wiring is employed.

Prototype of Sensor Element

Figure 14 shows the sensor element with 30 mm square and the twenty elements made for this research. At first, silicon rubber sheet with 1 mm thickness was adhered on both surfaces of PVDF. Next, a circular hole was drilled in the center of the silicon rubber, and the hole was filled with conductive rubber.

The conductive rubber was designed to contact the electrode of PVDF. Furthermore, aluminum plate was adhered on the silicon rubber on each side. The upside and downside electrode of PVDF are connected with the upper and lower aluminum plate respectively via conductive rubber. In order to assure the electric contact between the PVDF and the aluminum plate, the conductive rubber was pressed by screw. The reason of using 5 mm thick aluminum plate is that metal plate with certain thickness is necessary in order to prevent the bending deformation.

Each sensor element was placed on load cell and punched by hand hammer (with felt cap) several times. Figure 15 shows examples of the impact force waveforms. Solid curve is the sensor output and dashed curve is the load cell output. Eventually, sixteen elements among them were selected and used for the flexible sensor.

Figure 16 shows the output gradient of the chosen sensor elements. The horizontal axis is the peak impact force obtained by load cell and the vertical axis is the peak output of the sensor element. As shown in the figure, load-bearing capacity of a single sensor element is more than 1.5 kN. If the impact force is distributed to several sensor elements, then the load bearing capacity is more than 10 kN.

Prototype of Flexible Impact Force Sensor

Figure 17(a) shows a flexible sensor in which sixteen sensor elements are arranged in a 4 × 4 matrix. The sensor is covered with artificial leather.

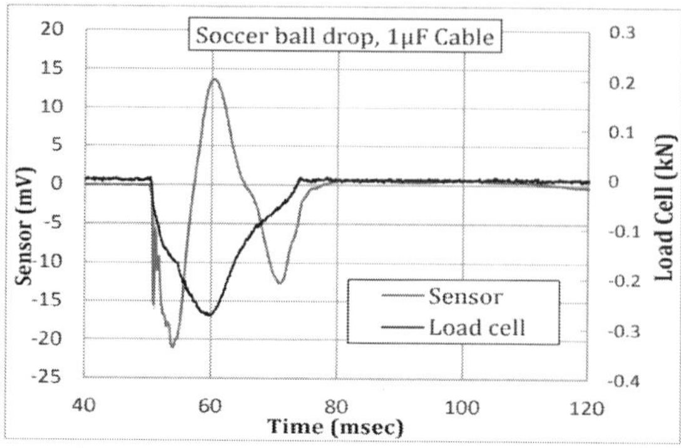

Figure 13. Impact force waveform in which vertical and lateral compression is mixed.

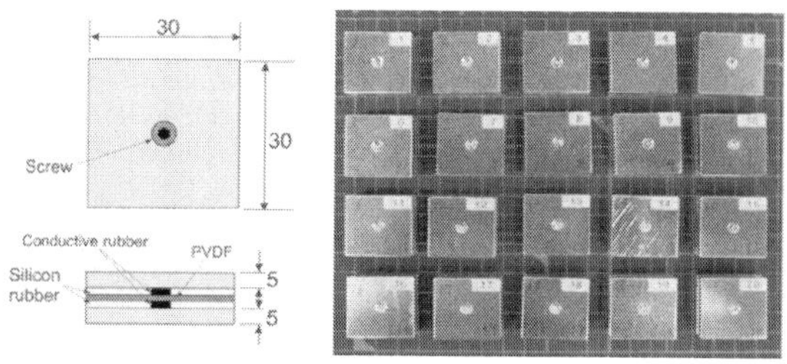

Figure 14. Configuration of the prototype sensor elements.

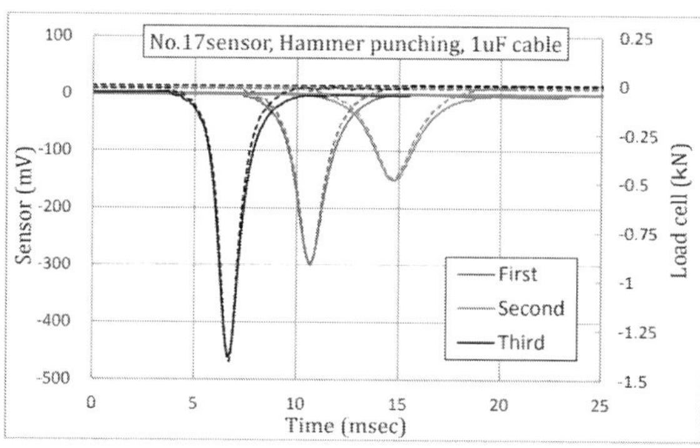

Figure 15. Example of impact force waveform of sensorelement (solid curve) and load cell (dashed curve).

Its surrounding and the vertical and horizontal column are finished by sewing. The total size is about 230 mm square. Figure 17(b) shows the sensor inserted into conductive cloth bag for static shielding. Green cable is the ground wire. Piezoelectric sensors are prone to be influenced by static electricity [9] [10] . In order to prevent the signal drift by static electricity, it is preferable to use in the state of electrostatic shielding. Figure 17(c) shows the state in holding by bending the sensor.

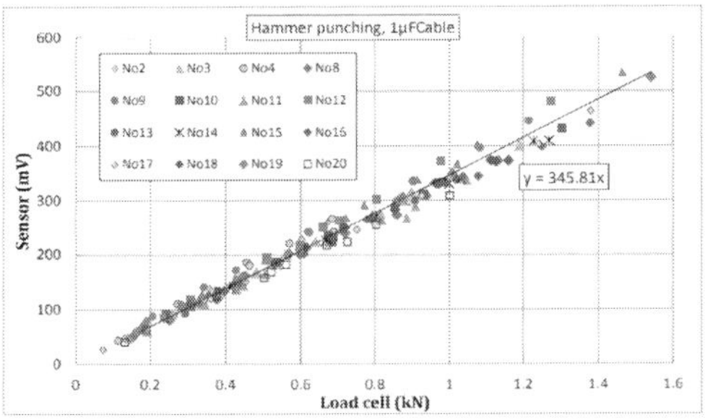

Figure 16. Relationship between sensor output and load cell output obtained by the hammer punching test.

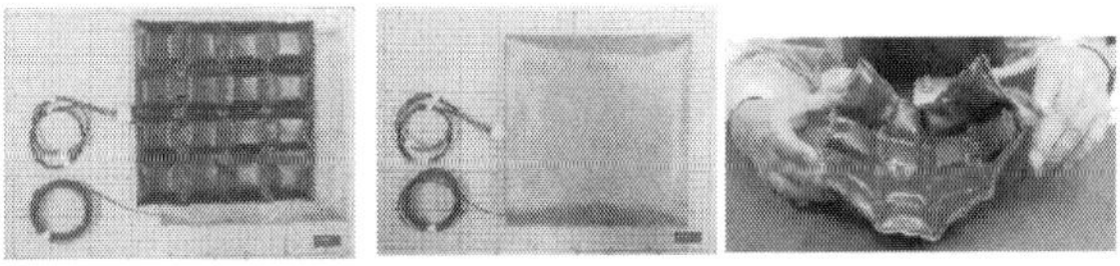

Figure 17. Prototype sensor. (a) Flexible sensor consisting of 4 × 4 sensor elements; (b) Conductive cloth bag; (c) Hold the sensor bend.

Even though the sensor element itself is a rigid body, the connecting portions of the sensor elements can keep the flexibility.

Figure 18 shows the cross section view of the flexible sensor. The upper aluminum plates of all sensor elements are in contact with the upper conductive cloth. Further, the lower aluminum plates of all sensor elements are in contact with the lower conductive cloth. In order to prevent short-circuiting of the upper and lower conductive cloth, bilayer deer skin is sandwiched in the thickness center of the sensor as the insulation sheet. After sewing the intermediate portion of the sensor element and element, each sensor element is in a state put into a pocket.

The electrical wires are connected to the lower and upper conductive cloth respectively. In other words, the output signal of the sensor element is obtained from the conductive cloth via upper and lower aluminum plate. Because there is no electrical wiring to connect each sensor element, there is no risk of electric disconnection.

It is obvious that the flexibility of the sensor increases as the increase of the number of sensor elements distributed. If the sensor element is assumed as "a plate placed on the elastic floor", the bending moment reduces along with the sensor dimension becomes smaller. Therefore, it is possible to reduce the sensor thickness as the size of sensor element is small.

EXPERIMENT

Comparison of Sensor Output and Load Cell Output

Figure 19 shows the photos of boxing punching test. First, thick acryl plate was fixed on the top of a load cell with 10 kN capacity. Sponge rubber sheet with 40 mm thickness was placed on it. Then, the flexible sensor inserted in the conductive cloth bag was placed on the sponge rubber and a student punched it by boxing glove. Since the sponge rubber is pretty thick, the central portion of the sensor is deformed concave by the punching.

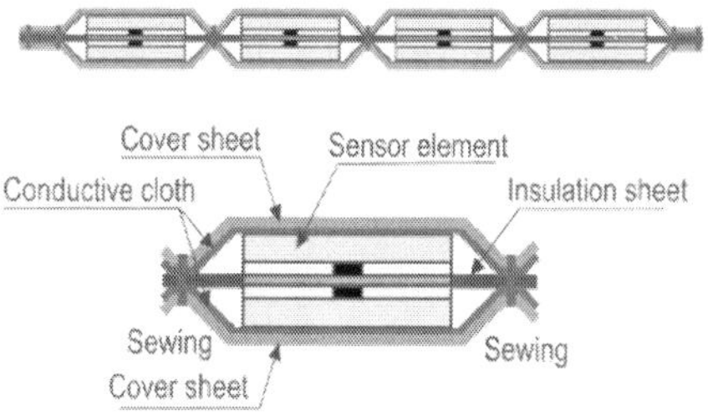

Figure 18. Schematic view of sensor cross-section. Cover sheet is artificial leather. Insulation sheet is bilayer deer skin.

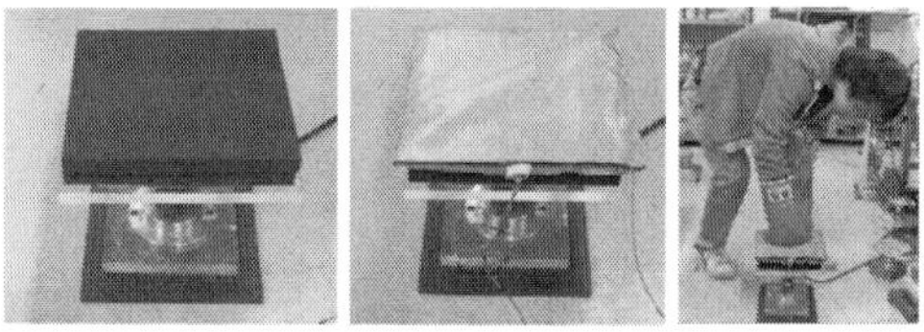

Figure 19. Impact test of flexible sensor on the load cell. (a) Sponge rubber (t = 40 mm) put on the load cell; (b) Sensor placed on the sponge rubber; (c) Punching test by boxing glove.

Also, similar experiment was conducted under the condition that the flexible sensor was placed directly on the acryl plate.

Figure 20 shows an example of impact force waveform measured in the above punching test. Solid and dashed curves indicate the experiments on the sponge rubber and the acryl plate respectively. The red curve is sensor output and the black curve is load cell output. The impact force waveform of the sensor is in good agreement with that of load cell.

Figure 21 shows the relationship between output peak of the sensor and output peak of the load cell. Regardless of the presence or absence of sponge rubber, the sensor output shows almost same gradient. This means that the sensor can measure the impact force in the flexible interface.

Figure 22 shows sponge sheet with 40 mm thickness overlaid on the sponge rubber. The total thickness of the bilayer is 80 mm. The sensor was placed on the sponge sheet and punched by boxing glove. In the figure, two examples of the impact force waveform are compared with that of the load cell output. As the sponge is so soft, the sensor was greatly deformed in a concave shape. There is somewhat difference in the waveform of sensor and load cell.

Figure 23 shows the sensor placed in bent condition on the wood semicircular column with diameter 150 mm. Some positions of the sensor were fixed with gum tape. Then, the sensor was punched by boxing glove. The relationship between the load cell output and the sensor output are indicated by black diamond. The red circle in the figure is the output gradient of the experiment shown in Figure 22. These two gradients are substantially coincident. Moreover, this output gradient is in good agreement with that in Figure 21.

Punching Test of the Sensor Installed on Concrete Wall

Figure 24 shows the punching test by boxing glove on concrete wall. Firstly, sponge rubber sheet (40 mm thickness) was pasted on the concrete wall. The sensor was placed on the sponge rubber sheet and the several parts are fixed on the wall with gum tape.

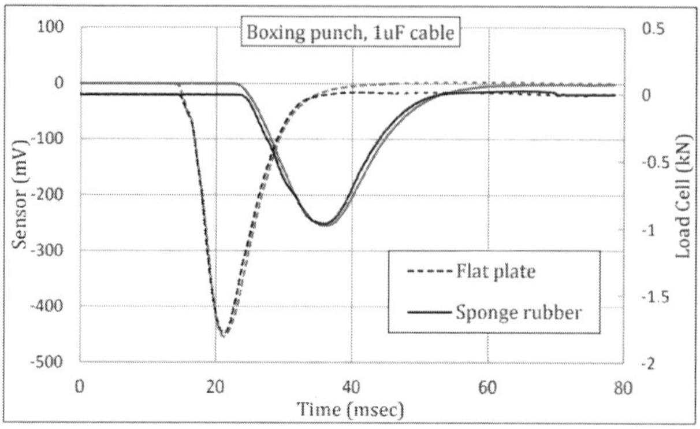

Figure 20. Impact force waveforms of the sensor and the load cell. Red curve is sensor, black curve is load cell.

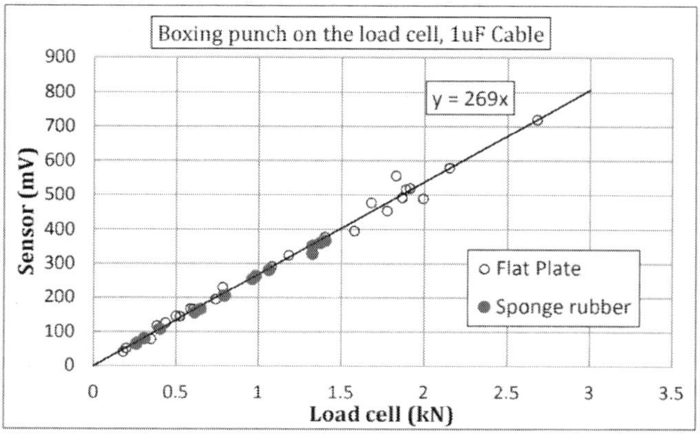

Figure 21. Comparison of the peak output between sensor and load cell.

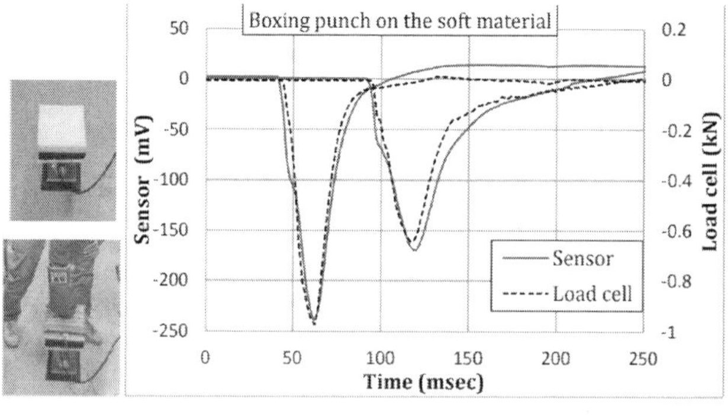

Figure 22. Impact force waveform of sensor and load cell. Sponge is overlaid on top of the sponge rubber.

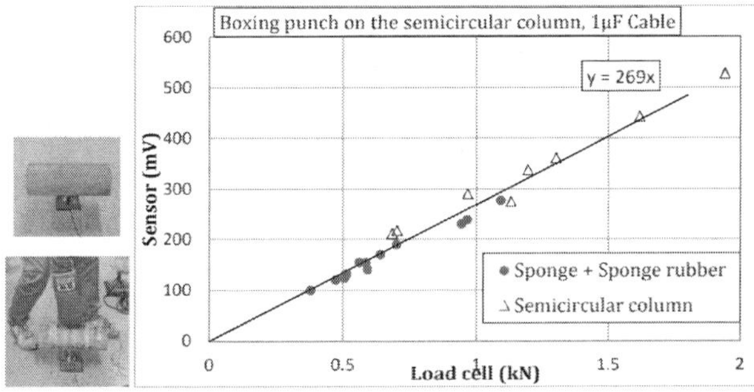

Figure 23. Relationship between sensor output and load cell output. Black diamond: semicircular column. Red circle: sponge & sponge rubber sheet.

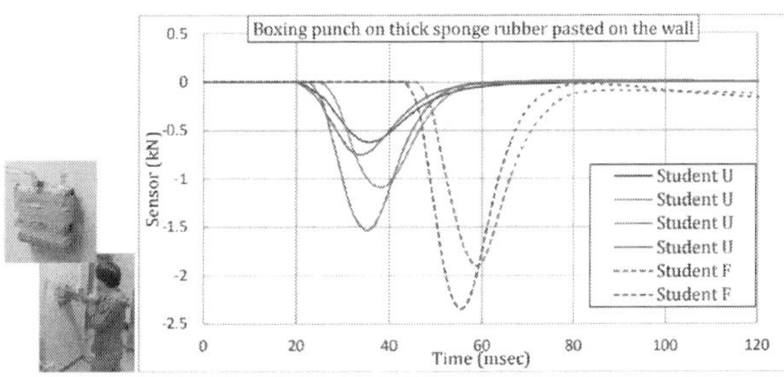

Figure 24. Boxing punching of the sensor installed on the 40 mm thick sponge rubber pasted on concrete wall.

Four impact force waveforms weakly punched by student "U" and two waveforms strongly punched by student "F" are shown in the figure. The peak impact force by the strong punching is around 2 kN. The contact time is about 30 - 40 msec.

Punching Test of the Sensor Installed on Sandbag

Figure 25 shows the experiment of boxing punching sensor was placed on the surface of a sandbag (soft sandbag) of gymnasium and fixed with gum tape. Compared with the punching on the concrete wall, the sandbag is softer and a concave deformation is induced on sandbag surface by punching.

Figure 26 shows the impact force waveforms. The peak impact force ranges 0.7 kN and 0.9 kN. The contact time is about 70 msec and this is longer than that on the wall punching. In the figure, it is seen that the gradient of impact force becomes gentle from the middle of unloading. This reason is considered as follows. The sandbag swings rearward by the punching.

Therefore, the forward speed of the glove persists for a while even after passing the peak load. Since the pull-back of the glove is delayed, the contact between glove and sandbag is continued for a while. The difference of the gradient of unloading segment is clear when compared to the waveform of wall punching in Figure 24.

IMPROVEMENT OF THE SENSOR FLEXIBILITY

Sensor elements described above is 30 mm square and 12 mm thickness. To improve the flexibility of the sensor, it is necessary to increase the distribution density by reducing size and thickness of the sensor element. Figure 27(a) shows slightly small sized sensor elements with 25 mm square and 8 mm thickness. Figure 27(b) is a flexible sensor made from the six elements.

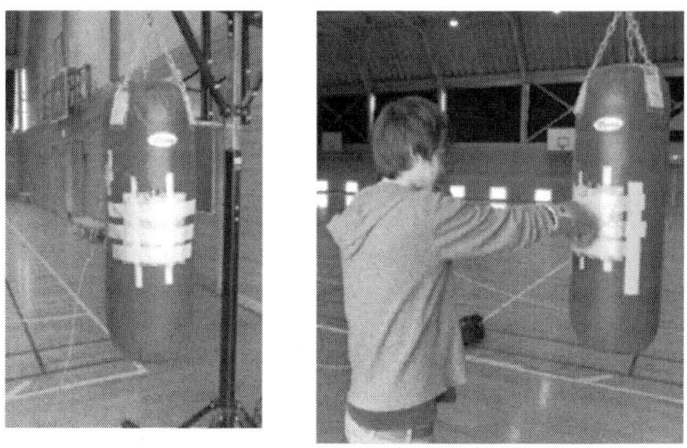

Figure 25. Boxing punching of the sensor installed on the sandbag.

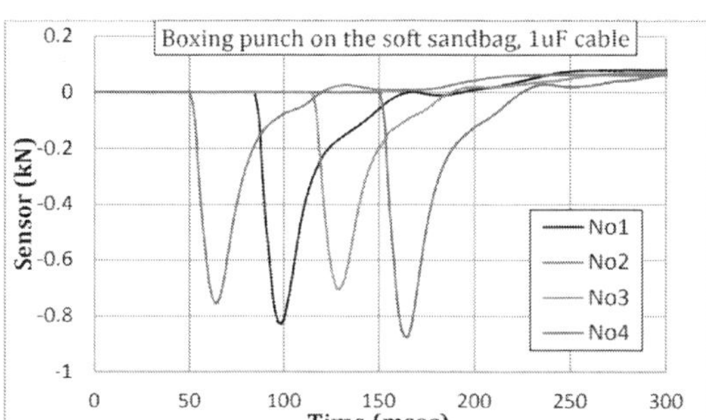

Figure 26. Impact force waveform of the sensor installed on the soft sandbag surface and punched by boxing glove.

Figure 28(a) shows the sensor fixed to the instep of soccer shoes with cords and rubber bands. Figure 28(b) shows the impact force waveforms of instep kick of soccer ball. The strength of the kick is the extent to path to nearby players. The peak impact force is about 1 kN and the impact time is about 12 - 15 msec. This contact time is close to the value shown in the reference [11].

From the experience, it seems possible to miniaturize the element size to 15 mm square and 5 mm thickness. When the element becomes smaller than that, however, reconsideration of the manufacturing process is necessary. In spite of the rough distribution density of the sensor elements, the sandbag sensor was not much discomfort at the time of use. On the other hand, in the soccer shoe sensor, the sensor weight and the lack of flexibility were felt uncomfortable. “Flexibility” and “light weight” are thought to be more important for the sensors installed on human body side.

CONCLUSIONS

This study is aimed to develop a flexible sensor which can be used in the measurement of impact force in flexible interface. The boxing punching is chosen as the measurement object. The results obtained are as follows.

Figure 27. (a) Sensor elements of 25 mm square and 8 mm thickness; (b) Flexible sensor made fromsix sensor elements.

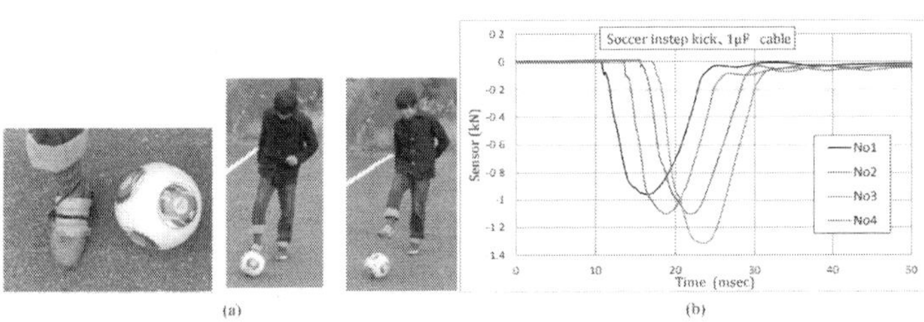

Figure 28. (a) Set up of the sensor on soccer shoe; (b) Impact force waveforms of instep kick of soccer ball.

1) A flexible sensor in which sixteen sensor elements are arranged in a 4 × 4 matrix is developed. Each sensor element is a rigid plate structure. The sensor elements are arranged at regular intervals. Both entire surfaces of the sensor elements are covered with artificial leather, and its surrounding and the vertical and horizontal columns are finished by sewing. Flexibility of the sensor is achieved by the flexible leather connections between adjacent sensor elements;
2) Sum of the output signal of each sensor element is collected through the conductive cloth. Since there is no electrical wiring inside the sensor, the sensor structure is hard to break by the impact force which is realized;
3) Boxing punching experiments were carried out by installing the sensor on concrete wall and sandbag. The flexible sensor showed a good performance in the experiments;
4) The proposed sensor is inadequate distribution number of sensor elements, but the proposed sensor includes mechanisms to achieve a flexible sensor with high accuracy. By increasing the distribution density of the sensor element, it can approach to the practical sensor.

Flexibility, thinness, lightness, toughness and accuracy are the keys of flexible impact force sensor. Overcoming of these problems is essential for the realization of the sensor that can be installed on human body or robot.

ACKNOWLEDGEMENTS

Chen Liu, Yukio Fujimoto, Yoshikazu Tanaka The authors wish to thank Tadashi Uesugi, Takahiro Namikawa, Atsushi Maeda and Takuya Okada of the Graduate School of Engineering, Hiroshima University for their assistance in experiments. This work was partially supported by the Venture Business Laboratory of Hiroshima University.

REFERENCES

1. Ishikawa, M. and Shimojo, M. (1982) A Method for Measuring the Center Position of a Two Dimensional Distributed Load Using Pressure-Conductive Rubber. Transactions of the Society of Instrument and Control Engineers, 18, 730-735.
2. Mei, T., Li, W.J., Ge, Y., Chen, Y., Ni, L. and Chan, M.H. (2000) An Integrated MEMS Three-Dimensional Tactile Sensor with Large Force Range. Sensors and Actuators, 80, 155-162. http://dx.doi.org/10.1016/S0924-4247(99)00261-7
3. Lee, H.K., Chang, S.I. and Yoon, E. (2006) A Flexible Polymer Tactile Sensor: Fabrication and Modular Expandability for Large Area Deployment. Journal of Micro Electromechanical Systems, 15, 1681-1686. http://dx.doi.org/10.1109/JMEMS.2006.886021
4. Donati, M., Vitiello, N., De Rossi, S.M.M., Lenzi, T., Crea, S., Persichetti, A., Giovacchini, F., Koopman, B., Porobnik, J., Munih, M. and Carrozza, M.C. (2013) A Flexible Technology for the Distributed Measurement of Interaction Pressure. Sensors, 13, 1021-1045. http://dx.doi.org/10.3390/s130101021
5. Lee, M.H. and Nicholls, H.R. (1999) Review Article Tactile sensing for Mechatronics—A State of the Art Survey. Mechatronics, 9, 1-31. http://dx.doi.org/10.1016/S0957-4158(98)00045-2
6. Lumelsky, V.J., Shur, M.S. and Wagner, S. (2001) Sensitive Skin. IEEE Sensors Journal, 1, 41-51. http://dx.doi.org/10.1109/JSEN.2001.923586
7. Walilko, T.J., Viano, D.C. and Bir, C.A. (2005) Biomechanics of the Head for Olympic Boxer Punches to the Face. British Journal of Sports Medicine, 39, 710-719. http://dx.doi.org/10.1136/bjsm.2004.014126
8. Fujimoto, Y. and Setyanto, A.T. (2007) Sheet Type Impact Force Sensor by the Use of Piezoelectric Film. Transactions of the Japan Society of Mechanical Engineers, Series C, 73, 184-191. (in Japanese) http://dx.doi.org/10.1299/kikaic.73.184
9. Fujimoto, Y., Shintaku, E., Tanaka, Y. and Fujiyoshi, J. (2012) Pad Type Compressive Force Sensor Suitable for High-Speed Impact Force Measurement. Transactions of the Japan Society of Mechanical Engineers, Series C, 78, 2438-2449. (in Japanese)
10. Fujimoto, Y., Liu, C., Uesugi, T., Tanaka, Y. and Shintaku, E. (2013) Pipe Surface Sensor for Impulsive Force Measurement. Transactions of the Japan Society of Mechanical Engineers, Series C, 79, 113-123.

(in Japanese) http://dx.doi.org/10.1299/kikaic.79.1949

11. Shinkai, H., Nunome, H., Ikegami, Y. and Isokawa, M. (2008) Ball-Foot Interaction in Impact Phase of Instep Soccer Kick. In: Reilly, T. and Korkusuz, F., Eds., Science and Football VI, The Proceedings of the 6th World Congress on Science and Football, Routledge, Abingdon, 41-46.

CHAPTER
11

A Sensing and Robot Navigation of Hybrid Sensor Network

Shuncai Yao[1], Jindong Tan[2], Hongxia Pan[1]

[1]Department of Electric Engineering, North University of China, Taiyuan, China

[2]Department of Electrical Computer Engineering, Michigan Technological University, Houghton, USA

ABSTRACT

Traditional sensor network and robot navigation are based on the map of detecting fields available in advance. The optimal algorithms are explored to solve the energy saving, shortest path problems, etc. However, in practical

Citation:- S. Yao, J. Tan and H. Pan, "A Sensing and Robot Navigation of Hybrid Sensor Network," Wireless Sensor Network, Vol. 2, No. 4, 2010, pp. 267-273. doi: 10.4236/wsn.2010.24037.

environment, there are many fields, whose map is difficult to get, and need to detect. This paper explores a kind of ad-hoc navigation algorithm based on the hybrid sensor network without the prior map. The system of navigation is composed of static nodes and mobile nodes. The static nodes monitor events occurring and broadcast. In the system, a kind of cluster broadcast method is adopted to determine the robot localization. The mobile nodes detect the adversary or dangerous fields and broadcast warning message. Robot gets the message and follows ad-hoc routine to arrive the events occurring place. In the whole process, energy saving has taken into account. The algorithms of nodes and robot are given in this paper. The simu- late and practical results are available as well.

Keywords: Robot Navigation, Hybrid Sensor Network, Routine Search

INTRODUCTION

Sensor network is an advance technology in recent years. There have been many papers in this field. Most of these are concerned in static sensor network or mobile sensor network separately. The communication and related technology have been studied. The static sensor network consists of the fixed nodes and adopts the ad-hoc communication. This kind of system can monitor the environment and send the messages to the base station. Energy would be saved, but the system can hardly do any- thing better if the whole condition is changed or the adversary available. In mobile sensor network, the nodes can adjust their places according the algorithms, which are based on the changing condition. However, at the same time, the energy would be cost much. It is well know that is a serious problem for sensor networks.

In navigation, most of these two systems always adopt a method that is based on mapping. That is to say, the whole environment map is known in advance. In spite that the optimal algorithms have been adopted in these systems, it is not so suitable for the genuine conditions yet.

In this paper, we explore a hybrid sensor network. It is composed of both static sensor nodes and mobile sensor nodes. The static sensor nodes are deployed regularly, and detect the events of environment. While, the mobile sensor nodes detect the adversary or dangerous field. When these fields are found the mobile nodes will stop at the edge of adversary or dangerous fields and send a message to robot. Finally, there will be more nodes at the edge of adversary or dangerous fields to protect the robot not to approach. The method is a compromise between the energy saving and flexible of whole system. Based on this scheme, non-mapping navigation comes into reality. It is proved that it is a better scheme in energy saving and flexibility through the simulate experiment at the end of this paper.

RELATED WORK

In a static sensor network, the quantity of nodes is an important problem. The more nodes means the more coverage of whole area and more accurate of navigation. Lots of research has focused on the sensor localization [1], structural monitoring [2] or protocols [3]. Recently, researchers have begun to explore the method, which low down the quantity. And, it is taken into account that the adversary and dangerous fields available in environment.

Since these methods adopt the skeleton graphs and optimal algorithm, the quantity of nodes and the price of communication have been slowed down.

The quantity of nodes is not such a pivotal problem in mobile sensor network because its flexibility. The problem of energy costs is tried to low down by sorts of algorithm based on the passive algorithm [4, 5] or statistic-learning algorithm. On the other hand, these algorithms are complicated. They are not so easy to come into realities and the use more capacity of nodes memory.

The sensor network of this paper develops the hybrid sensor network and adopts a kind of cluster navigation. At first, the static sensor nodes are deployed as regular as possible in the whole environment. That is to say, the localization of static sensor nodes should be arranged at the correct place, but we need not care it so much. These static sensor nodes may local in the adversary or dangerous fields, but it does not matter. We can take measure to solve the problem. The sensing program can be downloaded in these sensor nodes. They can sense the change of environment area that we care. The mobile sensor nodes, actually they are the small mobile robots, but they just do the dangerous fields detecting work and nothing else. Navigated robots are not these nodes. Actually, there is another robot being navigated. In Figure 1, the static sensor nodes are deployed in the whole area. They have sensing and communication abilities. Actually, these nodes are tmote Sky [6] made by Moteiv Corporation. They are ultra-low power wireless module for use in sensor networks, monitoring applications, and rapid application prototyping. Tmote Sky leverages industry standards like USB and IEEE 802.15.4 to interoperate seamlessly with other devices.

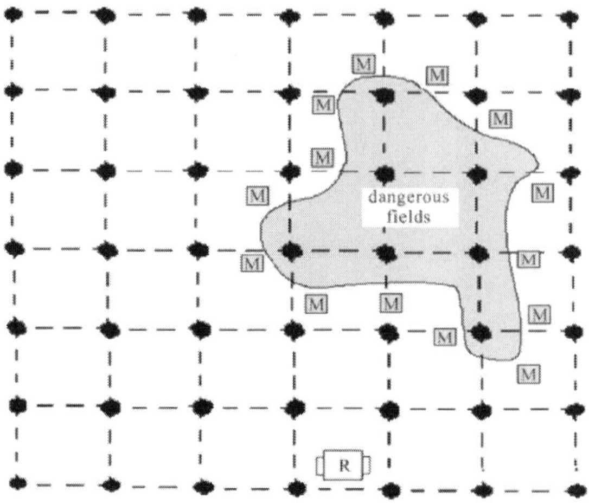

Figure 1. The structure of system.

The mobile sensor nodes can gather around the edge of adversary or dangerous. The robot is in a waiting status. The mobile nodes are four-wheel cart, which carry the Mica nodes. The Mica wireless platform serves as a foundation for the emerging possibilities. With sensing, communication, and I/O capabilities, Mica can simultaneously act as a data router, sensor interface, and control point. Mica's flexible design serves as a building block for creating efficient application-specific protocols. And Mica does not require use of predefined protocols [7].

We have done many works both on the communication of static sensors and moving of mobile sensor nodes. This is the base that we can explore the scheme.

THE DEPLOYMENT OF STATIC SENSOR NODES AND COMMUNICATION ALGORITHM

The static sensor nodes deploy all the detecting area. In view of the energy saving, we adopt a kind of street map and skeleton graph [8]. This scheme constructs a reduced graph with as few as possible working nodes instead of the whole area covering static sensor nodes. All the nodes stand at the key points of the detecting area. They are the working nodes. That does not mean we do not need large number of static nodes. If the quantity of static nodes is not enough, the navigation can become very difficult, we discuss this problem at navigation. The nodes, which are not on the key points, fall in sleep. That means these sleeping nodes are working at the low power state. When the system needs them, they will be taken up. These sleeping nodes are in working state, and afford robot navigation. That is to say, there is large number of nodes in system, but, the working nodes are not so many and energy is saved sequentially. On the other hand, the flexibility and accuracy of system can be maintained. Those nodes, which are not in adversary fields and not on the point, sleep unless the robot navigation needs them (We will discuss in Section 5.).

Except these non-key point nodes, there are still some nodes falls into sleep. They are the nodes lie in the adversary or dangerous fields. They can keep the sleeping status through the all process. Actually, they have quitted events monitor. Therefore, the street map and skeleton graph can be described as Figure 2.

Given such preliminary works, we modify the static nodes algorithm to arrange the nodes places, monitor events and broadcast the message as shown in Algorithm 1.

Algorithm 1. Deploy of static sensor nodes

```
while TRUE do
detect the location
if location! = dangerous fields & location! = key point then
send navigation requirement
else
node keeps sleep & quit monitor
end if
if navigation response! = TRUE then
node keeps sleep
else
broadcast navigation message
end if
end while
```

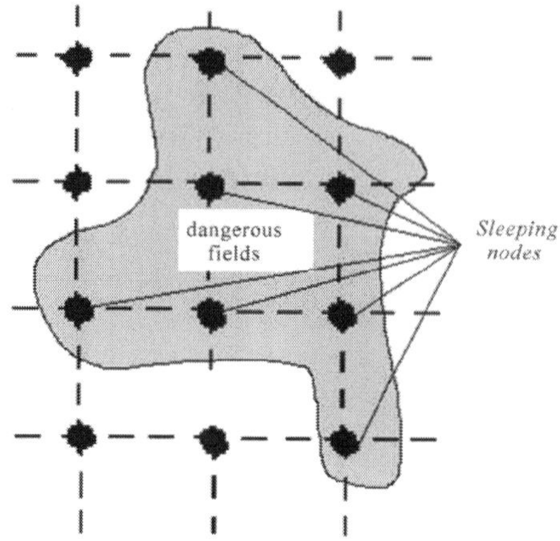

Figure 2. The diagram of on point nodes and the sleeping nodes. These nodes which is in the dangerous fields are put into sleeping status.

Once there is an event occurs, the monitoring message can be sent by ad-hoc module to find the robot. In order to save the energy, we set a threshold for the event. Therefore, the working static nodes are not always at "working status".

THE MOBILE SENSOR NODES

The mobile sensor nodes afford the task that detect adversary or dangerous for there is no prior knowledge about the field or map. Actually, mobile sensor nodes are the little robots. The working style of mobile nodes adopts a kind of active learning [9, 10].

Since the map of detected field is unknown, the mobile nodes scan the whole detecting area in raster fashion.

A penalized estimate can be constructed. It is $\hat{f}^c$. When they find these area or fields, they do the estimate of the area by using average smooth. Then, in this section, the boundary of dangerous area can be located roughly. Consequently, the mobile nodes calculate to smooth these detected regions and get refine localization by average algorithm. A complexity estimate will be got, just like the first coarse scan. It is r $\hat{f}^r$. Finally, the estimate of the fields can be generated by coarse scan and refine localization.

The mean square error can be described as follow, by using Craig-Bernstein inequality [11]:

$$E\left[\left\|\hat{f}-f\right\|^2\right] \le K_1 \left(\frac{\log n''}{n''}\right)^{\frac{1}{d}} 2^{-J} + K_2^{J(d-1)} \frac{\log n}{n} \quad (1)$$

$\hat{f}$ is calculated estimate value that conclude the coarse estimate $\hat{f^c}$ and refine estimate $\hat{f^r}$. K_1 and K_2 are the constants larger than 0.

Figure 3 shows the process about mobile sensor nodes detecting and their final localization.

Once the localization of adversary or dangerous regions are estimated, the mobile can communicated each other, move and stand around the edge of these regions according average criterion. From then on, the mobile sensor nodes work as static nodes and send warning message. Algorithm 2 has shown the mobile nodes motion and positioning method.

Algorithm 2 Mobile nodes define dangerous fields

1: while only received the static nodes' signal do

2: Doing raster scan

3: Collect the sample

4: Launch estimate algorithm

5: if routine = BLOCKED then

6: if other mobile nodes! = stay then

7: stay as a static node

8: sending the warning message

9: else

10: record the sample

11: end if

12: else

13: raster scan

14: end if

15: end while

Thus, the mobile nodes can gather around the edge of adversary or dangerous. According the adaptive statistic algorithm, the mobile nodes can find the adversary fields as soon as possible. The path length detect can be de- scribed as [12]:

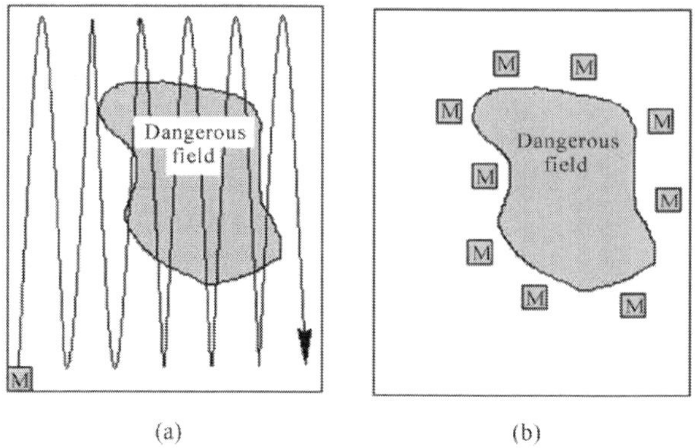

Figure 3. The mobile nodes (a) Raster fashion moving; (b) Mobile nodes location.

$$l \sim O(n^{\frac{d-1}{(d-1)^2+d}}) \tag{2}$$

where l is the path length, ~ means the order of polynomial, and d is the dimensions of function. And the time can be estimated as:

$$t \sim O(\varepsilon^{-(d-1+\frac{1}{d})}) \tag{3}$$

After the algorithms of static sensor nodes and mobile nodes are designed, we can go into the next step. That is the navigation of robot.

NAVIGATION OF ROBOT

In navigation, the robot is Pioneer II -AT four-wheel drive. The robot also has a variety of expansion power and I/O ports for attachment and close integration of additional sensors and other accessories. Expansion includes an addressable I/O bus for up to 16 devices, two RS-232 serial ports, eight digital I/O ports, five A/D ports, PSU controllers and more—all accessible through a common application interface to the robot server software, P2OS [13]. Therefore, we use the Mica as a communicate de- vice and serial ports send dictate to the robot.

Most navigation is based on the detect fields map, which is available in advance. So, the location of robot can be decided by coordinate of the map

[14]. And the optimal navigation algorithm can be explored. There have been many algorithms in these fields. But they can hardly adapt the non-mapping condition. That is to say, if there is a novel environment or a region that its condition is difficult to get to know, what these algorithms do is detect the whole region and acquire the whole information of the region. Only after the information of region is known, can the optimal algorithm can be adopted. Therefore, the energy may be cost in acquiring information. In order to solve the problem, we explore a kind of non-mapping navigation algorithm to guide the robot to arrive the event occurring place.

Generally speaking, navigation can be described as 3 questions. That is where I am, where I should go, and how to arrive [15]. We adopt a kind of dynamic cluster navigation algorithm. It can decide the location of robot and use both static nodes and mobile nodes to guide the robot to arrive the event place.

1). Where I am (Decide the location of robot)

Since what we discuss is no prior field map in this paper; the first thing is the location of robot should be decided. It is the important step.

In order to acquire the location of robot, we adopt a cluster-navigating algorithm. It depends on the ad-hoc relay communication. The static nodes work as event heads is 1, and so do as other nodes in the next cluster until the robot receive the message [16]. Robot receives the hierarchy of the static nodes. Therefore, the robot can decide approximately how long the distance between event and itself not precise. It is enough. And the robot should distinct the message of event or relay. It can decide itself whether arriving the place or not. At the same time, robot should also distinct the static or mobile nodes, so it can decide it is close to danger fields or not.

If the static nodes are too sparse to decide the robot in a field, it is difficult for robot to make a decision its direction to move. The robot cannot decide where to go. The system must take measure to make up. Those nodes, which are not on the key point, should be taken up the sleeping nodes around them and form a new cluster in order to inform the robot.

Figure 4 shows the process of detect the robot location. The detail algorithm of finding robot location is showed as follow:

Algorithm 3 The Transmission of Event 1: while robot! = arrived do

2: select working static nodes as cluster

3: if the nodes is too sparse then

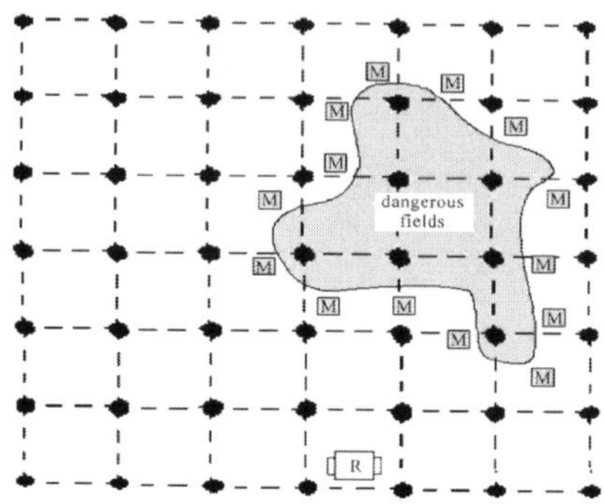

Figure 4. The process of finding robot by using cluster communication.

4: wake up the sleeping nodes

5: else

6: static nodes work as cluster

7: end if

8: if event = TRUE then

9: hierarchy = 0

10: broadcast message

11: else

12: if received message = TURE then

13: hierarchy < - hierarchy+1 14: broadcast

15: end if

16: end if

17: end while

2). Where should I go

In a large scope, robot can move to the direction whose hierarchy is always lowing. When robot arrives in the range of cluster, the static nodes should shut down the message broadcasting and its hierarchy minus one.

While in the cluster, robot always finds the leader of this cluster.

3). How to arrive

The robot can follow a routine, which the nodes' hierarchy is lowing unless it receives the warning message of the mobile nodes. When it receives warning message of the mobile nodes, it have to use the smooth routine in order to pass the danger field until the warning message vanish.

Finally, the robot will arrive the place where events occur when it receive the event's source message.

The distance from robot to dangerous fields can be calculated by potential [17]:

$$\phi(x, y) = \frac{1}{(x^2 + y^2)^{\frac{1}{2}}} \equiv \frac{1}{R^{\beta}}, \quad \beta > 0 \tag{4}$$

where R is the Euclidean distance from the edge to the mobile nods. In this paper, it can be shown as:

$$\phi_i = \sqrt{(x_i - x_e)^2 + (y_i - y_e)^2} \tag{5}$$

(x_e, y_e) is the location of dangerous fields. The energy of mobile nodes can be defined as follow:

$$V_i = \frac{1}{2}\sum_{i=1}^{m} k_i (\| \phi_i \|)^2 + \frac{1}{2} k_v \| v_i \|^2 \tag{6}$$

Therefore, we can conclude the algorithm of robot navigation. There are still some handicaps (they are the points not area) in the robot routine, but there is sonar in the robot. The robot can avoid these handicap points. It is the algorithm 4 shown as follow:

Algorithm 4 Navigation of Robot

1: While event message received = TURE Do

2: if arrived! = TURE then

3: if mobile warning message! = TURE then

4: use sonar (or communicate with the static nodes) follow navigation routine

5: else

6: adopt the smooth routine 7: end if

8: else

9: send arrive message

10: inform the event node turn off requirement 11: end if

12: end while

Since there is no map in advance, what the robot does is receive the message and do probing motion. Thus, the time to arrive the event is not shortest and navigation routine is not the optimal routine. It is the cost of none prior map available and the lower cost of both static and mobile nodes. On the other hand, there are no so many working nodes in this system, the energy cost of nodes can be saved.

4). How to wake up the sleeping nodes

When the static nodes, which are on the key point, is too sparse to accomplish the navigation task, it is necessary to wake up the sleeping nodes.

At first, the cluster head, the static nodes on the key point, acquires the message and sends the message to next closed cluster head. If the closed cluster head can receive the message, it will send back a message to the sending cluster head. That means that the statistic nodes are not too sparse there. The message can relay to the next static nodes. If the closed cluster head cannot receive the message, it cannot send back the message to the sending cluster head. It means that the statistic nodes are too sparse there. That is to say, the sleeping nodes in the cluster have to be taken up in order to relay the message

The construction of the information deliver is shown in Figure 5

The static nodes in this system adopt tmote Sky [6] made by Moteiv Corporation. They are ultra-low power wireless module for use in sensor networks, monitoring applications, and rapid application prototyping. In an open field, the communication distance of tmote nodes can reach 100 meters.

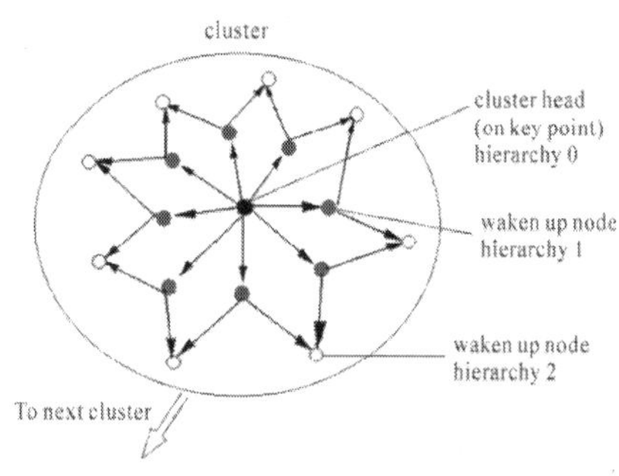

Figure 5. The construction of information deliver in a cluster.

But in this system, the communication distance of tmote nodes is restrained to 10 meters in or- der to saving energy. In a cluster, there is one tmote node working as the cluster and the other nodes working as sleeping nodes. The event occurs, such as the temperature is over the threshold, the cluster head sends the message to the next cluster head. If the next cluster head receives the message, it sends back a message to the former cluster.

Consequently, message can be relayed on. If the sending cluster head does not receive the backing message of the next cluster head, the sending cluster head will send a signal to the sleeping nodes in its cluster, in order to wake up these sleeping nodes. That is to say, the low power costs' nodes (sleeping nodes) will transfer from the low power costs' state (sleeping) to working state (waken up). Once the sleeping nodes are waken up, they can afford the message delay events.

After the works have been accomplished, we simulate the sensing and navigation of hybrid sensor network by using. Figure 6 and Figure 7 is result of simulating experiments.

It can be seen that the energy saving of mobile work is better than the static network. While, the energy saving of hybrid network is much better than the mono-static and mono-mobile network.

The practical experiments are performed with the setup shown in Figure 8. A sequence process of navigation is shown from (a) to (d). An emergent event occurs by a node (the temperature is over the threshold), the robot (Pioneer II -AT) should find the node and arrive the place in time. Part (a) shows the initial experiments situation. There are 100 tmote nodes (static nodes) and 8 small robots with the tmote nodes on them (mobile nodes) in a 25m2 (5m×5m) filed. The lawn simulates the dangerous fields, and the scope can be detected by moisture sensor on tmote nodes.

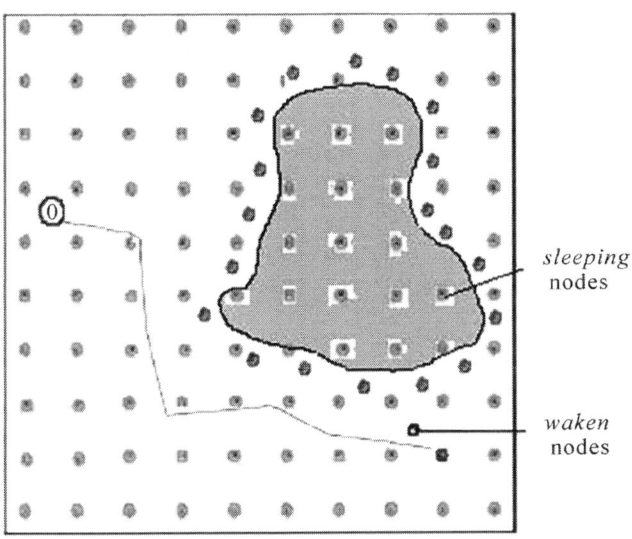

Figure 6. The result of simulating and the trajectory of robot.

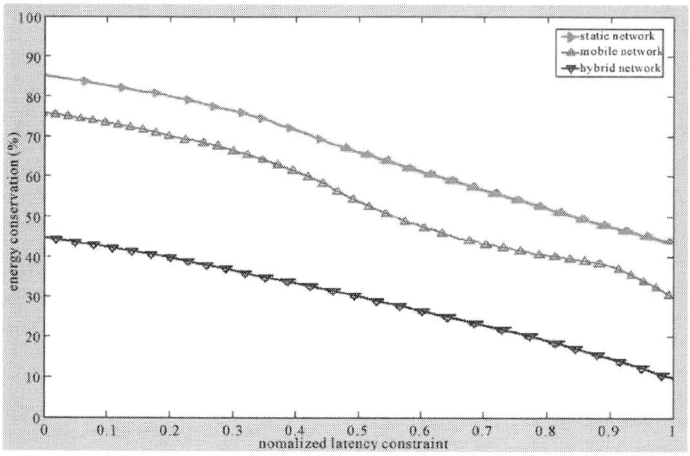

Figure 7. The energy conservation of different network.

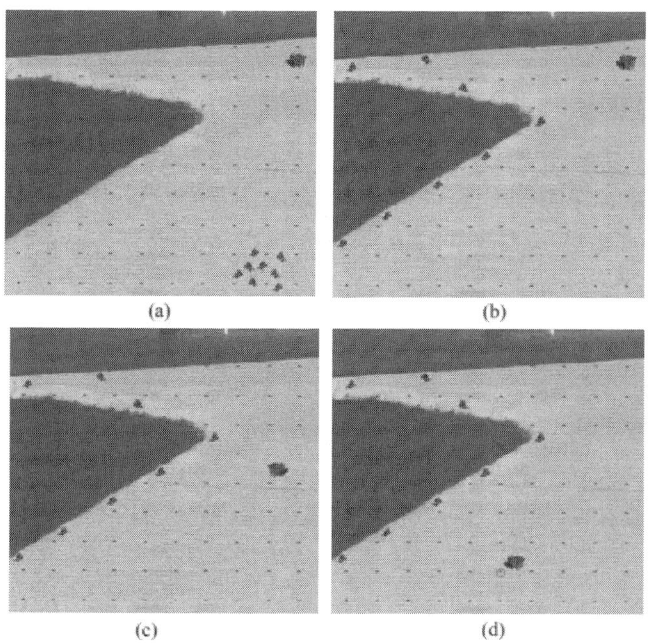

Figure 8. The practical experiments of the navigation.

Therefore, the range of dangerous fields can be defined by them. Consequently, part (b) shows the dangerous fields are defined. Part (c) shows the process of navigation, and part (d) shows the robot arrive the emergency place (red circle) finally.

Despite the difficulties with the dangerous fields' definition, the navigation is successfully performed many times. Although arriving the final location of the robot is not so quickly (about 3 minutes), the robot avoids the danger fields and being guided by this hybrid sensor network system indeed. We plan to conduct further re- search to get higher speed of this navigation system.

CONCLUSIONS

This implementation demonstrated robot guidance by a hybrid sensor network. The energy saving in sensor net- work is an important problem. We take it as the primary problem in this system. The skeleton of adversary or dangerous fields can be outlined by using the mobile sensor nodes and the robot navigation comes into reali- ties on the ad-hoc method. In spite of the less optimal routine, it is an effective scheme when there is no map of the whole detecting fields in advance.

However, the precision of the navigation system greatly depends on the robot hardware. Navigation can be problematic if environmental factors are not so ideal, such as electrical cables, higher power electrical equipment and steel structures exist. Although it is possible to compensate for these effects in a known environment, a search situation may not permit elaborate offline calibration. In addition, the response speed of whole system should be raised higher. We are going to do further re- search on them.

ACKNOWLEDGMENT

We would like to give special thanks to Dr. Jindong Tan, the professor of Michigan Technological University in

U.S.A. for his providing of experimental equipment and environment. The same thanks go to Qi An, who helped

me with studies of the project and sensor network,

TinyOS.

REFERENCES

1. D. Moore, J. Leonard, D. Rus and S. Teller, "Robust Distributed Networklocalization with Noisy Range Meas- urements," Proceedings of the 2nd International Confer- ence on Embedded networked sensor systems, Baltimore, 2004, pp. 50-61.
2. R. Szewczyk, A. Mainwaring, J. Polastre and D. Culler, "An Analysis of a Large Scale Habitat Monitoring Ap- plication," Proceedings of the 2nd International Confer- ence on Embedded networked sensor systems, Baltimore, 2004, pp. 214-226.
3. J. Polastre, J. Hill and D. Culler, "Versatile Low Power Media Access for Wireless Sensor Networks," Proceed- ings of the 2nd International Conference on Embedded networked sensor systems, Baltimore, 2004, pp. 95-107.
4. M. Batalin, M. Rahimi, Y. Yu, D. Liu, A. Kansal, G. Sukhatme, W. Kaiser, M. Hansen, G. Pottie, M. Srivastava and D. Estrin, "Call and Response: Experiments in Sampling the Environment," Proceedings of ACM Sensor Systems, Los Angeles, 2004.
5. M. Rahimi, R. Pon, W. J. Kaiser, G. S. Sukhatme, D. Estrin and M. Srivastava, "Adaptive Sampling for Environmental Robotics," Proceedings of IEEE International Conference on Robotics and Automation, Los Angeles, 2004.
6. http://www.moteiv.com
7. http://www.crossbow.com
8. R. Willett, A. Martin and R. Nowak, "Backcasting: Adaptive Sampling for Sensor Networks," Proceedings of IEEE International Conference on Information process- ing in sensor networks, New York, 2004, pp. 124-133.
9. R. Castro, R. Willett and R. Nowak, "Faster Rates in Re- gression via Active Learning," Proceedings of Neural In- formation Processing Systems, 2005. http://homepages.cae. wisc.edu/ rcastro/ ECE-05-3.pdf
10. M. Horstein, "Sequential Decoding Using Noiseless Feed- back," IEEE Transactions Information Theory, Vol. 9, No. 3, 1963, pp. 136-143.
11. J. Borenstein and H. R. Everett, "Navigating Mobile Ro- bots: Sensors and Techniques," John Wiley & Sons, 1992.
12. M. Burnashev and K. S. Zigangirov, "An Interval Esti- mation Problem for Controlled Observations," Problems in Information Transmission, Vol. 10, 1974, pp. 223-231.

13. O. Khatib, "Real-Time Obstacle Avoidance for Manipu- lators and Mobile Robots," The International Journal of Robotics Research, Vol. 5, No. 1, 1986, pp. 90-98.
14. J. Cortes, S. Martinez, T. Karatas and F. Bullo, "Cover- age Control for Mobile Sensing Networks," IEEE Trans- actions on Robotics and Automation, New Orleans, April 2004, pp. 243-255.
15. A. Korostelev, "On Minimax Rates of Convergence in Image Models under Sequential Design," Statistics and Probability Letters, Vol. 43, 1999, pp. 369-375.
16. G. Golubev and B. Levit, "Sequential Recovery of Ana- lytic Periodic Edged in the Binary Image Models," Mathe- matical Methods of Statistics, Vol. 12, 2003, pp. 95-115.
17. R. Castro, R. Willett and R. Nowak, "Faster Rates in Re- gression via Active Earning," Proceedings of Neural In- formation Processing Systems, Vancouver, 2005. http:// homepages.cae.wisc.edu/rcastro/ ECE-05-3. Pdf

CHAPTER

12

A Study on Vehicle Detection and Tracking Using Wireless Sensor Networks

G. Padmavathi[1], D. Shanmugapriya[2], M. Kalaivani[1]

[1]Department of Computer Science, Avinashilingam University for Women, Coimbatore, Tamil Nadu, India [2]Department of Information Technology, Avinashilingam University for Women, Coimbatore, Tamil Nadu, India

ABSTRACT

Wireless Sensor network (WSN) is an emerging technology and has great potential to be employed in critical situations. The development of wireless sensor networks was originally motivated by military applications like battlefield surveillance. However, Wireless Sensor Networks are also used in many areas such as Industrial, Civilian, Health, Habitat Monitoring,

Citation:- G. Padmavathi, D. Shanmugapriya and M. Kalaivani, "A Study on Vehicle Detection and Tracking Using Wireless Sensor Networks," Wireless Sensor Network, Vol. 2 No. 2, 2010, pp. 173-185. doi: 10.4236/wsn.2010.22023.

Environmental, Military, Home and Office application areas. Detection and tracking of targets (eg. animal, vehicle) as it moves through a sensor network has become an increasingly important application for sensor networks. The key advantage of WSN is that the network can be deployed on the fly and can operate unattended, without the need for any pre-existing infrastructure and with little maintenance. The system will estimate and track the target based on the spatial differences of the target signal strength detected by the sensors at different locations. Magnetic and acoustic sensors and the signals captured by these sensors are of present interest in the study. The system is made up of three components for detecting and tracking the moving objects. The first component consists of inexpensive off-the shelf wireless sensor devices, such as MicaZ motes, capable of measuring acoustic and magnetic signals generated by vehicles. The second component is responsible for the data aggregation. The third component of the system is responsible for data fusion algorithms. This paper inspects the sensors available in the market and its strengths and weakness and also some of the vehicle detection and tracking algorithms and their classification. This work focuses the overview of each algorithm for detection and tracking and compares them based on evaluation parameters.

Keywords: Wireless Sensor Networks, Acoustic and Magnetic Sensors, Acoustic and Magnetic Signals, Detection and Tracking Algorithms

INTRODUCTION

The Wireless Sensor Networks comprise of relatively inexpensive sensor nodes capable of collecting, processing, storing and transferring information from one node to another. These nodes are able to autonomously form a network through which sensor readings can be propagated. Since the sensor nodes have some intelligence, data can be processed as it flows through the network. Sensing devices will be able to monitor a wide variety of ambient conditions: temperature, pressure, humidity, soil makeup, vehicular movement, noise levels, lighting conditions, the presence or absence of certain kinds of objects, mechanical stress levels on attached objects and so on. These devices will also be equipped with significant processing, memory and wireless communication capabilities. Emerging low-level and low-power wireless communication protocols can be used to networks the sensors. This capability will add a new dimension to the capabilities of sensors. Sensors will be able to coordinate among themselves on a higher-level sensing task. The sensors can be deployed in any facility or area, which has to be sensed in three main types. It can either be 1) triangular sensor deployment, 2) square sensor deployment and 3) irregular sensor deployment [1]. These deployments are depicted in Figure 1.

Networking inexpensive sensors can revolutionize information gathering in a variety of situations.

A sensor node usually consists of four sub-systems:

- • A computing subsystem: In a sensor node, the microprocessor (microcontroller unit, MCU) is responsible for functions such as control of sensors and execution of communication protocols.

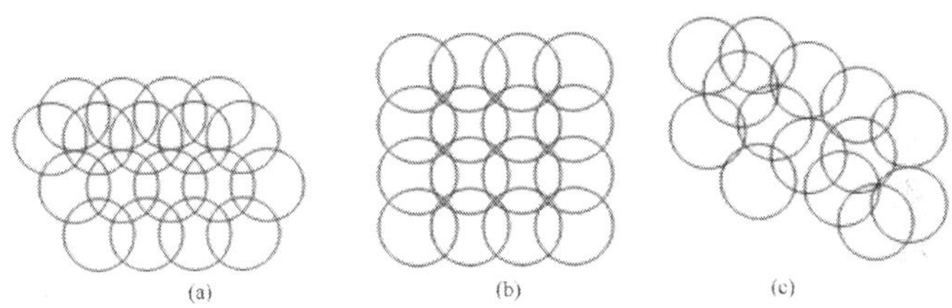

Figure 1. (a) Triangular; (b) Square; (c) Irregular deployments.

- A communication subsystem: This comprises of short range radios used to communicate with neighbouring nodes and the outside world. These devices operate under the Transmit, Receive, Idle and Sleep modes having various levels of energy consumption.
- A sensing subsystem: Low power components can help to significantly reduce power consumption. Since this subsystem (sensors and actuators) is responsible for the sharing of information between the sensor network and the outside world.
- A power supply subsystem: It consists of a battery which supplies power to the node.

Characteristics of WSN

Some of the unique characteristics of a WSN include:

- Limited power they can harvest or store.
- Ability to withstand harsh environmental conditions
- Ability to cope with node failures
- Mobility of nodes
- Dynamic network topology

- Communication failures
- Heterogeneity of nodes large scale of deployment
- Unattended operation

Applications

The various sensor network applications include,

1) Military Applications
- Monitoring Friendly Forces, Equipments and Ammunition
- Battlefield Surveillance
- Targeting
- Battle Damage Assessment
- Nuclear, Biological and Chemical Attack Detection

2) Environmental Applications
- Forest Fire Detection
- Flood Detection
- Monitoring through Internet
- Monitoring Biodiversity

3) Habitat Monitoring Applications
- Habitat Monitoring
- Great Duck Island System

4) Health Applications
- Tele-monitoring of Physical Data
- Tracking and Monitoring Doctors and Patients in a Hospital
- Drug Administration in Hospitals

5) Home and office Applications
- Smart Homes
- Managing Inventory Control
- Home automation
- Environmental control in office buildings

6) Other Applications
- Monitoring Nuclear Reactor.
- Target Tracking
- Suspicious Individual detection: Interactive museums

- Fire Fighters Problem (First Responders Problem) The major contribution of this paper includes classification of sensors available and vehicle detection and tracking algorithms using Wireless Sensor Networks.

The paper is organized as follows: Section 2 gives the classification of sensors used for vehicle detection and mainly concentrates on the acoustic and magnetic sensors. Section 3 gives the gist about the acoustic and magnetic signals acquired from the sensors. Section 4 discusses the various algorithms used in vehicle detection and tracking, analysis about these algorithms is given in Section 5 and summary of these algorithms given in Section 6 followed by the conclusion.

CLASSIFICATION OF SENSORS USED IN VEHICLE DETECTION

There are a wide variety of sensors available in market today [2]. Table 1 shows the variety of current sensor technologies and compares the strengths and weaknesses with respect to installation, parameters measured, and performance in bad weather, variable lighting, and changeable traffic flow.

Many over-roadway sensors are compact and mounted above or the side of the roadway, making installation and maintenance relatively easy.

Some sensor installation and maintenance applications may require the closing of the roadway to normal traffic to ensure the safety of the installer and motorist.

All the sensors listed here operate under day and night conditions. Sensors are broadly classified as Intrusive and Non-Intrusive sensors.

Table 1. Strengths and weaknesses of commercially available sensor technologies.

Technology	Strengths	Weaknesses
Inductive loop	Flexible design to satisfy large variety of applications. Mature, well understood technology. Large experience base. Provides basic traffic parameters (e.g., volume, presence, occupancy, speed, headway, and gap). Insensitive to inclement weather such as rain, fog, and snow. Provides best accuracy for count data as compared with other commonly used techniques. Common standard for obtaining accurate occupancy measurements. High frequency excitation models provide classification data.	Installation requires pavement cut. Improper installation decreases pavement life. Installation and maintenance require lane closure. Wire loops subject to stresses of traffic and temperature. Multiple loops usually required to monitor a location. Detection accuracy may decrease when design requires detection of a large variety of vehicle classes.
Magnetometer (two-axis fluxgate magnetometer)	Less susceptible than loops to stresses of traffic. Insensitive to inclement weather such as snow, rain, and fog. Some models transmit data over wireless radio frequency (RF) link.	Installation requires pavement cut. Improper installation decreases pavement life. Installation and maintenance require lane closure. Models with small detection zones require multiple units for full lane detection.
Magnetic (induction or search coil magnetometer)	Can be used where loops are not feasible (e.g., bridge decks). Some models are installed under roadway without need for pavement cuts. However, boring under roadway is required. Insensitive to inclement weather such as snow, rain, and fog. Less susceptible than loops to stresses of traffic.	Installation requires pavement cut or boring under roadway. Cannot detect stopped vehicles unless special sensor layouts and signal processing software are used.
Microwave radar	Typically insensitive to inclement weather at the relatively short ranges encountered in traffic management applications. Direct measurement of speed. Multiple lane operation available.	Continuous wave (CW) Doppler sensors cannot detect stopped vehicles
Active infrared (laser radar)	Transmits multiple beams for accurate measurement of vehicle position, speed, and class. Multiple lane operation available.	Operation may be affected by fog when visibility is less than ≈ 20 feet (ft) (6 m) or blowing snow is present. Installation and maintenance, including periodic lens cleaning, require lane closure
Passive infrared	Multizone passive sensors measure speed.	Passive sensor may have reduced vehicle sensitivity in heavy rain, snow and dense fog. Some models not recommended for presence detection.
Ultrasonic	Multiple lane operation available Capable of over height vehicle detection. Large Japanese experience base.	Environmental conditions such as temperature change and extreme air turbulence can affect performance. Temperature compensation is built into some models. Large pulse repetition periods may degrade occupancy measurement on freeways with vehicles travelling at moderate to high speeds.
Acoustic	Passive detection. Insensitive to precipitation. Multiple lane operation available in some models.	Cold temperatures may affect vehicle count accuracy. Specific models are not recommended with slow-moving vehicles in stop-and-go traffic.
Video image processor	Monitors multiple lanes and multiple detection zones/lane. Easy to add and modify detection zones. Rich array of data available. Provides wide-area detection when information gathered at one camera location can be linked to another.	Installation and maintenance, including periodic lens cleaning, require lane closure when camera is mounted over roadway (lane closure may not be required when camera is mounted at side of roadway). Performance affected by inclement weather such as fog, rain, and snow; vehicle shadows; vehicle projection into adjacent lanes; day-to-night transition; vehicle/road contrast; and water, salt grime, icicles, and cobwebs on camera lens. Reliable night-time signal actuation requires street lighting. Requires 30- to 50-ft (9- to 15-m) camera mounting height (in a side-mounting configuration) for optimum presence detection and speed measurement. Some models are susceptible to camera motion caused by strong winds or vibration of camera mounting structure. Generally cost effective when many detection zones within the camera field of view or specialized data are required.

Intrusive Sensors

Intrusive sensors are those that need to be installed under the pavement, in saw-cuts or holes on the roads. Popular intrusive sensors include inductive loops, magnetometers, micro loop probes, pneumatic road tubes, piezoelectric cables and other weigh-in-motion sensors. The main advantage of these sensors is their high accuracy for vehicle detection while the drawbacks include the disruption of traffic for installation and repair, resulting in high installation and maintenance cost.

Non-Intrusive Sensors

To overcome the disadvantage of intrusive sensors, non-intrusive sensors are developed eg, above ground vehicle detection sensors. Above ground sensors can be mounted above the lane of traffic or on the side of a roadway where they can view multiple lanes of the traffic at angles perpendicular to or at a slanting angle to the flow direction. Technologies used in aboveground sensors include video image processing (VIP), microwave radar, laser radar, passive infrared, ultrasonic, passive acoustic array, and combinations of these sensor technologies. However, these non-intrusive sensors tend to be large size and power hunger.

Smart-Dust Sensor Node-Hardware Platform

Smart-Dust is one of the potential sensor nodes which can be used in the future vehicle detection system. In Smart-Dust sensor node, essential components for vehicle detection (processor, memory, sensor and radio) could be integrated together as small as a quarter through MEMS technology. Together with its low power design [3], the Smart-Dust sensor node is suitable for implementing the vehicle detection sensor networks. Figure 2 shows the different generations of Smart Dust sensor nodes (Motes). The left picture is a 1st generation smart dust sensor node "Rene Mote". From left to right, the right picture shows the "MICA Mote" (2nd generation), "MICA2 Mote" (3rd generation) and "MICA2-Dot Mote (3rd generation)". Smart-Dust sensor node is designed by EECS department in UC Berkeley and Intel [4] using modular component approach and it consists of two major components: mother board and sensor board. Thus, different sensor boards could be attached to the same mother board for different applications. Thus, Smart

Figure 2. Dust family.

Dust sensor node could potentially be used in a wide range of applications such as vehicle detection, enemy monitoring in the battlefield, temperature measurement in a building, environmental monitoring etc.

The basic components of MICA mote shown in Figure 3 belongs to the Smart-Dust family. The components are listed in Table 2. The mother board consists of an Atmel 90LS8535 processor, 512KB SRAM, 8KB Flash RAM and a RF transceiver for wireless communication. The Sensor board consists of a 10-bit analog to digital converter, a Magnetometer (Honeywell HMC1002), a temperature sensor, a photo camera and an accelerometer sensor.

For vehicle detection system, the sensors used are the magnetometer and acoustic sensors. Next, the basic operating principles of magnetometer and acoustic sensors are will be reviewed.

Acoustic Sensors

The acoustic sensor in the Smart Dust sensor node is a condenser type microphone. The schematic for a typical condenser acoustic sensor is shown in Figure 4. It includes a stretched metal diaphragm that forms one plate of a capacitor. A metal disk placed close to the diaphragm acts as a backplate. A stable DC voltage is applied to the plates through a high resistance to keep electrical charges on the plates. When a sound field excites the diaphragm, the capacitances between the two plates vary according to the variation in the sound pressure. The change in the capacitance generates an AC output proportional to the sound pressure, which shows the ultra low-frequency pressure variation. A high-frequency voltage (carrier) is applied across the plates and the acoustic sensor output signal is the modulated carrier. The photo in the right of Figure 4 shows the Panasonic

Figure 3. MICA mote.

Table 2. Components of smart dust mote.

Mother Board	Sensor Board
Atmel 90LS8535 processor (clocked at 4 MHz)	10-bit analog to digital converter
RF Monolithics transceiver (916.50 MHz)	Magnetometer(Honeywell HMC1002)
	Microphone (Panasonic WM-62A)
512KB SRAM, 8KB Flash RAM	Temperature Sensor
	Photo Camera
	Accelerometer Sensor

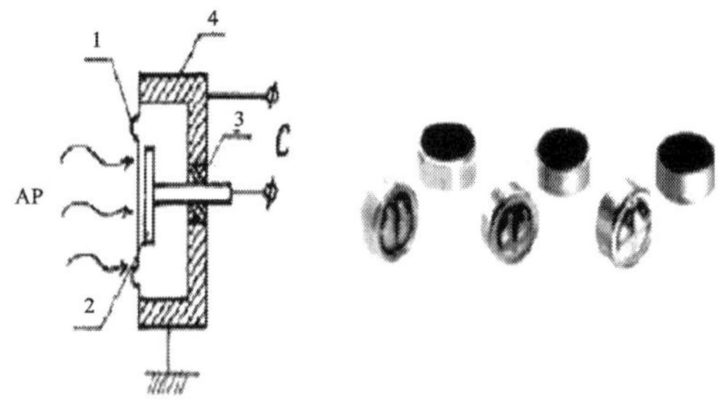

Figure 4. Condenser microphone. AP-the acoustic pressure; C-the variable capacitance; 1-the metal diaphragm; 2-the metal disk; 3-the insulator; 4-the case.

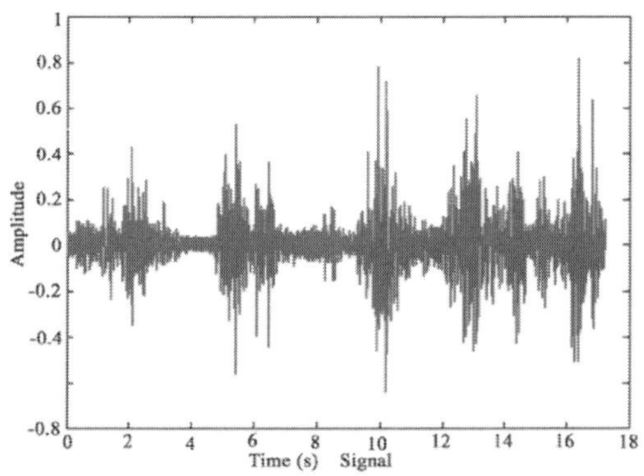

Figure 5. Waveforms of acoustic signals emitted from truck.

WM-62A condenser microphones used in Smart Dust Motes. Figure 5 shows a typical vehicle acoustic signal waveforms.

Magnetic Sensors

Magnetic sensors differ from most other detectors in that they do not directly measure the physical property of interest. Devices that monitor properties such as temperature, pressure, strain, or flow provides an output that directly reports the desired parameter (Figure 6). Magnetic sensors, on the other hand, detect changes, or disturbances, in magnetic fields that have been created or modified, and from them derive information on properties such as direction, presence, rotation, angle, or electrical currents. The output signal of these sensors requires some signal processing for translation into the desired parameter. Although magnetic detectors are somewhat more difficult to use, they do provide accurate and reliable data without physical contact.

The Honeywell HMC1002 magnetometer on the MICA sensor board is a magnetoresistive sensor. Honeywell magnetic sensors and magnetometers offer complete magnetic field sensing solutions that are highly accurate, and allow for easy integration for virtually any application. Figure 7 shows the honeywell magnetic sensors.

The Anisotropic Magneto Resistive (AMR) sensor is one type that has a wide Earth's field sensing range and can sense both the strength and direction of the Earth field. The AMR sensor is made of a nickel-iron (Permalloy) thin film deposited on a silicon wafer and patterned as a resistive strip. The strip resistance changes about 2%-3% when a magnetic field is applied. Typically, four of these resistive strips are connected in a wheatstone bridge configuration so that both magnitude and direction of a field along a single axis can be measured. The key benefit of AMR sensors is that they can be bulk manufactures on silicon wafers mounted in commercial integrated circuit packages. Figure 8 shows a typical vehicle magnetic signal waveforms.

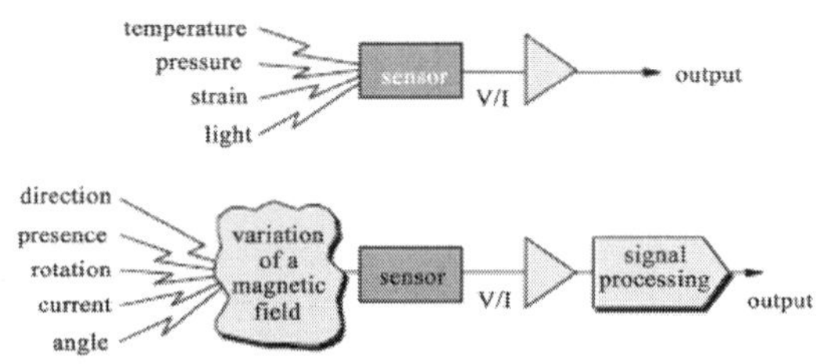

Figure 6. Magnetic sensor parameters.

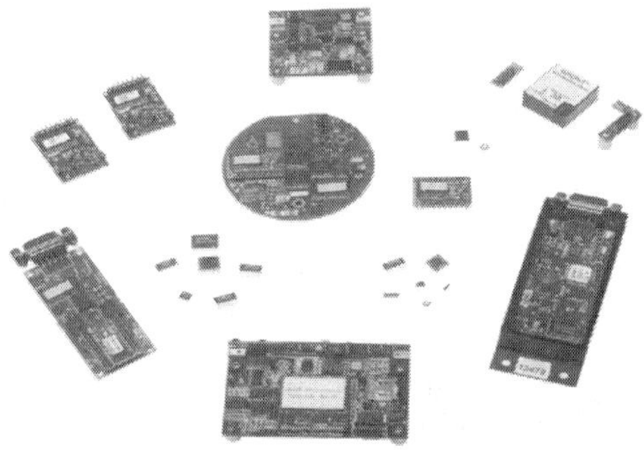

Figure 7. Magnetic sensors.

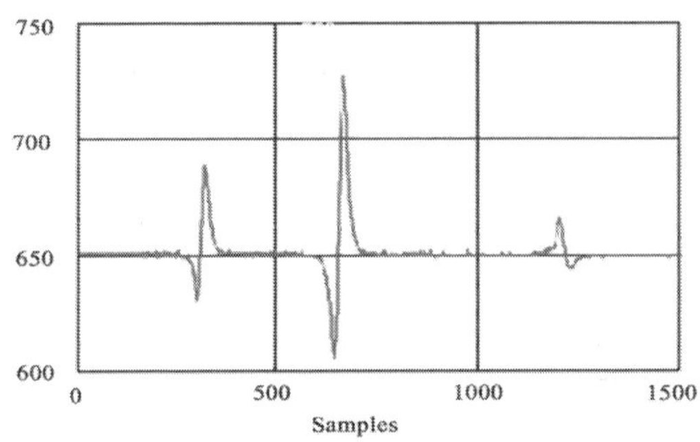

Figure 8. Waveforms of magnetic signals.

Magnetic Field

A magnetic field is a vector field that surrounds magnets and electric currents, and is detected by the force it exerts on moving electric charges and on magnetic materials. When placed in a magnetic field, magnetic dipoles tend to align their axes parallel to the magnetic field. Magnetic fields also have their own energy with an energy density proportional to the square of the field intensity.

Magnetic Data

Magnetic data is the term used for data that is acquired from magnetic (as opposed to optical) motion capture systems. A central magnet is used to create a field in which sensors can determine their position and orientation as they move about in the field. Raw magnetic data has no hierarchy information; the sensors do not know where they are relative to the other sensors, and know their position in the magnetic field. The file formats for magnetic motion capture data reflect this. In this respect, they are somewhat similar to the BVA file format.

The problem associated with magnetometer vehicle detection is similar to the acoustic sensors but the magnetic signals are much cleaner than acoustic signals. The magnetometers available today can sense magnetic fields within the earth's field-below 1 gauss.

They can be used for detecting the vehicles, which are ferrous objects that disturb the earth's field. The earth's field provides a uniform magnetic field over wide area in the scale of kilometers and even a ferrous object can create a local disturbance in this field.

This local field disturbance can be sensed by the magnetometers for vehicle detection. Figure 9 shows the disturbance of earth's magnetic field by a car. After presenting the basic principles of sensors, the characteristics of the measured acoustic and magnetic signals will be studied and algorithms are proposed for reliable low cost vehicle detection.

SIGNALS FOR VEHICLE DETECTION AND TRACKING

The signals taken into consideration for study in this proposal are acoustic and magnetic signals.

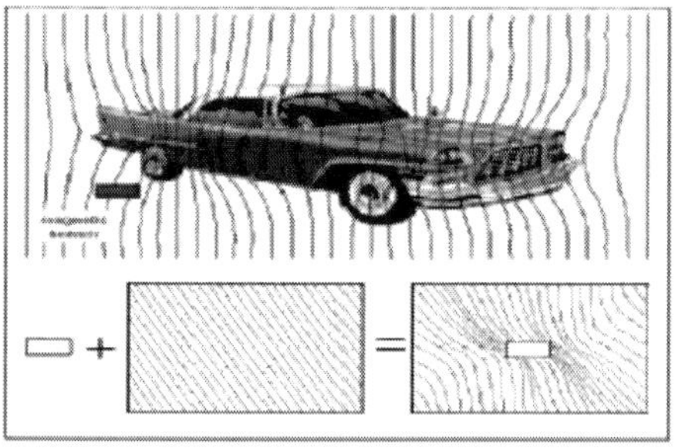

Figure 9. Earth's magnetic field disturbed by a car.

Acoustic Signals

The acoustic signature is made up of a number of individual elements. These include:

- Machinery noise: noise generated by a ships engines, propeller shafts, fuel pumps, air conditioning systems, etc.
- Cavitation noise: noise generated by the creation of gas bubbles by the turning of a ship's propellers.
- Hydrodynamic noise: noise generated by the movement of water displaced by the hull of a moving vessel.

These emissions depend on a hull's dimensions, the installed machinery and ship's displacement. Therefore, different ship classes will have different combinations of acoustic signals that together form a unique signature.

Sonar operating in passive mode can detect acoustic signals radiated by invisible submarines and these signals are used to target attacks.

Magnetic Signals

Magnetic Signals are changes that happen in the magnetic field caused by the movement of ferrous objects.

ALGORITHMS FOR VEHICLE DETECTION AND TRACKING

There are a wide variety of algorithms generally available for detection and Tracking of target. Figure 10 shows the classification of detection and tracking algorithms.

Target Detection Background

The basic idea behind a target detection algorithm is that sensor nodes are deployed within an area and the nodes determine whether or not a target has entered or is presently within the area. Nodes usually determine if a target is present by detecting a change in some sort of signal, whether it is acoustic, light, temperature or some other type [5]. Since the strength of this signal will scatter throughout the sensor network the nodes need to collaboratively work together in order to determine if a target has been detected or not. Several of these collaborative detection algorithms are discussed in the following section:

Basic Fusion

This algorithm is presented first since it has an extremely simple design and implementation and serves as an ideal algorithm to use as the foundation for explaining other more complex detection algorithms. This algorithm collects the captured data from each of the nodes in a network. Then, to remove any invalid data provided by faulty nodes, the largest and smallest data values are dropped. The remaining data is then averaged together. If this average exceeds a predetermined threshold, then the system is notified that a target has been detected. This algorithm has two versions in which the data provided by each node differs. The data is either the original data collected by the node or it is a detection decision already made by the individual nodes using their original data [5].

Hybrid

This algorithm is similar to the basic fusion algorithm. In that a detection decision is based on data collected by numerous nodes throughout the network. However, this algorithm differs in that either a yes/no target present bit decision is sent or the collected data is sent. In this algorithm, each node is given two threshold values for a previously determined detection signal type, such as sound. One of these threshold values determines that if the data collected by the node exceeds this value then a target has been detected. Correspondingly, the other threshold is a minimum in that if the data is below, then no target is present. If either of these two thresholds is met then a bit message indicating the presence or lack of a target is transmitted. However, if neither of these thresholds is exceeded then the raw data collected by the node is passed on to the base station where it is used with the data of other nodes to determine if a target is present or not [18].

Predicate Clustering

In this algorithm nodes are divided into groups called clusters in which each group has a cluster head that all sub nodes of the cluster report their data to. Typically, one cluster is formed around a single target with multiple clusters being in the network if multiple targets are present. Cluster heads are chosen by all nodes within the area of a detected target "electing" to become the head by broadcasting their intent to become a cluster head. The node with the peak sensor reading is chosen as the cluster head [6]. Once cluster heads are chosen, the remaining nodes choose group to join by running a decision predicate on their captured sensor data and the data of their neighbours. One particularly interesting feature of this algorithm is that a cluster head can be responsible for multiple targets. When it is detected, multiple targets are going to enter the same cluster, both clusters provide the target information to the "merged" cluster head that is now responsible for tracking both targets.

Grids

In this approach the sensor network is broken up into "virtual grids" where each grid contains points. For each grid point, it is determined which nodes' sensing area covers the grid point and then these nodes create a schedule so that they all collaboratively cover this grid point at alternating times. So at any point of time't' at least one node's sensing area covers grid point 'x'. Since each node covers multiple grid points, each node will contain several sensing schedules to keep in sync so that at any point of time all grid points are covered by at least one node. In order to do this, all nodes first determine the schedules for each grid point, then schedule all their individual point schedules together so that each node is not constantly waking and sleeping instead sleeps a little and then wakes for a while. Once each node determines its own individual schedule, it must be correlated with those of its neighbours so that as few as possible nodes are awake at any point of time but all grid points are covered at all times. Once the schedules are completed for each node, the nodes start to follow their schedule, going through states of being asleep and awake in order to detect a new target [7].

Wave

The Wave protocol has the same goal in mind as that of the Grid algorithm, detecting targets while conserving energy by not having all the nodes on all the time. The basic idea of the wave algorithm is to wake up a series of nodes at certain times so that only nodes that are awake at one point of time are in one concentrated area. Then, over time, this section moves over the network by putting nodes to sleep and waking new ones. This gives the appearance of an awake section of nodes flowing over the entire network like a wave [8].

This algorithm proposes three types of waves:

- Two straight lines of nodes are awake at the same time, with the lines being parallel to each other. The two lines start on opposite sides of the network from each other and gradually work towards each other to meet in the centre of the network.
- A wide line of nodes are awake at the same time and this line moves across the network from one edge to the opposite edge.
- Waves of nodes are awake and the wave traverses from one edge of the network to the other. This approach differs from the wide line in that the boundary shape is curvy or wave like instead of

being a straight line boundary. The wave is created by an awake sensor waking all nodes within its communication range on the side of it that the wave needs transition too. This approach also differs from the previous two in that it uses radio communication to wake up the next nodes to start sensing instead of time syncing the movement of wake/sleep nodes across the network.

Like the other algorithms, when a target is detected, the node(s) pass the information on for use.

Patrolling

Patrolling algorithms are similar to the wave algorithms in that only a series of nodes are awake at one point of time. This algorithm just chooses which nodes to wake differently. This can be done on an on-demand basis or a coverage-orientated basis. The difference between the two is that on-demand is some predetermined path that is currently interesting whereas coverage-orientated just repeatedly watches an area [9].

The algorithm for both approaches is the same. First, a patrol is defined as a set of information containing the patrol speed, duration, iteration period and path. Once this information is collected, a Patrol Host node is selected, which is usually the first node on the patrol path. The duty of the Patrol Host is to periodically transmit the patrol data along with the current patrol time. The idea is that over the time the patrol virtually traverses down the specified path. As the patrol advances, nodes that are closest to the patrol path wake up and collect information. Nodes determine if they are near the patrol path by using the patrol information that the Patrol Hosts broadcast. Additionally, as the patrol moves, new Patrol Hosts must be elected. As the patrol virtually moves out of the current patrol host's range, a new patrol host is selected on the path which then takes over the patrol host duties and the previous patrol host can now sleep. This process of waking nodes and handing off patrol host duties continues until the end of the path is reached. Once the path end is reached, the patrol is reversed in order to traverse the path back to the original starting point. This process continues according to the duration and iteration parameters are set in the original patrol parameters [9].

Mobile Nodes

Until this point, all the target detection algorithms discussed so far have assumed that the sensor nodes remain stationary. This section looks at a few target detection algorithms in which the nodes are mobile within the network. Even though these algorithms allow nodes to be mobile, their goal remains the same as the previously mentioned target detection algorithms. These algorithms are still trying to detect targets in a timely and efficient manner while trying to cover as much of the network as possible using as little energy as possible. The theory behind allowing nodes to be mobile is that they may now detect a target that would have otherwise gone undetected in a stationary node network [10].

Past Detections

Like other algorithms, the network in this algorithm is divided into cell areas. Each cell is unique in that it maintains the current state and past history of targets that have been in the cell. This includes the location in which the target was first detected within in the cell. Using this data, the system calculates the areas of the cell that are most likely to contain a target based on the locations of previously detected targets [11].

Once these locations are determined, each node is assigned an area to watch and a priority. These priorities are used in the "coordination mechanism" part of the algorithm. The coordination part of the algorithm is responsible for altering the path nodes so that their observation paths overlap as little as possible. For example, the node that watches the area where targets are the most likely to occur is given the highest priority of the system and its path is not altered. A second node, whose priority is slightly less than the first node, whose path also slightly intersects that of the first node has its path altered so that the two paths will not cross. This path altering continues throughout the system from highest priority down to lowest priority until all node paths are altered to prevent too much overlapping of observation areas [11].

Collaborative Coverage

This algorithm is little different in that the paths the nodes traverse are not coordinated in any way. Each node is allowed to create and follow its own arbitrary path. While the node is travelling on this path at different locations and times, it collects data and saves. The sensed data are saved with time and location details [12]. While the nodes move across their paths they share their collected data with those nodes they come into radio communication with. Data is usually shared on a request basis. Each node is provided a coverage threshold and confidence threshold. Actual target detection in this system is determined by examining the data taken at several points that are close to each other. If the combined data collected by this group of close points exceeds some predetermined threshold then it is determined that a target is detected at this area. Note that the value of this detection threshold will determine how sensitive/ insensitive the system is at detecting targets.

Bayesian Estimation Track Before Detect (TBD) Estimator

Integrated tracking and detection, based on unthresholded measurements, also referred to as Track Before Detect (TBD) is a hard nonlinear and non-Gaussian dynamical estimation and detection problem. However, it is a technique that enables the user to track and detect targets that would be extremely hard to track and detect, if possible at all with "classical" methods. TBD enables one to be better able to detect and track weak, stealthy or dim targets in noise and clutter. The particle filters have shown to be very useful in the implementation of TBD algorithms.In TBD, the detection problem is done using the track output over multiple scans. The detection decision will be made at the end of the processing chain, i.e., when all information has been used and integrated over time [15].

Although it is called track before detect, the tracking and detection processes occur simultaneously. In this way, the energy of a (weak) target is integrated and correlated over time and position. The concept will lead to a better performance when detecting and tracking weak targets. TBD also implicitly solves the problem with data association.

Target Tracking Background

Target tracking algorithms usually focus on the aspect of the sensor nodes' interaction with a target after the target has already been detected within the area the sensor nodes cover. Once the object has been detected, the nodes collect information and then use one of many different types of algorithms to calculate the current location of the object relative to the sensor nodes' locations. From here, it is the goal of the sensor network to track the object as it moves through the network. This may or may not involve predicting the next location of the object when it moves. In order to forewarn those nodes it will be heading towards to prepare to capture data [14]. Several of these tracking algorithms are discussed in the following section:

Simple Triangulation

This algorithm is an extremely simple design and implementation and serves as an ideal algorithm to use as the foundation for explaining other more complex tracking algorithms. The whole goal of this algorithm [14] is to provide a simple algorithm that uses simple computation in order to calculate an object's current location and predict where it is headed and notify those nodes near the predicted next location of the object.

This algorithm first assumes that all nodes in the network are localized to a common reference point and can detect and estimate the distance to a target using signal strength [14]. When a node detects an object within its range it broadcasts a Target Detected message.

This message contains the location of the sensor node and the distance to the target. All nodes that hear this message store its data in their local memory. When a node that has detected the target hears two other

Target Detected messages from two other nodes it performs triangulation on the three coordinates to calculate the location of the target. (Note that this means that more than one node may perform this calculation for the same target at the same time.)

This node then continues to project the trajectory of the target. When the estimated target trajectory has been calculated, all nodes that are within some distanced' perpendicular to the target's trajectory are sent a Warning message to alert them that the target is headed towards them. These newly awoken nodes then track the object as it enters their area and repeat the Target Detected and Warning message sending process.

Clusters

The cluster target tracking algorithm has been widely discussed through many research papers. In this section, the basic idea of target tracking using clusters is discussed, followed by an overview of the variances in each of the different cluster algorithm research papers.

Basic Cluster Algorithm

The basic algorithm for tracking an object using clusters is as follows [20]:

- Some (or all) of the nodes in a cluster detect the object and report their data to a cluster head. Note that each cluster has only one cluster head.
- The cluster head node uses all the target detection information from the sensor nodes to estimate the target's location.
- The cluster head uses the calculated target location and past

locations of the target to predict the next location of the target.

- Those sensors around the predicted location are woken up to form a new cluster (if not already in one) to detect the target.
- When the target is detected in this new cluster, the previous cluster's nodes are all put into a sleep state. This new cluster then continues the cluster tracking algorithm.

Hierarchical Super nodes

This algorithm deviates from the basic algorithm in that the cluster heads (called super nodes) have a higher Communication range and more computational power [21]. These "super nodes" are distributed throughout the network and the otherwise normal nodes are assigned to super nodes. Clusters are not dynamically generated in this algorithm. Interestingly enough, super nodes do share target location information among each other, whereas regular sensor nodes do not.

Dynamic Clustering

Like the super node algorithm, this algorithm also assumes that cluster head nodes have more power than normal sensor nodes. However, sensor nodes are not assigned to clusters in this algorithm. Instead, they are invited to join a cluster by the cluster head. The cluster head does this by broadcasting a join message that includes the time and signature of the target the cluster head detected [19]. Those sensor nodes that have stored data that matches the data in the broadcast message respond to the broadcast by sending their captured data to the cluster head node. Interestingly, the cluster head only waits for a certain number of replies and when the required numbers of replies are received, the cluster head calculates the target's location. Unlike the super node algorithm, this algorithm has one active cluster head at a time. In other words, cluster heads do not work together [22].

Cell Collaboration

Like predicate clustering all nodes in this algorithm are of the same type. Nodes are formed into clusters (called cells) in which all nodes within the cells collaboratively decide when a target has entered their cell and if they should track it [22]. It is interesting to note that cells can have different sizes; the size of cells is determined by the observed velocity of the target. So cell size will increase for faster moving targets.

Probabilistic Localization

This algorithm deviates even further from the previously discussed algorithms. Cluster heads are special high powered nodes that know the location of every node within their cluster [23]. This algorithm takes advantage of the fact that the cluster head knows these locations. When sensor nodes detect a target, they send a very small notification message to their cluster head and store the target location data, time and other relevant data in their local memory. Upon receiving several notifications, the cluster head performs a probabilistic localization algorithm to determine which sensor nodes to query saved data from. In other words, the cluster head runs an algorithm that helps to predict which sensor nodes are closest to the target. Therefore, the cluster head has the best data to calculate an accurate estimation of the target location. When the cluster head determines which nodes to query, they are asked for the data they saved in their local memory and the cluster head uses this information to calculate the location of the target.

Distributes Predictive Tracking

This algorithm is very similar to the predicate localization algorithm in that cluster head knows the ID, location and energy level of each of the nodes within their cluster. However, not all nodes belong to a cluster. Nodes on the border of the network are not in a cluster and those are on sense at all times. Similarly, nodes next to the border are not in a cluster. Moreover the, cluster heads only choose 3 nodes from target tracking data. Cluster heads determine which 3 nodes to query based on their location in relation to the predicted target location calculated by a previous cluster [24]. When the cluster head has chosen which nodes to query, those nodes are woken up to prepare to detect the approachable target. This algorithm differs even further in that if a cluster head cannot find 3 suitable nodes within its own cluster, it can seek the help of neighbouring clusters and ask for one or more of their nodes to be turned on and report information.

Rooted Tree

Related to the clusters tracking algorithm is the idea of the rooted tree tracking algorithm. This algorithm differs somewhat in that instead of having multiple cluster heads, there is only one head node and it is referred to as the root. The root node is the node that is closest to the target.

When a new target location is predicted, if there is another node that is closer to that position than the current root then new node becomes the root. Other nodes around the root work together to form a tree in which their sensed data is collected and passed up through the tree (children to parents) until all data reaches the root [25]. When the root receives all the data it calculates the target position and predicts the new target position as mentioned in the basic target tracking algorithm. The tree itself is reconfigured on every root change. When a root change occurs, a "reconfigure" message is broadcast containing the location of the new root. When a node detects the "reconfigure" message, it detaches itself from the old tree and attaches itself to the new tree by recognizing its neighbour node that is closest to the new root node as its new parent. New data is then collected and passed up the tree for the new root to use.

Particle Filtering

The basic idea behind a particle filtering tracking algorithm is that numerous object state descriptions are saved that contain data necessary to calculate the target position at a certain time. These state descriptions are referred to as particles and each particle has its own weight. The weight of a particle determines how much the data it contains will contribute to the location estimation of an object. When new particles are created, the weights of the pre-existing particles are adjusted and then all the particles are used to calculate a new target location [26]. Eventually, particles with weights that are below a certain threshold are eliminated as duplicate particles.

Some of the particles filtering algorithms differ a little. Certain algorithms have all the particles stored in a central node and this node does all the target location processing [28]. Other algorithms distribute the stored particles across the network nodes.

ANALYSIS OF ALGORITHMS

This section analyses each target tracking algorithm mentioned with a focus on how it could be improved by combining it with one or more of the ideas presented in the target detection algorithms.

Simple Triangulation

The main problem with this algorithm lies in the fact that when a node has detected a target and hears two other nodes broadcast the same target detection, the nodes starts to perform the triangulation localization calculation. This means that at the same point of time there will be multiple nodes calculating the same target location. This is a waste of computation and energy resources. This algorithm needs to be improved so that when a node starts to calculate the location of the target, it first broadcasts its intention to calculate the location to all the other nodes, similar to the data sharing that is done in the collaborative coverage mobile detection algorithm. Then all the other nodes that would have performed this same calculation will not perform it once they hear the broadcasted intention of the first node.

Basic Cluster

The majority of the cluster algorithms had all the nodes on during their target detection phase. It would be better if they instead adopt a technique similar to that used in the wave detection algorithm to detect targets. Since each area of the network is already divided into clusters, each node within the cluster could take turns monitoring the area for the arrival of a target using one of the wave detection methods mentioned. This would increase the lifetime of the network.However, using one of the mobile detection algorithms is not recommended since the nodes have to register with the cluster head. Depending on the mobile detection algorithm used, the nodes would constantly be moving in and out of clusters. This would cause an increase in the amount of message communications with these nodes in order keep track of the nodes within the cluster. Mobile detection algorithm of this type can be applied to the situation only if the mobility of the nodes is limited and the node always stay within the same cluster.Additionally, the cluster algorithms could benefit from the basic fusion detection algorithm. Once a target has been detected and all the information is passed to the cluster head the cluster head could use the technique used in the basic fusion algorithm to eliminate any data provide by faulty nodes. This would help make the network less vulnerable to faulty nodes.

Dynamic Clustering

The biggest problem with this algorithm is that it will not work with mobile nodes. The network calculates the sensor node's locations at start-up. Since these locations need to be known by the cluster head and are calculated only once, this prevents any nodes from moving or even being added to the network. This problem could be easily fixed by periodically refreshing the locations of the sensor nodes, although this will decrease the network's lifetime. So it is better to use a detection algorithm and constantly refreshing the nodes' location isn't an optimal solution is better in this case.

Cell Collaboration

This clustering algorithm is already perfectly setup to use the grid detection algorithm for detection of targets. Since the nodes in this algorithm are already divided into cells, the basic structure to use the grid algorithm is already in place. All that needs to be done is to add the grid algorithm for the detection part. Once, this is done the algorithm would be more efficient since the nodes would be alternating sleep schedules with each other in order to detect the target instead of all being awake at the same time.

Probabilistic Localization

This algorithm already sounds like that it has incorporated parts of the hybrid detection algorithm in it. When a target is detected, this algorithm sends a small message to the cluster head. Similarly, when a target is detected in the hybrid algorithm a small yes/no bit message is sent to the decision maker in the network, or cluster head.

Then the cluster head uses this data to determine which nodes to query for further data. In determining which nodes to query, it would be a good idea to use an averaging technique similar to that used in the basic fusion detection algorithm in order to eliminate the impact of any data provided by a faulty node. The similarities to the hybrid algorithm continue. When the cluster head determines which nodes to query, the nodes respond back to the cluster head with more detailed information.

However, this algorithm should not be combined with a mobile node detection algorithm because of the way it decides which nodes to query. In this algorithm, the cluster heads create probability tables in order to determine which sensors to query [23]. The use of a table is acceptable until nodes move outside or are added to the system. When this happens, the probability table must be updated. Therefore, the use of a mobile detection scheme would cause a lot of additional computational overhead since these tables are to be constantly refreshed.

Distributed Predictive Tracking

This algorithm is unique in that it is already doing its own target detecting. Instead of having the border nodes being on at all times to detect nodes it would be beneficial to have these nodes adopt a wave or patrol like technique. This way, they would still detect targets but would be consuming a considerably smaller amount of energy.

Rooted Tree

One of the basic problems with the rooted tree technique is that it doesn't provide any fault tolerance facts. In order to make this algorithm less vulnerable to problems caused by faulty nodes, it should use an averaging technique similar to that used in the basic fusion detection algorithm.

Additionally, this algorithm encounters problems when the velocity of the target is very high. As the velocity of the target increases so does the number of times in which the tree must be reconfigured [25]. This will cause problems when a threshold velocity of the target is reached in which the tree can no longer be generated quickly enough to keep up with the moving target. This problem could be eliminated by altering this algorithm to work more like the patrolling algorithm. Instead of constantly rebuilding the tree, the tree is treated as a path that can have steps (or nodes) along the path added or deleted from it. This way when the target moves, just add a new node to the path and remove any one that is too far away to contribute any longer. The basic nodes needed to collect data are still members of the path and new ones are added. It saves an enormous amount of time and message communication needed to periodically rebuild the tree structure each time.

Particle Filtering

In order to track targets, this algorithm saves an enormous amount of data in the form of particles. Over time, these particles take up a large amount of memory space. The amount of memory actually used could be decreased if this algorithm is to adopt the idea behind the TBD algorithm. Bayesian Estimation Track Before Detect (TBD) based on particle filtering gives high performance and hybrid detection algorithm stops recording data once the data it currently contains exceeds a threshold to know whether a target is present or not. The particle filtering algorithm could use the same principle. Once enough particle data is collected to exceed some threshold, the node would stop collecting data. This would not only decrease the amount of memory used but also decrease the amount of work to be done to save energy.

Evaluation of Target Detection and Tracking Algorithms

The best algorithm is established based on the evaluation criteria. The parameters used for evaluation are time, energy, propagation delay, vulnerability, memory used, stability and scalability. Particle Filtering based on the Bayesian TBD estimator algorithm that sounds the best for vehicle detection and tracking in WSNs. Table 4 shows the evaluation summary.

SUMMARY OF ALGORITHMS

It is observed that, one of the best target detection and tracking algorithms is the combination of the Bayesian Estimation Track before Detect (TBD) Estimator and particle filtering algorithm. The fact that the TBD covers the network area in a short amount of time and enables nodes to sleep at times saves a huge amount of energy. This enables a target to be detected, without too much delay, in an energy efficient manner.

The particle filtering algorithm is also on the right track since the pre-existing particles are adjusted when the new particles are created and hence particles with weights that are below a certain threshold are eliminated as duplicate particles, which also conserves energy. Additionally, the use of several central localization calculating points instead of one helps to reduce the vulnerability of the system to an attacker. It eliminates the one point of failure problem. It also shares the burden of energy usage that one node would face across several nodes.

Particle filtering approach allows all the nodes to share the energy burden. Hence particle filtering algorithm combined with Bayesian Estimation Track before Detect (TBD) Estimator algorithm can be used for target detection and tracking.

CONCLUSIONS AND FUTURE DIRECTIONS

In this paper an attempt is made to gather the information about the unauthorized vehicle detection and tracking in the battlefield surveillance and also to survey the sensors that are widely available for vehicle detection. The effort is also set to survey and evaluate the detection and tracking algorithms intended for target (eg vehicle or animal) detection and tracking using Wireless Sensor Networks. An overview of each algorithm type was presented. The combinations of target detection algorithms with target tracking algorithms are also discussed. The various combinations of algorithms have been classified and each category is evaluated according to the identified criteria. Particle Filtering based on the Bayesian TBD estimator algorithm is an interesting one for target detection and tracking using WSNs. It has good resource efficiency, propagation delay is minimum. It also increases scalability and particle based approach increases stability. Real time experiments are required in order to conclude whether a Particle Filtering based on Bayesian TBD estimator algorithm will work in a practical scenario. This paper aims to give a brief overview of sensors and algorithms used for Vehicle detection and tracking which directs the future researchers to show new directions used in battlefield surveillance or in any place where human monitoring is not possible.

ACKNOWLEDGMENT

The authors would like to thank the Armament Research Board (ARMREB-DRDO) for supporting this Research project by funding.

REFERENCES

1. Y. C. Tseng, S. P. Kuo, H. W. Lee, and C. F. Huang, "Location tracking in a wireless sensor network by mobile agents and its data fusion strategies," Lecture notes in computer science, Springer Link, pp. 2–3, 2003.
2. E. Y. Luz and A. Mimbela, "Summary of vehicle detection and surveillance technologies used in intelligent transportation systems," The Vehicle Detector Clearinghouse, Southwest Technology Development Institute (SWTDI) at New Mexico State University (NMSU), Fall 2007. http://www.nmsu.edu/traffic/.
3. S. Coleri, M. Ergen, and T. J. Koo, "Lifetime analysis of a sensor network with hybrid automata modelling," Processings of ACM International Workshop on Wireless Sensor Networks and Applications (Atlanta, GA), pp. 98–104, 2002.
4. D. Jiagen, S. Y. Cheung, C. W. Tan, and V. Pravin, "Signal processing of sensor node data for vehicle detection," IEEE Intelligent Transportation Systems, pp. 70–75, 2004.
5. C. Thomas, K. S. Kewel, and R. Parameswaran, "Fault tolerance in collaborative sensor networks for target detection," IEEE Transactions on Computers, pp. 320–333, 2004. http://citeseer.ist.psu.edu/clouqueur03fault.html.
6. F. Qing, Z. Feng, and L. Guibas, "Lightweight sensing and communication protocols for target enumeration and aggregation," Proceedings of the 4th ACM international symposium on Mobile ad hoc networking & computing, pp. 165–176, 2003.
7. Y. Ting, H. Tian, and A. S. John, "Differentiated surveillance for sensor networks," Proceedings of the 1st international conference on Embedded networked sensor systems, pp. 51–62, 2003. http://www.cs.virginia.edu/papers/ diff_surveillance_sn_p51-yan.pdf.
8. S. S. Ren, Q. Li, H. N. Wang, and X. D. Zhang, "Design and analysis of wave sensing scheduling protocols for object-tracking applications," Lecture notes in computer science, SpringerLink, pp. 228–243, 2005. http://www.cs. wm. edu/~liqun/paper/dcoss05.pdf.
9. C. Gui and P. Mohapatra, "Virtual patrol: a new power conservation design for surveillance using sensor networks," Information Processing in Sensor Networks, pp. 246–253, 2005.
10. B. Y. Liu, P. Brass, O. Dousse, P. Nain, and D. Towsley, "Mobility

improves coverage of sensor networks. international symposium on mobile ad hoc networking & computing," pp. 300–108, 2005.
11. M. K. Krishna, H. Hexmoor, and S. Sogani, "A t-step ahead constrained optimal target detection algorithm for a multi sensor surveillance system," IEEE Intelligent Robots and Systems, pp. 357–362 2005. http://arxiv.org/ ftp/cs/papers/0505/0505045.pdf.
12. K. C. Wang and P. Ramanathan, "Collaborative sensing using sensor of uncoordinated mobility," Lecture notes in computer science, SpringerLink, pp. 293–306, 2005. http: //www.ece.wisc.edu/~wander/ papers/dcoss05_wang.pdf
13. E. B. Ermis and V. Saligrama, "Adaptive statistical sampling methods for decentralized estimation and detection of localized phenomena," ACM'05. http://iss.bu.edu/ srv/Publications/Manuscript-final.pdf.
14. R. Gupta and S. R. Das, "Tracking moving targets in a smart sensor network," Vehicular Technology Conference, pp. 3035–3039, 2003. http://www.cs.sunysb.edu/ ~samir/Pubs/VTC03-long.pdf.
15. S. J. Davey, M. G. Rutten, and B. Cheung, "A com parison of detection performance for several track-bef ore-detect algorithms," EURASIP Journal on Advance s in Signal Processing, 2008, http://www.hindawi.com /journals/asp/2008/428036.html.
16. B. Tatiana, H. Wen, K. Salil, R. Branko, G. Neil, B. Travis, R. Mark, and J. Sanjay, "Wireless sensor networks for battlefield surveillance," Land Warfare Conference, 2006.
17. L. A. Klein, M. K. Mills, and D. R. P. Gibson, "Traffic detector handbook," Operations and Intelligent Transportation Systems Research, 2006. http://www.tfhrc.gov/ its/pubs/06108/index.htm.
18. L. Yu, L. Yuan, G. Qu, and A. Ephremides, "Energydriven detection scheme with guaranteed accuracy" Information Processing in Sensor Networks, pp. 284–291, 2006.
19. W. P. Chen, J. C. Hou, and L. Sha, "Dynamic clustering for acoustic target tracking in wireless sensor networks," 11th IEEE International Conference on Network Protocols (ICNP'03), November 2003.
20. D. Li, K. D. Wong, Y. H. Hu, and A. M. Sayeed, "Detection, classification and tracking of targets in distributed sensor networks" IEEE Signal Processing Magazine, pp. 1163–117, 2002.
21. S. Oh and S. Sastry, "A hierarchical multiple-target tracking algorithm for sensor networks," IEEE Robotics and Automation," pp. 2197–2202,

2005. http://www.eecs. berkeley.edu/~sho/papers/icra05_hmttsn.pdf.

22. R. B. Richard, R. Parameswaran, and M. S. Akbar, "Distributed target classification and tracking in sensor networks," In Proceedings of the IEEE, Vol. 91, No. 8, pp. 1163–1171, 2003. http://www-net.cs.umass.edu/cs791_ sensornets/papers/brooks.pdf.
23. Y. Zou, and K. Chakrabarty, "Target localization based on energy considerations in distributed sensor networks," 1st IEEE Workshop Sensor Network Protocols and Applications (SNPA'03), pp. 51–58, 2003.
24. H. Yang and B. Sikdar. "A protocol for tracking mobile targets using sensor networks, sensor network protocols and applications," pp. 71–81, 2003. http://networks.ecse. rpi.edu/~bsikdar/papers/hua_icc03.pdf.
25. W. S. Zhang and G. H. Cao, "Optimizing tree reconfiguration for mobile target tracking in sensor networks," Infocom, pp. 2434–2445, 2004.
26. G. Ing, "Distributed particle filters for object tracking in sensor networks," Proceedings of the 3rd international symposium on Information processing in sensor networks pp. 99–107, 2004.

CHAPTER

13

Deployment of Pre-Industrial Autonomous Microbe Sensor in Saudi Arabia's Injection Seawater System

Mohammed A. Al-Moniee[1], Xiangyang Zhu[1], Rikke Markfoged[2], Aabdullah H. Al-Wadei[1], Poul L. Pedersen[2], Anders K. Tuxen[2], Fuad I. Al-Nuwaiser[1], Lone Tang[2], Tinna Staghøj Roesen[2], Thomas Lundgaard[2]

[1]Saudi Arabian Oil Company (Saudi Aramco), Dhahran, Saudi Arabia

[2]Danish Technological Institute (DTI), Aarhus, Denmark

ABSTRACT

Microbial growth in water injection systems can lead to many problems, including biofouling, water quality deterioration, injectivity loss, microbial corrosion, and reservoir formation damage. Monitoring of microbial

Citation:- A. Al-Moniee, M. , Zhu, X. , Markfoged, R., Al-Wadei, A. , Pedersen, P., Tuxen, A., Al-Nuwaiser, F., Tang, L., Staghøj Roesen, T. and Lundgaard, T. (2018), "Deployment of Pre-Industrial Autonomous Microbe Sensor in Saudi Arabia's Injection Seawater System". Journal of Sensor Technology, 8, 1-17. doi: 10.4236/jst.2018.81001.

activities is required in any mitigation strategy, enabling operators to apply and adjust countermeasures properly and in due time. In this study, the pre-industrial autonomous microbe sensor (AMS) was constructed with technical improvements from the prototype for increased sensitivity, durability, robustness, and maintainability. The pre-industrial AMS was lab validated, field proven, and deployed at critical locations of seawater injection network for automated detection of microorganisms under the Saudi Arabia's harsh environment. An excellent correlation between AMS measurement data (fluorescence count) and actual count of microbial cell number under microscope was established (coefficient of determination, R^2 > **0.99) for converting AMS fluorescence count to cell numbers (cell mL^{-1}**) in the injection seawater. The pre-industrial AMS only required monthly maintenance with solutions refill, and was able to cope with hot summer months even without protection in an air-conditioned shelter. The study team recommended wider deployment of the online AMS for real-time monitoring of bacteria numbers in the various strategic locations in Saudi Aramco's complex seawater injection network, as an integral component of pipeline corrosion and leak mitigation program.

Keywords: Automated Monitoring, Autonomous Microbe Sensor, Microbial Sensor, DNA Staining, Injection Seawater

INTRODUCTION

Saudi Arabian Oil Company (Saudi Aramco) has the largest seawater injection system in the world. The growth of microorganisms and the formation of biofilm on the pipeline's inner surfaces and process equipment lead to an array of challenges, including biofouling, water quality deterioration, infectivity loss, reservoir plugging, and microbiologically influenced corrosion (MIC) [1] [2] . A critical cornerstone in an effective mitigation strategy is the routine surveillance of the system microbiology, enabling operators to apply and adjust microbiological countermeasures accordingly.

The microbial number and activity are traditionally monitored with conventional growth methods that require manual sampling and handling. Monitoring the microbial activity with conventional methods in this large system, especially at the remote locations, presents a long-standing challenge to Saudi Aramco. In an attempt to ensure the injection water quality and maintain the integrity of injection pipelines network, Saudi Aramco and the Danish Technological Institute (DTI) embarked on a three-phase project with an overall objective of developing and deploying an online sensing technology for automatic and real-time monitoring of microbial numbers in Saudi Aramco's large injection seawater system.

In the feasibility phase study [3] [4], five single-analyte methods were evaluated in the laboratory setup for the suitability of automation for detection of microbial activity in the Saudi Aramco injection seawater. Staining of cells with DNA binding fluorescent dyes (SYBR Green), followed by quantification of fluorescence signals, was identified as a reliable single-analyte method for automated, online determination of microbial cell numbers in the injection seawater system.

In the subsequent phase, an autonomous microbe sensor (AMS) prototype was constructed, tested, and optimized in the laboratory, followed by a validation in the field for automated detection of microorganisms in the harsh Saudi Arabia desert environment [5]. In the field validation, the AMS prototype was able to monitor and follow the general microbial status in the system, including detection of periods with increased microbial growth or decreased microbial numbers following biocide injection. The infield AMS prototype detection limit was 10^5 cells/ml.

The objective of the current phase study was to implement the technical improvements identified in the prototype field validation [5], construct two pre-industrial AMS devices, validate and deploy the pre-industrial AMS in Saudi Aramco's seawater injection network. The technical improvements aimed to increase the sensitivity, durability, robustness, and maintainability, while reducing the uncertainties related to AMS operation and function.

TECHNICAL IMPROVEMENTS OF PRE-INDUSTRIAL AMS

The pre-industrial AMS is an instrument that provides an online and real-time detection of total microbial numbers in seawater pipelines. It is provided with a flow cell that measures the fluorescence intensity of microorganisms in the seawater sample after mixing with DNA staining dye SYBR Green [6] [7] [8] . The sensor contains all necessary ancillary systems to sample seawater and fluorescence dye, inject them in precise volumes, mix them and dispose them after measurement, to a waste container. The microbial number is derived from the fluorescence intensity through a calibration procedure that can be checked periodically. The calibration procedures involve the fluorescence intensity measurement with AMS, the temperature during measurement and the direct microscopy counting of known concentrations of bacteria dilutions after staining with 4, 6-diamindino-2-phenylindole (DAPI) [9] [10] [11] .

The areas for technical improvement and optimization identified in the field testing of the AMS prototype [5] have been incorporated in the design and construction of the pre-industrial AMS. Some of the improvement areas were:

1) Optimization of cleaning procedures: The internal AMS cleaning procedures were optimized to minimize the risk of bio fouling and iron precipitation. Alkaline (NaOH) and acid (oxalic acid) cleaning procedures were incorporated into the measurement cycle to prevent the biofouling and formation of iron precipitations in liquid handling components.
2) Incorporation of an internal DNA control: The internal DNA control was used to validate the AMS functionality and compensate the temperature effect on fluorescence counts.
3) Optimization of fluorescence detection procedure and software: The sampling procedure and software were optimized in the pre-industrial AMS to minimize the scattering of data points caused by uneven cell distribution.
4) Sensitivity improvement: The detection limit of pre-industrial sensor was improved by optimizing hardware components (spectrophotometer, optical fiber, and flow cell) and microbial staining and measurement procedures.
5) Incorporation of temperature correction factor: The temperature inside the AMS was measured simultaneously with the fluorescence measurements. The inside temperature and the fluorescence count of internal DNA control were used to develop a temperature compensation factor to compensate the temperature effect on the fluorescence dye.
6) Redesigned user interface: The new software interface, layout and data output allowed data export and automatic plotting and display in remote computer.

DESIGN AND CONSTRUCTION OF PRE-INDUSTRIAL AMS

The design of pre-industrial AMS has incorporated the technical improvement areas identified from AMS prototype. Design changes from the prototype to the pre-industrial AMS are summarized in Table 1.

The pre-industrial AMS consists of three separate modules: hardware module, consumables module, and power module:

1) Hardware module: includes a custom-made flow cell for achieving flow during quantification, a heating chamber for improved staining conditions and two dedicated pumps.

Table 1. Design changes from the prototype to the pre-industrial AMS.

Part	Prototype AMS	Pre-industrial AMS
Unit layout	Single box	Three modules: hardware, consumables, and battery
Internal DNA control	N/A	Synthetic DNA in ethyleneiaminetetraacetic acid (EDTA) and ascorbate buffer
Sample staining	40 min at ambient temperature	13 min at 60°C
Flow cell	Commercially available flow cell	Custom-built flow cell with minimal volume and large numerical aperture
Measurement	Sample is static when illuminated	Sample flows slowly when illuminated to avoid bleaching of dye
Dye dispensing	Fixed volume	Dedicated syringe pump for dye
Lower limit of field detection	~10^5	Low limit of detection, LLOD
		– AMS 1: 1.9×10^4; AMS 2: 1.2×10^4
		Low limit of quantification, LLOQ
		– AMS 1: 5.0×10^4; AMS 2: 3.2×10^4
Cleaning	NaOH	NaOH and oxalic acid
Power	AC mains	Battery, charged either by AC mains or solar panels
Remote communication	Team Viewer through Saudi Aramco intranet when AMS is on	Permanent RS-485 connection. Data retrieval and AMS control through dedicated software on remote computer

Pump A is for dispensing water sample and cleaning solutions; while Pump B is dedicated for SYBR Green dye distribution.

2) Consumables module: includes four containers for water, oxalic acid, sodium hydroxide and waste.

3) Power module: A 24 V DC battery pack was chosen as the power supply. The batteries are connected to three solar panels or 230 V AC mains power.

The modular design of AMS improves the equipment functionality and durability in the challenging Saudi Arabia dessert environment and operation conditions of seawater injection system [4]. Each of these three modules are mounted in a stainless-steel cabinet and may be stacked with the battery module at the bottom, the consumables module in the middle and the hardware module at the top (Figure 1). The AMS is designed to run on batteries that are charged by solar panels or 230 V mains power. The AMS device installed in the field is connected to the remote computer through a MODBUS protocol using RS-485 cable. The Lab View application [12] on the remote computer controls the sensor operation, equipment shut down following measurements, and data retrieval.

PERFORMANCE ASSESSMENT

Lab validation

After construction of two pre-industrial AMS units, the AMS was thoroughly tested and validated in the laboratories at DTI premises. The repeatability and functionality of the units were tested with respect to procedures as well as software. The major improvements in lab validation and their effect on AMS performance are summarized in Table 2.

After implementation of above improvements on AMS units, a correlation of fluorescence signals from AMS and direct counting of microbial cell number under microscope after DAPI staining was established (Table 3).

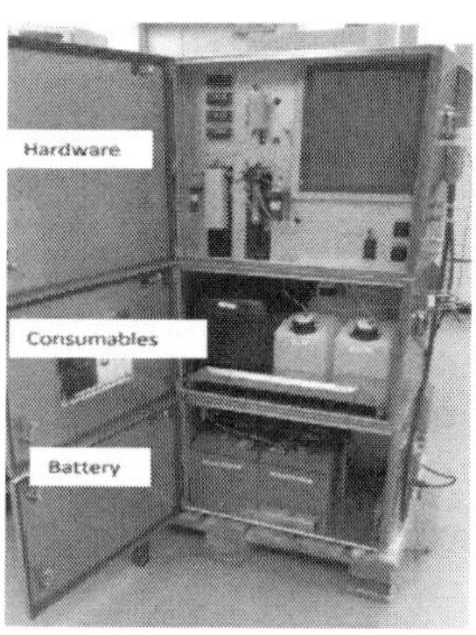

Figure 1. Hardware module, consumables module, and power module of pre-industrial AMS.

Table 2. Major improvements incorporated during the lab validation.

Improvement	Effect
Decrease of Pump A rate to 3 mL · min^{-1}	Eliminate air from surroundings into syringe A
New loading procedure for internal DNA control	Improved mixing of internal control (IC) DNA and Milli-Q water
Increased total volume of diluted internal DNA control to sample volume	Improved loading of IC DNA
New dye loading procedure. Dead space in Pump B valve used as dye dosing volume	Accurate dye dosing
Selection of incubation conditions of 13 min at 60°C	Improved staining of cells with SYBR Green I
No air is moved through flow cell	Eliminate introduction of air bubbles in flow cell
Flow direction of flow cell changed	Remove potential air bubbles in flow cell
New flow rate through flow cell	Optimal flow rate for maximum reading
Increased rate of dispensing and aspirating with Pump A for all non-critical steps	Minimize total AMS measurement cycle time to 75 min
Incorporated data export function	For automatic plot of raw and calibrated readings on graph in Excel

The correlation is performed using a series of dilutions of sulfate-reducing bacteria (SRB) enriched from Saudi Aramco injection seawater and diluted in phosphate buffer (pH 8, 5.5% NaCl).

The correlation is required to convert the data obtained by the AMS, in terms of fluorescence counts, into cell count per volume (cells mL^{-1}).

The calibrations show that the AMS units are linear in a wide range of at least two orders of magnitude, with a laboratory detection limit of around 10^3.

From the performed calibrations, the AMS units were accepted for field installation, calibration, and validation.

At the end of lab validation, the optimal AMS measurement settings were determined and implemented in both AMS units (Table 4).

Table 3. Laboratory calibration data and equation for AMS1 and AMS2.

	AMS1		AMS2	
	Fluorescence count (AMS)	Cells mL^{-1}	Fluorescence count (AMS)	Cells mL^{-1}
Sterile filtered system water (blank)	154 ± 7, n = 3		98 ± 3, n = 3	
LLOD	167	6.0×10^3	104	8.8×10^2
LLOQ	188	1.6×10^4	114	2.3×10^3
Regression equation	473X − 72,788		146X − 14,265	
Regression, R^2	0.9902		0.9490	
Linear range	6.4×10^3 - 5.0×10^6		1.1×10^4 - 2.4×10^6	

LLOD: Lower Limit of Detection; LLOQ: Lower Limit of Quantification.

Table 4. Final settings for one full AMS measurement cycle.

Improvement	Effect
Volume of sample	4 ml
Volume of IC DNA	4 ml (0.1 ml concentrated IC discarded at first, 0.2 ml IC in 3.8 ml sterile H_2O)
SYBR Green I	Diluted 1:20 in Dimethyl sulfoxide (DMSO). Volume equals to dead space in Pump B (20 **µL**)
Incubation of sample	13 min at 60°C
Incubation of IC	Less than one minute at ambient temperature
Flow rate and time	3.5 $\mu L \cdot s^{-1}$ **of 180 sec and 10 sec**
Measurement wavelength interval	520 - 530 nm
Fluorescence count	Average of readings every one sec for 180 and 10 sec
Background reading (auto-fluorescence)	To be subtracted from sample reading. Fluorescence reading of sample without dye, for correction of any auto-fluorescence of the sample water. Background reading is not subtracted from internal control readings.
Cleaning	Sterile H_2O and air between sample and IC measurement. Acid, alkaline, sterile water and air at the end of a cycle. No air through the flow cell.
Time consumption	75 min for one full cycle, including washing/cleaning

Field installation and Calibration

After laboratory validation at DTI premises, both AMS units were shipped to Saudi Aramco, and tested for functionality and stability after transportation. After functionality check, AMS1 was connected to the side-stream of a 60" shipping line at an injection facility and AMS2 was installed at an injection plant further downstream of AMS1, connecting to a side-stream of 40" seawater pipeline. The seawater flow continuously passes AMS through a bypass connection to the disposal when AMS is not doing the measurement. During the measurement cycle, a fixed volume of seawater sample is taken into the AMS for staining and measurement of microbial cell number. During the field testing and validation, the sensors were programmed to take two measurements a day, with a monthly maintenance for solutions refill. The data were retrieved by the operators in the Control Room.

The fluorescence intensity from a stained sample depends on the amount of DNA present in the sample, which in turn, depends on the cell size and the metabolic status of the cell, as an actively dividing cell contains more DNA than a dying cell [13] [14] [15] . Therefore, the fluorescence intensity from the AMS was re-calibrated in Saudi Aramco with active aerobic heterotrophic bacteria enriched from injection system seawater. A 3-point field calibration was performed on AMS1 and AMS2 using the bacteria dilution series prepared with sterile filtered system water.

The calibration correlates the cell number (cell mL^{-1}) of the system water determined by DAPI staining and counting to AMS fluorescence counts from dilution series of the sample measured in the field. A typical DAPI staining for counting microbial cell number under microscope is shown in Figure 2. The results of 3-point field calibration between cell number and fluorescence count is shown in Figure 3.

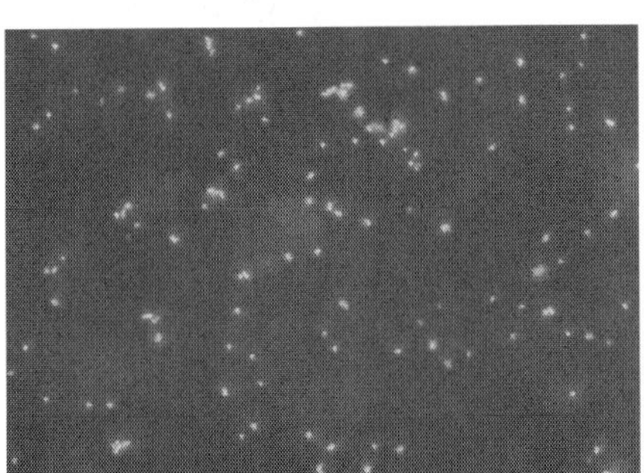

Figure 2. DAPI staining and counting under microscope.

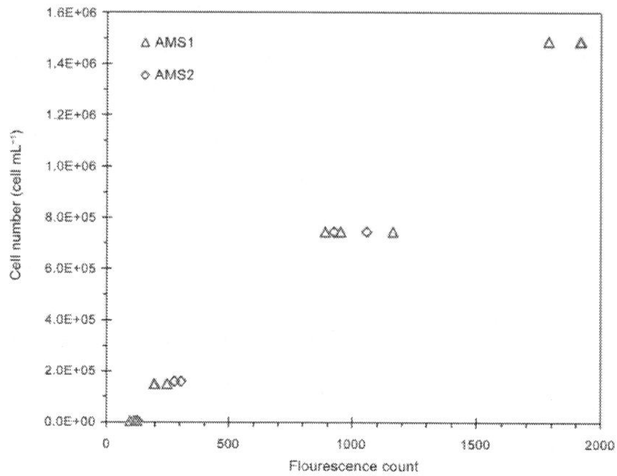

Figure 3. Relation between cell number and fluorescence count of AMS1 (blue) and AMS2 (red) using enrichment culture from the system, diluted in sterile filtered system water. All AMS readings are performed with an AMS interior temperature of 30°C.

A linear regression between AMS fluorescence measurement and DAPI count for converting AMS fluorescence count to cell numbers in the sample is summarized in Table 5. Compared to the calibration performed in the laboratory, these data show that the fluorescence count relative to the cell number is lower in the field, and therefore the regression equation in the lab at 25°C is not recommended for field trial. The field regression equations for each AMS unit were loaded in the specific AMS units, and used for the data conversion from fluorescence count to cell number.

Temperature effect on Fluorescence Intensity

During the field validation of the AMS units, a tendency of higher fluorescence counts with lower AMS inside temperatures has been observed. A full correlation of all internal DNA control readings shows an exponential decrease in fluorescence with increasing temperature, which fits well with an exponential function with both a vertical and horizontal offset [Figure 4 and Equation (1)].

$$f(x) = A \times \exp\left[B\left(x - x_0\right)\right] + C \tag{1}$$

where the fitted parameters are $A = 2.4 \times 10^4\ s^{-1}$, $B = -0.10°C^{-1}$, $x_0 = 4.16°C$, $C = 2.49 \times 10^3\ s^{-1}$, and x = AMS inside temperature (°C).

The function f(x) between temperature and fluorescence intensity was used to temperature compensate internal control (IC) readings. The compensation is done by multiplying the IC reading with the ratio f(30)/f(T), where 30 is the reference temperature, and T is the AMS inside temperature of the given data point.

Table 5. Field calibration data and equation for AMS1 and AMS2.

	AMS1		AMS2	
	Fluorescence count (AMS)	Cells mL^{-1}	Fluorescence count (AMS)	Cells mL^{-1}
Sterile filtered system water (blank)	114 ± 12, n = 3		123 ± 7, n = 3	
LLOD	136	1.9×10^4	137	1.2×10^4
LLOQ	173	5.0×10^4	160	3.2×10^4
Regression equation	843X – 96,020		878X – 107,935	
Regression, R^2	0.9939		0.9940	

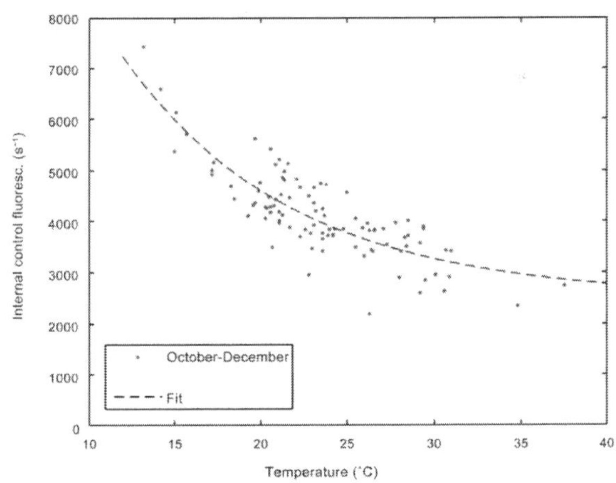

Figure 4. Internal control DNA fluorescence readings versus temperature inside AMS during field readings. October-December data have been fitted with an exponential function with both a vertical and horizontal offset.

The temperature effect on fluorescence intensity is likely related to the molecular structure of SYBR Green I, as the supplier of SYBR Green (Invitrogen) explained that "the temperature effect is likely caused by less rigid bonds in the fluorescent molecule as the atoms become excited and move around." Therefore, the observed temperature response is a direct consequence of the reduced SYBR Green fluorescence at elevated temperatures. In this case, the fluorescence intensity of stained IC DNA or stained cells are equally affected by temperature.

An IC measurement is performed for each AMS run with equal DNA content for each run. As the IC and system water samples are affected equally by temperature, the IC can, besides validating the performance of the AMS unit, also be used for an internal temperature compensation. Figure 5 and Figure 6 showed the raw fluorescence readings and readings after temperature compensation from IC DNA and system seawater, respectively.

Other possible causes for the temperature dependency (melting of oligonucleotides DNA and heating of spectrometer) have been considered unlikely. Internal control DNA (a synthetic 55-base pair oligonucleotides with hairpin structure) was tested for the stability at a wide range of temperature. The synthetic DNA in EDTA and ascorbate buffer was found to be stable from 20°C to 80 °C for at least 28 days when stored in a sealed container.

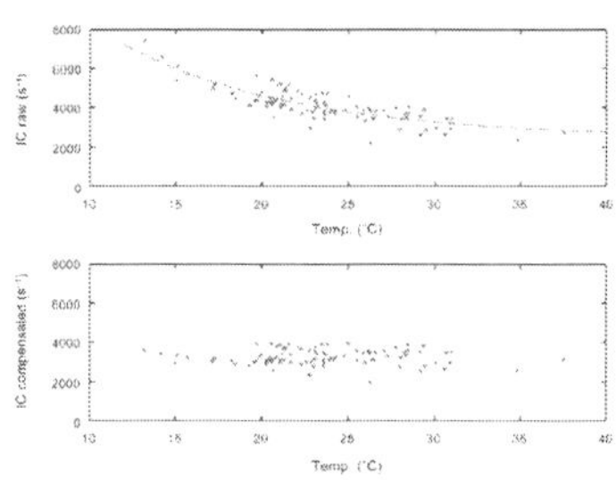

Figure 5. Internal control DNA fluorescence readings (October-December)-raw (top) and temperature compensated (bottom).

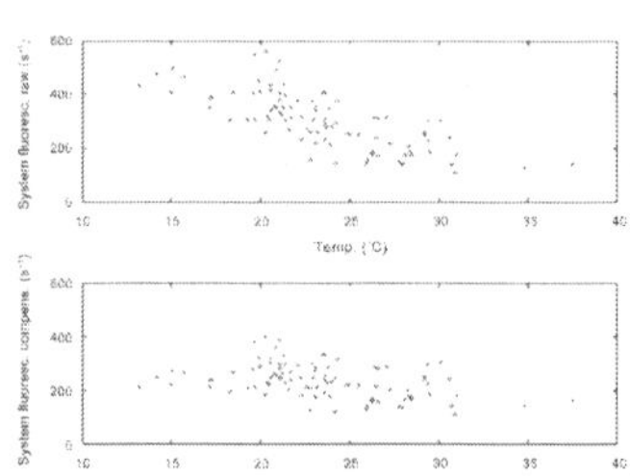

Figure 6. System seawater fluorescence readings (October-December)-raw (top) and temperature compensated (bottom).

Melting of the double stranded (ds) Oligo DNA (internal control) would result in a lower binding between the SYBR molecule and the DNA, as SYBR Green I only binds to ds DNA [6] [16] . The oligo DNA as well as the internal control buffer have been designed to withstand high temperatures, and based on calculations from the Oligo DNA supplier (Integrated DNA Technologies), the melting temperature (T_m, 50%) is 73°C for the conditions provided from the internal control buffer solution (Figure 7).

The spectrometer must be cooled for proper readings, and cooling has been included in the design of the AMS. To test whether the AMS spectrometer is affected by AMS inside temperature, fluorescein was injected manually into the flow cell at high and low temperatures (40°C and 25°C, respectively). Fluorescein is a dye that emits fluorescent light with excitation and emission spectra comparable to those of SYBR Green when bound to DNA [17]. The effect of AMS inside temperature on spectrometer readings of fluorescein at two concentrations was not detected (data not shown).

Field measurements

The AMS1 measurement data, from October 2016 to February 2017, on the fluorescence count of the IC DNA and microbial cell number in the seawater are plotted in Figure 8 and Figure 9, respectively. Figure 8 shows that the IC readings are relatively stable in this 4-month period, after temperature compensation of fluorescence intensity, verifying the stability of both IC DNA and SYBR Green dye, and the functionality of AMS in the field. Figure 9 shows that the majority of the measurements of seawater cell number are above the detection limit. When system water fluorescence count has been temperature compensated, the cell numbers from winter are in the range comparable to fall cell numbers.

Temperature effect on fluorescence intensity is determined to be significant when SYBR Green I dye is used to stain IC DNA and cell DNA in seawater samples. As a result, temperature compensation for all fluorescence counts of both IC DNA and microbial cells is required. The temperature compensation factor was incorporated in the software for converting the raw data to temperature compensated data for data exporting and plotting.

LESSONS LEARNED

During the field testing and validation, a number of technical challenges have been either encountered or identified, and the solutions have been implemented to address these challenges (Table 6). Among the challenges, the two main threats are the pressure regulator clogging and leakage from the fluid handling system, which are discussed in detail in the following text, to illustrate how the technical challenges have been resolved.

The pre-industrial AMS is equipped with a restrictor to reduce the side-stream pressure from 10 bars to less than 1 bar before water enters the AMS system. The restrictor is equipped with a pressure gauge to monitor inlet pressure. Since the seawater pipeline pressure is much higher, it is important that another pressure regulator is installed on the side-stream to reduce the pressure to about 10 bars for prevention of damage to AMS hardware.

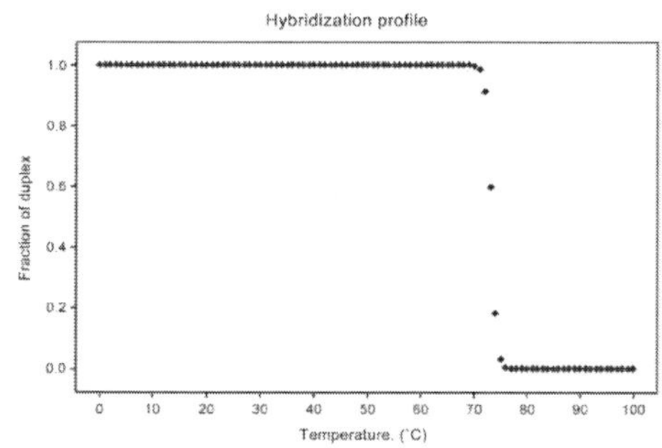

Figure 7. Hybridization profile of the Oligo DNA used as internal control. The total DNA is intact up to 70°C, and the $T_{m\ 50\%}$ is 73°C.

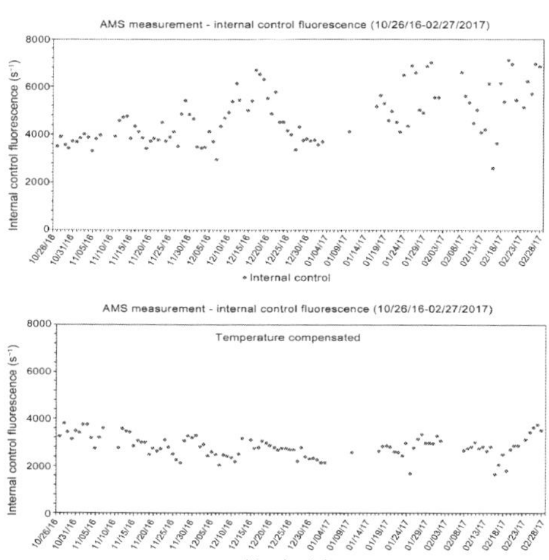

Figure 8. AMS1 autonomous sampling and measurement for fluorescence signals of IC DNA from October 2016 to February 2017-raw (top) and temperature compensated (bottom).

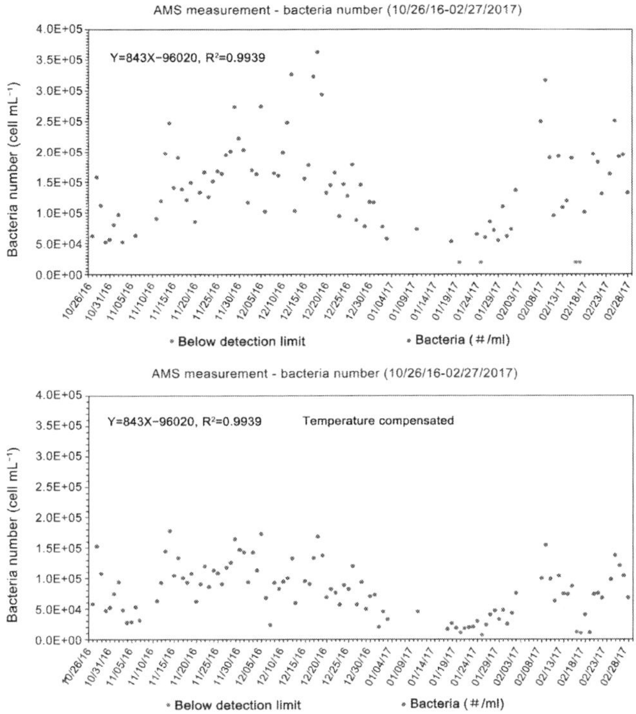

Figure 9. AMS1 autonomous sampling and measurement for microbial cell number from October 2016 to February 2017-raw (top) and temperature compensated (bottom).

Due to the harsh conditions in the seawater environment, the needle type pressure regulator tends to be easily clogged. To avoid clogging, the ball valve type or spring type pressure regulator is required for both side-stream and AMS. After the right type of pressure regulator is selected and installed, the regulator clogging and potential AMS damage due to high pressure are eliminated.

AMS liquid handling system consists of two syringe pumps, two multichannel valves, and many tubings to transport various liquids around the AMS system for measurement, cleaning, and disposal. Loose fittings, wear and tear, and corrosion have been noticed during long-term field trials. Replacement of original Delrin fittings with PEEK fittings has solved the corrosion problem caused by oxalic acid and reduced the leaking occurrence. Occasional tightening and close monitoring of the fittings are required when there is large ambient temperature fluctuation.

Table 6. Lessons learned from field test of the AMS.

Element name	Issue	Lessons Learned	Action Required
Reagents and solutions			
SYBR dye	Experienced inaccurate volume of dye collected by syringe.	Solvent DMSO in the dye will freeze below 12°C.	Consider insulating SYBR vial or enclosing AMS modules within an air conditioned shelter.
Reaction and measuring system			
Flow cell	Biocide entering flow cell.	Need to coordinate with corrosion engineer for biocide injection time.	Programmed AMS measurement time around the estimated biocide arrival time.
Liquid handling system			
Pressure regulator	Experienced clogging of pressure regulator inside AMS.	Select right type of regulator.	Replaced original needle type pressure regulator with a ball valve to reduce the pressure. Increased sampling tubing size from 1/8" to 6 mm.
Pressure regulator	Experienced damage of pressure regulator inside AMS.	Avoid overpressure of side-stream.	Installed a pressure regulator on side-stream to reduce the pressure to below 10 bars before connecting side-stream to AMS. Installed a restrictor with a pressure gauge to monitor inlet pressure.
Corrosion	Experienced corrosion of fittings.	Delrin fittings is prone to oxalic acid corrosion.	Replaced all Delrin fittings with PEEK fittings.
Leakage	Experienced leakage and precipitates around fittings.	Loosening of fittings due to large Temperature fluctuation.	Equipped all fittings with lock nut. Tighten fittings, if necessary.
Electronics and power system			
Solar panel	Experienced lots of dust collection on solar panels.	Required monthly cleaning.	Consider installing solar panels at proper height for easy access during monthly maintenance.
Power socket	Power socket is not water proof.	Risk of water splash entering power socket.	Replaced power socket with built-in power cable.

For AMS field deployment, an air conditioned (AC) shelter is recommended to maintain a relatively stable temperature in AMS modules, which will have many beneficial effects: eliminate fitting loosening due to large temperature change, prevent SYBR Green dye freezing, prevent temperature-regulated AMS shutdown, and extend the life span of the AMS device.

In general, pre-industrial AMS sampling, conditioning, measurement sequence and method, solution and reagent refilling were found suitable for field application. When in operation, the AMS was subjected to routine maintenance from the operation manual (Table 7).

Table 7. Pre-industrial AMS routine maintenance schedule.

Maintenance items	Schedule
Check instrument inlet pressure	Monthly
Check instrument sample flow rate	Monthly
Inspect internal sample leak and blockage	Monthly
Clean solar panels	Monthly
Refill solutions and reagents	Monthly
Inspect tubings and fittings, and replace if necessary	Half-year service
Validate performance and recalibrate if necessary	Half-year service

CONCLUSION AND RECOMMENDATIONS

The online AMS, based on fluorescent DNA staining technology, is developed in three phases for real-time detection of microbial cell numbers in large seawater injection systems under Saudi Arabian conditions. Numerous technical improvements have been implemented in each developmental phase for increased sensitivity, durability, robustness, and maintainability, while reducing the uncertainties related to AMS operation and function. Two pre-industrial AMS units have been constructed and deployed at critical field locations for automated monitoring of microbial numbers in injection seawater system (salinity ~5.5%). The two units are currently operational, including the hot summer months in Saudi Arabia without AC shelters, requiring only minimal monthly intervention from field operators and engineers. The wider deployment at more strategic locations in large Saudi Aramco sea water injection network is planned. In addition, a renewed market study is on-going to assess the market opportunities, technology landscape, and commercialization strategy.

For long-term reliability and maintainability, it is recommended to house the AMS hardware and solution modules inside an AC shelter. AC shelter is expected to provide two major benefits. One is to prevent solvent DMSO of SYBR Green dye from freezing when temperature in Saudi Arabia winter drops to below 12°C. DMSO freezing affects the delivery of SYBR Green dye for bacteria staining, and hence the fluorescence signals. For hardware protection, AMS is programmed to shut down automatically when temperature is above 40°C, resulting in many missing measurements during hot Saudi Arabia summer. An AC shelter increases the flexibility of programming measurement time and frequency throughout the year. It is also recommended to use dual power supply (mains power and solar panel) for battery charge, whenever mains power is available at the installation site. This will ensure the sufficient battery charge to power AMS operation in case of reduced solar panel efficiency caused by adverse weather and environmental conditions (e.g., rain, clouds, dust, and sand storms). The cleaning of solar panels should be maintained even when the mains power is connected.

Calibration is sensitive to the specific water system in question, therefore specific calibration shall be performed during commissioning of each water system, to ensure the accurate conversion of fluorescence count to cell number in the sample. In addition, the performance of the AMS shall be validated after major maintenance and service, and re-calibrated if necessary.

ACKNOWLEDGMENTS

The authors would like to acknowledge the Saudi Arabian Oil Company (Saudi Aramco) and the Danish Technological Institute for granting permission to publish this paper.

REFERENCES

1. Ren, H., Wang, W., Liu, Y., Liu, S., Lou, L., Cheng, D., He, X., Zhou, X., Qiu, S., Fu, L., Liu, J. and Hu, B. (2015) Pyrosequencing Analysis of Bacterial Communities in Biofilms from Different Pipe Materials in a City Drinking Water Distribution System of East China. Applied Microbiology and Biotechnology, 99, 10713-10724. https://doi.org/10.1007/s00253-015-6885-6
2. Moniee, M.A., Juhler, S., Sorensen, K., Abeedi, F.N., Lundgaard, T. and Sanders, P.F. (2014) A Review of Saudi Aramco's Water Flooding System and Methods for Monitoring Microbial Activity. Proceedings 15th Middle East Corrosion Conference, Bahrain, 2-5 February 2014, 1-19.
3. Moniee, M.A., Juhler, S., Sorensen, K., Zhu, X.Y., Lundgaard, T., Abeedi, F.N. and Sanders, P.F. (2016) Laboratory-Scale Evaluation of Single Analyte Bacterial Monitoring Strategies in Water Injection Systems. Journal of Sensor Technology, 6, 11-26. https://doi.org/10.4236/jst.2016.62002
4. Moniee, M.A., Zhu, X.Y., Tang, L., Juhler, S., Nuwaiser, F.I., Sanders, P.F. and Abeedi, F.N. (2016) Optimization of DNA Staining Technology for Development of Autonomous Microbe Sensor for Injection Seawater Systems. Journal of Sensor Technology, 6, 27-45. https://doi.org/10.4236/jst.2016.63003
5. Moniee, M.A., Zhu, X.Y., Tang, L., Nuwaiser, F.I., Voigt, N.V., Sanders, P.F., Abeedi, F.N. and Habboubi, H.H. (2016) Validation of Autonomous Microbe Sensor Prototype for Monitoring of Microorganisms in Injection Seawater Systems. Journal of Sensor Technology, 6, 81-100. https://doi.org/10.4236/jst.2016.64007
6. Dragan, A.I., Pavlovic, R., McGivney, J.B., Casas-Finet, J.R., Bishop, E.S., Strouse, R.J., Schenerman, M.A. and Geddes, C.D. (2012) SYBR Green I: Fluorescence Properties and Interaction with DNA. Journal of Fluorescence, 22, 1189-1199. https://doi.org/10.1007/s10895-012-1059-8
7. Noble, R.T. and Fuhrman, J.A. (1998) Use of SYBR Green I for Rapid Epifluorescence Counts of Marine Viruses and Bacteria. Aquatic Microbial Ecology, 14, 113-118. https://doi.org/10.3354/ame014113
8. Shibata, A., Goto, Y., Saito, H., Kikuchi, T., Toda, T. and Taguchi, S. (2006) Comparison of SYBR Green I and SYBR Gold Stains for

Enumerating Bacteria and Viruses by Epifluorescence Microscopy. Aquatic Microbial Ecology, 43, 223-231. https://doi.org/10.3354/ame043223

9. Maruyama, A. and Sunamura, M. (2000) Simultaneous Direct Counting of Total and Specific Microbial Cells in Seawater, Using a Deep-Sea Microbe as Target. Applied Environmental Microbiology, 66, 2211-2215. https://doi.org/10.1128/AEM.66.5.2211-2215.2000
10. Kapuscinski, J. (1995) DAPI: A DNA-Specific Fluorescent Probe. Biotechnic & Histochemistry, 70, 220-233. https://doi.org/10.3109/10520299509108199
11. Tarnowski, B.I., Spinale, F.G. and Nicholson, J.H. (1991) DAPI as a Useful Stain for Nuclear Quantitation. Biotechnic & Histochemistry, 66, 296-302. https://doi.org/10.3109/10520299109109990
12. LabVIEW 2015, Version 15.0f2 (64-bit), National Instruments Corporation, Austin, Texas, US. http://www.ni.com/labview/
13. Gasol, J.M., Zweifel, U.L., Peters, F., Fuhrman, J.A. and Hagstro, A. (1999) Significance of Size and Nucleic Acid Content Heterogeneity as Measured by Flow Cytometry in Natural Planktonic Bacteria. Applied Environmental Microbiology, 65, 4475-4483.
14. Sgorbati, S., Barbesti, S., Citterio, S., Bestetti, G. and de Vecchi, R. (1996) Characterization of Number, DNA Content, Viability and Cell Size of Bacteria from Natural Environments using DAPI PI Dual Staining and Flow Cytometry, Minerva Biotechnology, 8, 9-15.
15. Gasol, J.M., del Giorgio, P.A., Massana, R. and Duarte, C.M. (1995) Active versus Inactive Bacteria: Size-Dependence in a Coastal Marine Plankton Community, Marine Ecology Progress Series, 128, 91-97. https://doi.org/10.3354/meps128091
16. Marie, D., Partensky, F., Jacquet, S. and Vaulot, D. (1997) Enumeration and Cell Cycle Analysis of Natural Populations of Marine Picoplankton by Flow Cytometry Using the Nucleic Acid Stain SYBR Green I. Applied Environmental Microbiology, 63, 186-193.
17. Thermo Scientific, FITC and TRITC, Instructions. https://assets.thermofisher.com/TFS-Assets/LSG/manuals/MAN0011224_FITC_TRITC_UG.pdf

CHAPTER

14

A Survey of Modeling and Control of Piezoelectric Actuators

Jingyang Peng, Xiongbiao Chen

Department of Mechanical Engineering, University of Saskatchewan, Saskatoon, Canada

ABSTRACT

Piezoelectric actuators (PEAs) have been widely used in microand nanopositioning applications due to their fine resolution, fast responses, and large actuating forces. However, the existence of nonlinearities such as hysteresis makes modeling and control of PEAs challenging. This paper reviews the recent achievements in modeling and control of piezoelectric actuators. Specifically, various methods for modeling linear and nonlinear

Citation:- J. Peng and X. Chen, "A Survey of Modeling and Control of Piezoelectric Actuators," Modern Mechanical Engineering, Vol. 3 No. 1, 2013, pp. 1-20. doi: 10.4236/mme.2013.31001.

behaviors of PEAs, including vibration dynamics, hysteresis, and creep, are examined; and the issues involved are identified. In the control of PEAs as applied to positioning, a review of various control schemes of both model-based and non-model-based is presented along with their limitations. The challenges associated with the control problem are also discussed. This paper is concluded with the emerging issues identified in modeling and control of PEAs for future research.

Keywords: Piezoelectric Actuators; Modeling; Control

INTRODUCTION

Piezoelectric actuators (PEAs) and PEA-driven positioning systems have been widely used in the fields of microand nanopositioning such as atomic force microscopes [1-3], adaptive optics [4], computer components [5], machine tools [6], aviation [7], internal combustion engines [8], micromanipulators [9], and synchrotron-based imaging systems [10] due to their high displacement resolution (sub nanometer) and large actuating force (typically a few hundreds of N). PEA-driven positioning systems have also been developed with various configurations, for example, 1-degree-of-freedom (1-DOF) positioning systems with flexure hinge mechanisms [11,12], stick-slip actuators [13], multiple PEAs-driven inchworms [14], or walking actuators [15]; multi-DOF positioning systems with series mechanism [16], parallel mechanism [17], or stick-and-clamping actuators [18].

In all positioning control applications, the hysteresis and creep effects of PEAs have shown to be able to significantly degrade the system performance and even system stability [19]. For improvement, models are desirable to represent these effects and on this basis, tracking controllers can be designed and implemented to achieve desired positioning requirements.

This paper reviews the modeling and control methods of PEAs in microand nanopositioning applications and some remaining issues to be solved, and is arranged as follows.

Firstly, the working principles and behaviors of PEAs are briefly introduced in Section 2. In Section 3 various models and modeling techniques for PEAs are examined, along with their limitation; and in Section 4 various control schemes for PEAs are reviewed with associated challenges. Section 5 concludes with emerging issues identified in modeling and control of PEAs for future research.

PIEZOELECTRIC ACTUATORS AND THEIR BEHAVIORS

Working Principles of Piezoelectric Actuators

PEAs utilize the converse piezoelectric effect of piezoelectric materials to generate displacement and force, i.e. a piece of piezoelectric material will be mechanically strained if subject to an electric field (by placing it into the electric field or applying voltages to its surfaces) [20]. This property is resulted from the microscopic structure of piezoelectric materials as explained in the following. Most piezoelectric materials used in PEAs, e.g. lead zirconate titanate, undergo a structural phase transition as its temperature drops through the so-called Curie temperature, during which their structurally and electrically symmetric cubic unit cells deform into structurally and electrically asymmetric tetragonal unit cells, resulting in spontaneous strain and polarization [21]. Groups of adjoining unit cells with uniformly oriented spontaneous polarization are called Weiss [20] or ferroelectric [21] domains. The directions of spontaneous polarization of different domains thus developed are random, so in this state the piezoelectric materials exhibit no overall piezoelectric behavior.

To fabricate PEAs from such piezoelectric materials, they are further exposed to a strong electric field (10^6 to 10^7 V/m) at a temperature just below the Curie temperature, forcing the directions of spontaneous polarization of the domains to align with the electric field. Such alignment can be approximately preserved after the removal of the electric field. This process is referred to as poling [21,20]. After poling, due to the approximate alignment of the spontaneous polarization of the domains, the deformations of the domains in the direction of their respective spontaneous polarization resulting form the application of an voltage to the piece of piezoelectric material (which generates an electric field that either enhance or suppress the spontaneous polarization of the domains, causing their dimension to change in the direction of the spontaneous polarization) can accumulate and causes an overall deformation or displacement. Thus the piece of piezoelectric material now possesses overall piezoelectric property including the converse piezoelectric effect and can be used as a PEA [20].

Behaviors of Piezoelectric Actuators

In microand nanopositioning applications, typical behaviors of PEAs concerned include hysteresis, creep, and vibration dynamics.

Hysteresis is the nonlinear dependence of a system not only on its current input but also on its past input. In PEAs, hysteresis exists in both the electric field (voltage)-polarization relationship and the electric field (voltage)-strain (deformation or displacement) relationship, with the latter being mostly concerned in microand nanopositioning, and it is caused by the nonlinearities in the converse piezoelectric effect of the unit cells and the switching and movement of domain walls [21]. According to [22], the relationship between the actuating force exerted on the PEA and the resulting displacement of the PEA is linear, as such the electric field (voltage)-strain (displacement) hysteresis can also be treated as the result of the voltage-internal actuating force hysteresis.

The strain (deformation or displacement)-electric field (voltage) hysteresis, which is usually also the input-output hysteresis relationship, of a typical PEA is shown in **Figure 1**(a). It can be seen in **Figure 1**(a) that the hysteresis trajectory of a PEA can be treated as being composed of three types of components: 1) the major loop which is the hysteresis loop that spans the whole input (voltage) range, 2) the minor loops which are the hysteresis loops that only span portions of the input range, and 3) the initial ascending curve which is the input/output trajectory traversed whenever the magnitude of the input applied to the PEA exceeds the maximum magnitude in the input history.

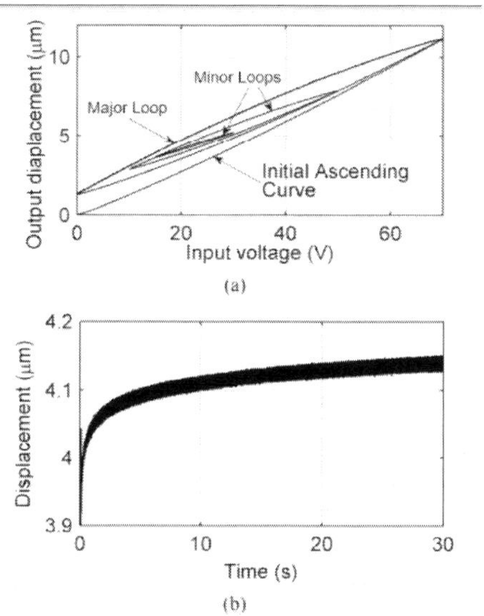

Figure 1. (a) Hysteresis of a PEA and (b) creep of a PEA subject to a 30 V step input.

Hysteresis in PEAs occurs in both relatively static operating conditions (i.e. with constant or slow varying input) and dynamic operating conditions (i.e. with fast varying input). If the influence of the rate of change of the input on the hysteresis can be ignored, then the hysteresis is referred to as rate-independent, otherwise ratedependent. The latter is usually treated as a combined effect of the rate-independent hysteresis and the vibration dynamics.

As hysteresis being the major nonlinearity of PEAs and possessing detrimental effects on the positioning accuracy and stability margins of feed-back control systems [23], compensation of hysteresis has always been a major concern in modeling and control of PEAs.

Creep is the slow variation in the PEA displacement that occurs without any accompanying change in the input voltage [24] as shown in **Figure 1**(b). It is caused by the same piezoelectric material properties as PEA hysteresis [20]. Being a slow and a small effect (on the order of 1% of the last displacement per time decade [20]), creep is sometimes neglected in closed loop and high frequency operations, e.g. in [25,26]. However, for slow and open-loop operations of PEAs, creep must be considered to avoid large positioning error [27].

Finally, PEAs also exhibit linear vibration dynamics, which is the dynamic relationship between the total force exerted on a PEA and the displacement of the PEA, and it resembles a distributed parameter system as the mass of the PEA is not concentrated at certain points [22]. However, if the positioning mechanism attached to the PEA, which is usually much more massive than the PEA, is also taken into account, the resulting overall vibration dynamics will approach that of a lumped parameter system [28].

MODELING OF PIEZOELECTRIC ACTUATORS

A large number of PEA models have been developed to mathematically represent the behaviors of PEAs mentioned in Section 2.2, and they can be generally classified into macroscopic models, which models a PEA as a whole, microscopic models, which models a PEA as a combination of a series of ferroelectric domains or discretized cells [21,29-39], and hybrid models, which combines the ideas behind these two categories has also been reported in [40]. Due to the requirement of using finite element methods, which is computationally expensive, to obtain the overall PEA behaviors, microscopic models and hybrid models are not suitable for the use in microand Nano positioning applications. As such, the following discussion only concerns macroscopic models.

The linear electromechanical model reported in [41] is an early example of macroscopic PEA models. But it can represent neither the nonlinear behaviors (hysteresis and creep) nor the linear vibration dynamics in PEAs due to its linear and static nature. To solve this, various submodels have been developed with each representing one or two of the linear/nonlinear behaviors of PEAs mentioned in Section 2.2, and then connected in appropriate manners to form (or used alone as) macroscopic PEA models. These sub-models, the methods of constructing macroscopic PEA models based on such sub-models, and the methods of inverting such sub-models for the use in model-based open-loop feedforward control of PEAs are reviewed as follows.

Hysteresis Sub-Models

Models for the Major Hysteresis Loop

At the early stage, hysteresis models were developed only to represent the major loop in the hysteresis trajectories of PEAs (refer to **Figure 1**). For example, in [42] polynomials were used to represent the major loop. The JilesAtherton hysteresis model, which was initially developed for representing ferromagnetic hysteresis [43] and afterwards adopted for the ferroelectric hysteresis in piezoelectric materials [44], is also limited to represent the major loop [45]. The inability of such models to represent minor loops in the hysteresis curves of PEAs limits their applications.

Rate-Independent Hysteresis Models

To model both the major loops and the minor loops in the hysteresis of PEAs without concerning the influence of the rate of chance of the input, in other words, to model the rate-independent hysteresis of PEAs, both existing and newly developed rate-independent hysteresis models have been adopted. Among which the Preisach hysteresis model [46] and its modifications [47-58], the PrandtlIshlinskii (PI) hysteresis model [25,59-63], and the Maxwell resistive capacitor (MRC) model [64-67] are among the most widely used. These rate-independent hysteresis models are reviewed as follows.

The Preisach hysteresis model represents the hysteresis by the combined effect of an infinite number of Preisach hysteresis operators $\delta_P(\alpha,\beta,u(t))$, as shown in **Figure 2**. Two parameters are used to characterize a Preisach hysteresis operator: the up switching value α and the down switching value

β, with $\alpha \geq \beta$. Each operator has two saturation values: 0 and 1, and its contribution to the model output is adjusted by $\mu(\alpha,\beta)$, referred to as the Preisach weighting function.

$$f(t)^{CPM} = \iint_{\alpha \geq \beta} \mu(\alpha,\beta)\delta_P(\alpha,\beta,u(t))\mathrm{d}\alpha\mathrm{d}\beta \qquad (1)$$

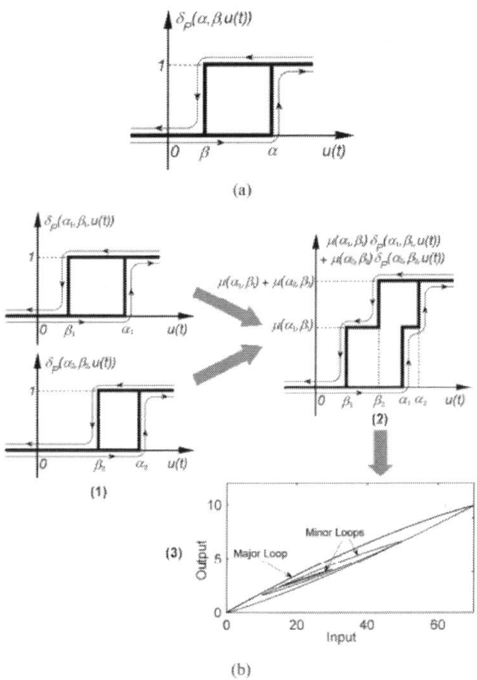

Figure 2. (a) Preisach hysteresis operator and (b) working principle of the Preisach hysteresis model: the sum of two hysteresis operators of different switching values as shown in (1) can represent a simple but unsmooth hysteresis loop as shown in (2); by using the sum of an infinite number of hysteresis operators of different switching values, smooth hysteresis loops including both major and minor loops can then be represented as shown in (3).

As such, the resulting Preisach hysteresis model is expressed as where $u(t)$ is the input and $f(t)^{CPM}$ is the output of the hysteretic system [47]. To reduce the computational effort due to the involvement of the double integration, alternative expressions of the Preisach hysteresis model that only involves arithmetic operations have also been derived from Equation (1) and the properties of the Preisach hysteresis operator [68,69].

In PI hysteresis models, hysteresis is represented by the combined effect of a finite number of play or backlash operators as shown in **Figure 3**. Following this idea, the expression of a PI hysteresis model involving n play operators is given as

$$f(t)^{\mathrm{PI}} = \sum_{i=1}^{n} \delta_{\mathrm{PI}}\left(w_i, r_i, u(t)\right) \tag{2}$$

where $f(t)^{\mathrm{PI}}$ is the model output; i denotes the index of the play operator; w_i and r_i are the weight parameter and the threshold parameter of the i-th play operator, respectively.

Similar to the PI hysteresis model, the MRC hysteresis model represents the hysteresis by the combined effect of a finite number of elasto-slide elements or operators as shown in **Figure 4** [64]. Each elasto-slide element or operator (**Figure 4**(a)(1)) is composed of 1) a mass sliding on a surface with a Coulomb friction μN, where μ is the friction coefficient and N is the normal force between the mass and the surface, and 2) a spring of stiffness k with one end connected to the mass whilst the displacement of the other end (the free end) $u(t)$ can be freely assigned and use as input to the elasto-slide element.

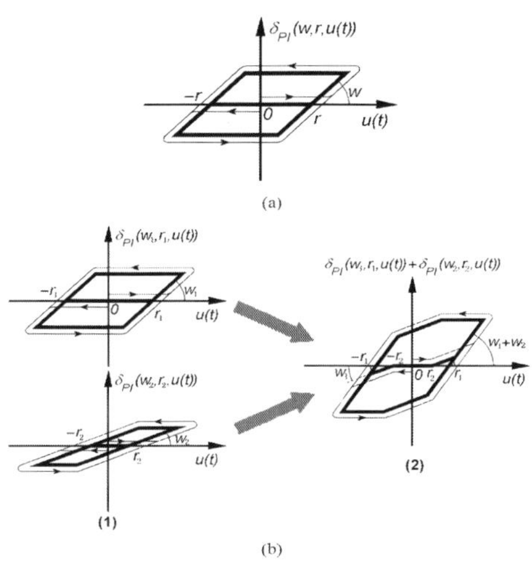

Figure 3. (a) Play operator and (b) working principle of the PI hysteresis model: the sum of two play operators of different parameters as shown in (1) form a simple PI hysteresis model as shown in (2).

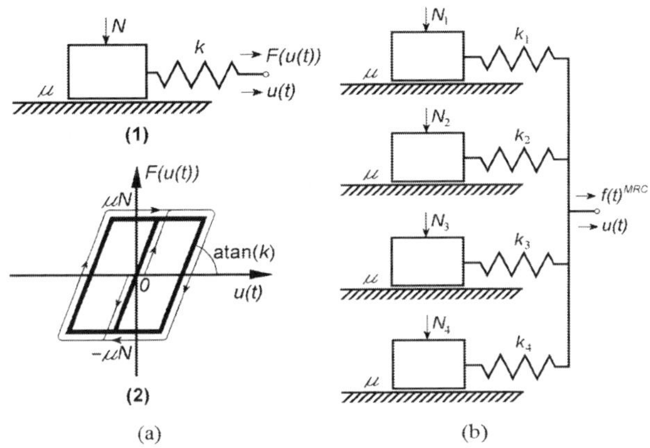

Figure 4. (a) An elasto-slide element and its input-output relationship; (b) The physical interpretation of the MRC hysteresis model [63].

As such, hysteresis relationship exists between the (input) displacement of the free end of the spring $u(t)$ and the resulting force in the spring $F(u(t))$, as can be seen in **Figure 4**(a)(2).

Connecting n elasto-slide elements as in **Figure 4**(b), the resulting MRC hysteresis model or the relationship between the input displacement of the free ends of the springs $u(t)$ and the total force experienced at the free ends of the springs $f(t)^{\mathrm{MRC}}$ can be expressed as

$$f(t)^{\mathrm{MRC}} = \sum_{i=1}^{n} F_i(u(t)) \tag{3}$$

where $F_i(u(t))$ is the force in the spring of the i-th elasto-slide element induced by $u(t)$.

Rate-Dependent Preisach and PI Hysteresis Models

The rate-independent hysteresis models discussed in Section 3.1.2 are only capable of representing the hysteretic behavior of a PEA in narrow (a few Hz) frequencies bands. In wide band applications where both hysteresis and vibration dynamics are significant, the rate-independent hysteresis models need either to be combined with vibration dynamics models or to be modified into rate-dependent hysteresis models to represent the behaviors of the PEA. In the literature, both the Preisach hysteresis model and the PI hysteresis model have been modified into rate-dependent hysteresis models.

In the rate-dependent Preisach hysteresis models, the Preisach weighting function $\mu(\alpha,\beta)$ are adjusted to account for the input-rate dependency by using rate-dependent multiplicative modifying factors [68,70] or neural network approaches [69]. In [68], the rate-dependent multiplicative modifying factor for adjusting $\mu(\alpha,\beta)$ is a function of 1) the average voltage input rate between two input extrema and 2) the variation of the voltage between two input extrema. Even though this model can effectively represent the PEA hysteresis up to 800 Hz, it requires the a priori knowledge of the input voltage waveform (at least to the next extremum), and thus being not applicable to conventional situations in which the future input voltage is unknown. To avoid this problem, in [70] a multiplicative modifying factor which depends only on the input-rate or the first derivative of the input voltage is used instead. However, the small differences between the input-rate at the input extrema at different input frequencies limits the model accuracy at higher input frequencies (the model is effective typically below 10 Hz). To achieve better accuracy (error of the model prediction being less than 5% of the maximum displacement of the PEA) in operations involving higher frequencies (~30 Hz), in [69] a neural network with rate-dependent output is used to account for the input-rate dependence of the Preisach weighting function. But the applicable frequency range remains low compared to the resonant frequencies of conventional PEAs (typically over 1 KHz).

Similarly, the rate-dependent PI hysteresis models, which are modified from the PI models by modeling the parameters as functions of the input rate, are also limited to represent low frequency (<30 Hz) PEA operations or the error becomes excessively high (over 10% of the peak-to-peak sinusoidal displacement of the PEA) [25,26].

So, in conventional applications where the a priori knowledge of the input voltage is not available, the rate dependent Preisach and the rate-dependent PI hysteresis models are both limited to represent PEA operations over frequency bands only slightly wider than the applicable frequency bands of the rate-independent models.

Inverse Preisach and PI Hysteresis Models

Inverse hysteresis models are essential to the model inversion-based feed forward hysteresis compensation technique for PEAs. The inverses of the two most widely used hysteresis models, i.e. the Preisach hysteresis model and the PI hysteresis model, have been established by employing different methods in the literature.

The Preisach hysteresis model cannot be inversed analytically; hence several methods of approximately inverting the Preisach hysteresis model have been developed instead [71]. Such approximate inverting methods can be classified into two categories.

The first category involves directly identifying the Preisach hysteresis model inversely (using the measured plant output as the input to the Preisach hysteresis model while using the plant input as the output to the Preisach hysteresis model) [1,72,73] or directly identifying the inverses of some intermediate functions that constitute the alternative expressions of the Preisach hysteresis model and then expressing the inverse Preisach hysteresis model based on them [56,71]. The advantage of this category of methods is that the computational effort of the inverse Preisach hysteresis model is the same as the Preisach hysteresis model, not more. However, in applications that also require the Preisach hysteresis model in additional to the inverse one, a different parameter identification procedure has to be performed (assume without using iterative methods) since the former cannot be analytically derived from the latter, and the two resulting models are not necessarily accurate inversions of each other due to the different model errors introduced in their respective parameter identification procedures.

The second category involves iterative-algorithms-based methods [71], which find the input to an identified Preisach hysteresis model iteratively until the output of the hysteresis model converges to the desired value. Such methods are developed based on the contraction mapping principle, which are firstly proposed in [74,75], and they have been successfully applied to inverse the Preisach hysteresis modeled hysteresis in a magnetostrictive actuator [71] and a PEA [76]. The advantage of such methods is that both the Preisach hysteresis model and the inverse one can be obtained via a single parameter identification procedure and the inversion can be highly accurate. However, the computational effort involved in the iterative-algorithms-based methods is generally much higher than that of calculating the Preisach hysteresis model output since the Preisach hysteresis model output needs to the calculated in each iteration.

The PI hysteresis model, in the contrary, can be inversed analytically, so good accuracy and low computational effort can be achieved simultaneously, making it much easier to implement than the inverse Preisach hysteresis models. The inversion algorithms for PI hysteresis models can be found, for example, in [63]. In [25,26] the inversion of the rate-dependent PI hysteresis model was also presented. In either case of rate-independent or rate-dependent, the PI hysteresis model inversion algorithms fail if the input frequency is so high that the rounding of the hysteresis loops due to phase lag introduced by other dynamics causes the hysteresis loading and unloading curves to be no longer monotonic. This problem was later solved in [77] by additional mappings to and back from a domain where the inversion of the rate-dependent PI hysteresis model always exists.

Other Hysteresis Models

Other hysteresis models, such as a first-order nonlinear differential equation model [78] and the Bouc-Wen hysteresis model [79-81] have also been used to model hysteresis in PEAs with promising results [22,82-84]. In [85], a non-linear auto-regressive moving average model with exogenous inputs (NARMAX) was employed to represent the hysteresis of a PEA. In [73] a hysteresis model was developed for PEAs by using a polynomial based linear mapping strategy. In [86], a hysteresis model based on a new hysteresis operator, which in turn is based on two hyperbola functions, are developed for PEAs along with its inverse model. Such hysteresis models have not seen wide application mainly due to their high complexity but limited improvements on accuracy as compared to other widely used hysteresis models discussed before.

Creep Sub-Models

Frequently Used Creep Models

The creep response in a PEA has a logarithmic shape over time, so it is sometimes referred to as the Log(t)- type creep, and the nonlinear creep model given by Equation (4) has been used to represent such creep behavior [87,88].

$$y(t) = y_0\left[1+\gamma \log_{10}\left(\frac{t}{t_0}\right)\right] \qquad (4)$$

where $y(t)$ is the creep model output, y_0 is the creep output at t_0 after the static input voltage is applied, and γ is a coefficient depending on the input voltage. The dependence of γ on the input voltage and the unsoundness of such a nonlinear creep model as $t \to 0$ and $t \to +\infty$ pose difficulties in its implementation [19].

To avoid the problem of unboundness encountered in the nonlinear creep model Equation (4) and to avoid the nonlinearity, a linear creep model was proposed as.

$$G_c(s) = 1 + \sum_{i=1}^{n_c} \frac{1}{c_i s + k_i} \quad (5)$$

where the creep transfer function $G_c(s)$ is represented by the combined effects of n_c spring (k_i)damper (c_i) systems [1,88]. Besides, an operator-based linear creep model was also reported in [60].

However, these linear creep models have linear and memory free equilibrium values, whist the actual equilibrium value of the creep effect of PEA exhibits hysteresis nonlinearity [89].

To further solve the problems encountered in both the nonlinear and linear creep model above, the PI hysteresis model design approach was adopted in [63] to develop an operator-based nonlinear creep model, whose output is a weighted sum of the values of a series of elementary creep operators.

Inverse Creep Models

Similar to the case of inverse hysteresis models, inverse creep models are essential in feed forward creep compensation.

While the analytic inversion of the nonlinear creep model was not mentioned in the literature, a method of feeding a voltage with an inverse Log(t)-type creep itself to a PEA to compensate for the creep in the output displacement was found to be effective [87].

On the other hand, the inversion of the linear creep model (Equation (5), not operator-based) is usually carried out together with the inversion of the vibration dynamics model since they can be combined into a single linear dynamics model.

An optimal inversion approach, which minimizes a quadratic cost function representing the weighted sum of the input voltage energy and the output displacement error energy [90], can be used to inverse the combined vibration dynamics and creep model in [1].

Linear creep models can also be combined with hysteresis models and then inverted, e.g. via iterative methods [88].

The inversion of the operator-based linear and nonlinear creep models together with other dynamics are reported in [60,63], respectively.

Vibration Dynamics Sub-Models

Frequently Used Vibration Dynamics Sub-Models

According to Section 2.2, for the consideration of accuracy, the vibration dynamics of a PEA should be modeled as a distributed linear system for a stand-alone PEA [22]. But for the convenience of practical uses, the vibration dynamics of a PEA and the attached positioning mechanism is often modeled as a whole as a single second-order system, as in [28,64,91,92], whilst higher order linear dynamical systems are only used in applications requiring higher level of accuracy [1]. Such lumped linear vibration dynamics models of PEAs are usually identified through fitting the measured frequency response of the PEA to that of a specific model structure by using a dynamic signal analyzer [1,93].

On the other hand, the axiomatic design method has also been used to find a higher order linear system representation of the vibration dynamics of a PEA with an attached positioning mechanism [94,95].

Inverse Vibration Dynamics Sub-Models

Inverse vibration dynamics are usually used in open-loop feed forward control of PEAs, especially in the case of controlling piezoelectric tube actuators as in [1]. Methods for inverting the vibration dynamics have been reviewed in Section 3.2.2.

Model Structures

With different sub-models representing hysteresis, creep, and vibration dynamics of PEAs defined, macroscopic PEA models are constructed by connecting such sub models according to certain model structures. There are two categories of model structures for macroscopic PEA models.

In the first category, the behavior of a PEA, when subject to a voltage input and an external loading force, is decoupled into several effects such as rate-independent hysteresis, vibration dynamics, and creep based on physical laws. Such effects are then represented by individual sub-models and such sub-models are connected in various manners to form a comprehensive model of the PEA.

The most well-known model structure of PEAs belonging to this category is the electromechanical model proposed in [64], which is further refined in [22]. **Figure 5** shows the schematic representation of this model.

There are 3 effects in **Figure 5**, namely 1) the hysteresis, H, between the charge in the PEA, q, and the resulting voltage, u_h; 2) the electromechanical transducer, T_{em}, converts the mechanical deformation of the PEA, y, into charge, (inverse piezoelectric effect), and converts the voltage, u_p, into an actuating force, F_p (piezoelectric effect); and 3) the vibration dynamics, M, relating the deformation of the PEA, y, to the internal actuating force, F_p, and the external load, F_e. Other symbols in **Figure 5** include u_{pea}, which is the voltage across the PEA, L, which is the original length of the PEA, and C, which is the capacitance of the PEA. The model can then be expressed as

$$u_{pea} = u_h + u_p \quad (6)$$

$$q = H(u_h) \quad (7)$$

$$q = Cu_p + q_p \quad (8)$$

$$q_p = T_{em} y \quad (9)$$

$$F_p = T_{em} u_p \quad (10)$$

$$y = M(F_p - F_e) \quad (11)$$

The hysteresis effect, H, and the vibration dynamics, M, can be modeled by different kinds of sub-models. Specifically, His usually represented by rate-independent hysteresis models, e.g. the MRC model was used in [64] whist the first-order nonlinear differential equation model, which was firstly proposed in [78], was employed in [22].

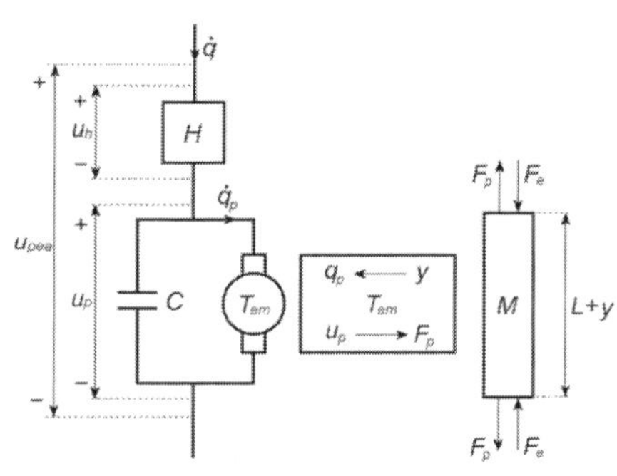

Figure 5. Schematic representation of a electromechanical model of a PEA [22].

The vibration dynamics, M, can be modeled by any models discussed in Section 2.1.3. It is noted that the use of this electromechanical model usually requires the measurements of the charge, q, in the PEA for model parameter estimation, which is difficult in practice. So this model is usually employed in applications that the relationship between u_{pea} and F_p, which is difficult to identify, can be approximated by a simple function (e.g. an affine function) with an error term as model uncertainties, such as in [91,92]. Also, creep is usually not considered in this electromechanical model [22,78] expect being treated as model uncertainties [91,92].

Another model structure proposed in [1] models the relationship between u_{pea} and F_p by a rate-independent hysteresis model directly, based on a conclusion drawn from the analysis in [64] that the rate-independent hysteresis of a PEA only exists between the input voltage and the resulting internal actuating force. This hysteresis sub-model is then cascaded to a vibration dynamic submodel and a creep sub-model to form the complete model of a PEA, as shown in **Figure 6**. Without the involvement of charge measurement for parameter estimation, this model structure is much easier to apply than the electromechanical model. Similar to the case of the electromechanical model, this model structure is highly flexible that each sub-model can be implemented in various ways, e.g. in [1], the Preisach hysteresis model was used for H and a fourth-order and a third-order linear dynamic systems are used to represent M and the creep effect, G_c, whist in [63], a modified PI hysteresis model is used for H and an operator based model of complex Log(t)- type creep for G_c. Also, due to the cascading connection between the sub-models, this model structure is very suitable for implementing control schemes with inverse nonlinearity and/or vibration dynamics compensation as in [1,63,93,96,97]. A similar model structure constructed using a bond-graph representation was also presented in [67].

In the second category, the different behaviors of a PEA are not decoupled. In such cases, rate-dependent hysteresis models such as the rate-dependent Preisach hysteresis model, the rate-dependent PI hysteresis model (Section 3.1.3), and the rate-dependent neural-networkbased hysteresis model [98] are used directly as models of the PEA. However, as mentioned in Section 3.1.3, such models are only accurate over small frequency ranges. And, as can be seen above, creep is usually neglected. These disadvantages seriously limit their usage.

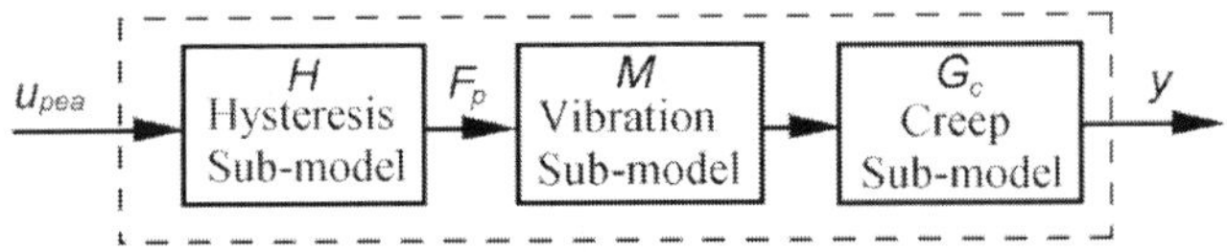

Figure 6. Model structure of a PEA as in [1].

CONTROL OF PEAS

Over the years, a number of control schemes have been reported for positioning control of PEAs. Due to the detrimental effects of the nonlinearities (especially hysteresis) of PEAs on positioning control performances in terms of positioning error and stability [19], the compensation or rejection of such nonlinear effects has been a major consideration in all of these control schemes. In the following, the typical control schemes for PEAs are reviewed.

Open-Loop Control Schemes

Open-loop control schemes are usually employed in applications in which position feedback are difficult to implement due to mechanical constraints, e.g. atomic force microscopes [1-3]. In such control schemes, inverse model of the PEA to be controlled is found and then cascaded to the PEA. The methods for finding such inverse models have been reviewed in Sections 3.1.4, 3.2.2, and 3.3.2. The inverse model generates an input voltage u_{pea} to the PEA according to the desired displacement y_d, such that the PEA subject to u_{pea} produces an output displacement of PEA y that follows y_d. **Figure 7** shows a typical open-loop control scheme for PEAs.

In some of the open-loop control schemes, all three effects in the PEA shown in **Figure 6** are compensated for by the inverse model of PEA to achieve best tracking accuracy (as the case in **Figure 7**). For example, in [1,67], inverse models of PEA representing hysteresis, vibration dynamics, and creep are used to compensate for these effects; whist in [93] a charge control technique was firstly employed to remove the hysteresis effect and then an inverse model was used to compensate for the linear dynamics.

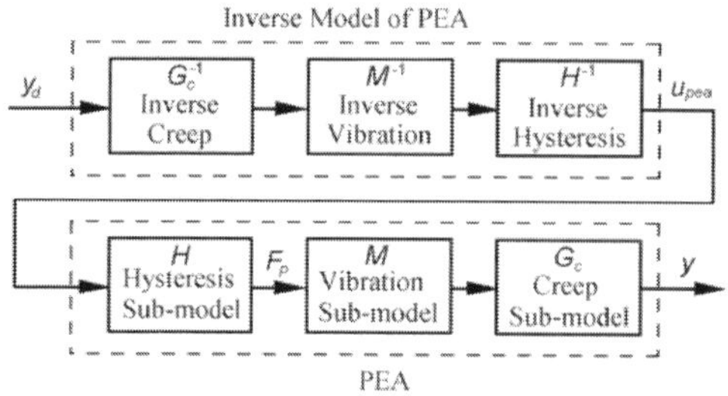

Figure 7. Open-loop control scheme for a PEA [1].

Other open-loop control schemes only focus on compensating for some of the three effects shown in **Figure 6**. For example, in [72,73,86,99], only hysteresis was compensated for by using the inverses of the Preisach hysteresis model, a hysteresis model developed by using a polynomial-based linear mapping strategy, a hysteresis model developed based on hyperbola-shaped hysteresis operators, and a hysteresis model developed in [100], respectively, since hysteresis is the most significant nonlinearity in PEAs. In [90,101], creep is also compensated for in additional to hysteresis by using operator-based inverse creep models and inverse PI hysteresis models to achieve better tracking performance. However, due to the negligence of the vibration dynamics in these schemes, their performances are only guaranteed in fixed frequency or narrow band operations unless inverse ratedependent hysteresis models are employed, e.g. the use of inverse rate-dependent PI hysteresis models in [25,26,77]. Nonetheless, the applicable frequency ranges (with acceptable tracking error, e.g. <5% of the maximum displacement of the PEA) of the schemes using inverse ratedependent hysteresis models are still relatively small (less than 50 Hz) compared with those also inverting the vibration dynamics (up to a few hundreds of Hz) [1,25,77].

In no load condition the open-loop control schemes have been shown to be highly effective in their applicable frequency ranges [1].

However, one major disadvantage of the open-loop control schemes is that their positioning performances are highly sensitive to unknown effects such as model errors, external loads, and changes in the dynamics of the PEA. To partially alleviate the problem, reference [65] proposed to use a disturbance-observer-based scheme in addition to the inverse MRC hysteresis model to compensate for the hysteresis and unknown effects simultaneously. Specifically, the estimated disturbance input to the PEA, which has the same effect as the aforementioned unknown effects on the PEA output, is subtracted from u_{pea}. However, the effectiveness of this scheme was not examined through experiments. In [102], the inverse hysteresis model is updated adaptively to account for the unknown effects. However, the influence of the unknown effects on the PEA output, though largely suppressed, can still be clearly observed in the experiment results. As such, the best solution to this problem remains introducing feedback into the control schemes, which is to be discussed in the following sections.

Feedback

Feedback control schemes as shown in **Figure 8** lead to strong suppression of the unknown effects including model errors, external loads, and changes in the dynamics of the PEA on the position control performances, hence they are widely used.

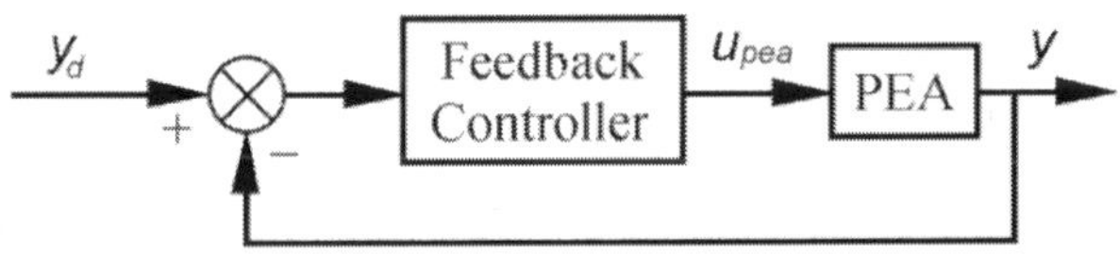

Figure 8. A feedback control scheme for PEAs [19].

In static or low frequency operations, classical control techniques such as PID control or using multiple integrators for tracking desired displacement profiles are widely used because of their simplicity and capability of eliminating steady state errors in such applications [20,103,104]. Various PID tuning techniques has been reported in PEA positioning control applications, e.g. by trial and error [20], by grey relational analysis [105], using an optimal linear quadratic regulation method [106], by a semiautomatic tuning technique [107], and by an automatic tuning technique [108]. However, in broadband operations with large system uncertainties including modeling errors, nonlinearities, external loads, etc., advanced control techniques are required because PID control is limited in bandwidth while dealing with uncertainties [109].

Among all of such advanced control techniques, sliding mode control has drawn a lot of attention in the recent years. This is due to the fact that sliding mode controllers can completely reject the effects of the so-call matched uncertainties or uncertainties in the input channel of the plant, resulting in strong robustness [110].

Specifically, in its basic form, the control signal generated by a sliding mode controller can be divided into two parts. The first part is discontinuous, referred to as the switching control, for compensating for the matched uncertainties while driving the plant trajectory to a prescribed geometric entity in the phase space (a space with each dimension corresponding to one state variable in the plant model), referred to as the sliding surface. The second part is continuous, referred to as the equivalent control, for keeping the plant trajectory on the sliding surface after the latter is reached. Since the sliding surface and hence the closed loop system dynamics after reaching the sliding surface can be freely designed, the closedloop system becomes insensitive to the matched uncertainties after the sliding surface is reached [110]. The major problem of implementing the sliding mode control techniques on PEAs is chattering and the need of the knowledge about the bounds of the matched uncertainties for designing the discontinuous control signal [110], which is to be discussed in more detail in Section 5.2.

Based on the model structure shown in **Figure 6**, hysteresis and creep in a PEA can be treated as (matched) uncertainties in the input channels of the vibration dynamics sub-model of the PEA (given that a linear creep model is used). As such, a sliding mode controller can be designed using the vibration dynamics sub-model as a nominal model of the PEA, and the effects of hysteresis and creep on the PEA output displacement can be strongly suppressed without knowing the exact expression of the hysteresis sub-model and the creep sub-model, thus largely reducing the efforts involved in hysteresis and creep modeling and compensation [95,111-114]. Various modifications of the basic sliding mode control technique have also been reported for controlling PEAs for improved performance. For instance, adaptive methods are combined with sliding mode control to remove the requirement of model parameter estimation or lead to better compensation for the nonlinearities as in [115-117]. And in [118] the nonlinearities of the positioning mechanism driven by a PEA in additional to those of the PEA itself were considered in sliding mode controller design.

Robust control techniques constitute another branch of widely used advanced control techniques for positioning control of PEAs. They try to find the control law via optimizing an objective function that incorporates the robustness objective, for example, minimizing the H_∞ norm [88,119-122] or the H_2 norm [76] of the transfer functions relating the disturbances to the plant output, thus minimizing the effects of the disturbances on the plant output and enhancing robustness. Robust control techniques can be combined with sliding mode control to suppress unmatched uncertainties (which the sliding mode control techniques cannot compensate for), while sliding mode control are considered to be superior in dealing with matched uncertainties because robust control techniques can only minimize the effects of the matched uncertainties on the plant output whilst sliding mode control techniques can completely reject such effects in theory [110].

Other advanced control techniques such as state feedback [123], optimal control [109,124], adaptive control [125-128], and neuron network methods [129-131] have also been applied to control PEAs.

Inverse sub-models of the nonlinearities of PEA are sometimes inserted between the feedback controller and the controller in **Figure 8** to linearize the PEA thus facilitating the use of linear feedback controllers. For example, an inverse Preisach hysteresis model is used to cancel out the hysteresis and then a linear H_2 controller is designed based on the remaining linear dynamics of a PEA in [76], while an inverse PI hysteresis model and a sliding mode controller are used in [113].

One problem with the feedback control schemes for PEAs is that the resulting closed-loop systems have low gain margins, which limit the use of high gain controllers. Such low gain margin is largely due to the quick phase loss near the first resonant peak in the frequency response of a PEA [23]. This problem can be alleviated by using a notch filter to reduce the gain around the first resonant peak [20,132,133].

Feedback with Feed forward

Feed forward is sometimes used to augment feedback controllers for nonlinearity compensation. A typical control scheme of this type is shown in **Figure 9**. An advantage of this scheme is that the low gain margin problem can be alleviated such that high frequency positioning performances can be improved over the feedback control schemes [124,134,135].

Recent researches adopting this scheme can be found, e.g. in [136] a hysteresis compensator based on the inverse Preisach hysteresis model was used as the feedforward controller to compensate for hysteresis whist a PID feedback controller was used to account for other effects, in [137] a inverse nonlinear differential equation hysteresis model was used as the feed forward controller and a PI controller was used as a feedback controller, and in [16] a inverse extended Coleman-Hodgdon model was used as the feed forward controller and a feedback controller designed via loop shaping techniques was employed.

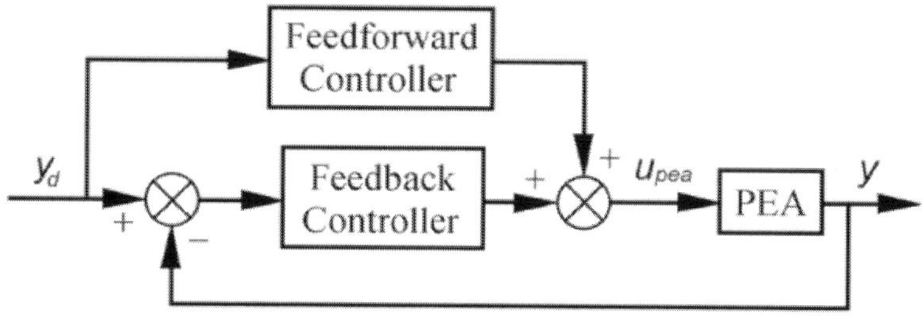

Figure 9. A feedback augmented with feedforward control scheme for PEAs [19].

Another control scheme involving feedback and inverse model feed forward is shown in **Figure 10** [97,138]. In this scheme, the feedback controller is a high gain controller used for nonlinear effect suppression such that the closed loop system has linear dynamics.

Then a feed forward controller, which is the inverse linear dynamics of the closed loop system, is applied to make the output y follow the desired output y_d. This control scheme can be seen as being evolved from the open-loop control scheme shown in **Figure 7**. The rationale behind this scheme is that the accurate modeling and inversion of the nonlinear effects such as hysteresis are complicated whist the inversion of a linear plant is relatively easy to compute [139]. However the low gain margin problem with the feedback loop exists in this scheme [19]. Also, similar to the open-loop control schemes, this scheme is sensitive to the disturbances acting outside the closed-loop system.

Disturbance Observer Based Schemes

Recently, a technique call disturbance observer is used for the compensation of hysteresis and other nonlinear/uncertain effects in PEAs. In this scheme, a PEA is modeled as a linear dynamic system $G(s)$, so when subject to the same input voltage u_{pea}, the output of the PEA y and that of the corresponding $G(s)$ are different due to hysteresis and other nonlinear/uncertain effects. The disturbance observer utilizes u_{pea} and y to estimate an additional input voltage u_d such that if $u_{pea}+u_d$ is applied to $G(s)$, the output of $G(s)$ is the same asy. u_d is referred to as an input disturbance to $G(s)$ that represents the hysteresis and other nonlinear/uncertain effects.

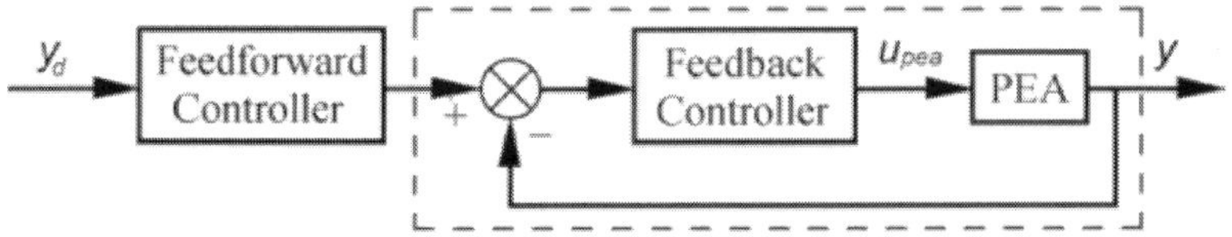

Figure 10. A feedback controller is used for nonlinear suppression whist a feed forward controller is the inverse model of the closed loop system for canceling out the dynamics of the closed loop system, which can be treated as linear [19].

As such, to compensate for hysteresis and other nonlinear/uncertain effects in the PEA, one only needs to subtract u_d from u_{pea} before u_{pea} being applied to the PEA. A control scheme with such a disturbance observer is shown in **Figure 11**. This scheme has been shown to be effective in both step [140] and high frequency tracking (200 Hz sinusoidal) [141] operations.

EMERGING ISSUES IN PEA MODELING AND CONTROL

Based on the reviews in the previous sections, some emerging issues regarding the modeling and control of PEAs and PEA-driven positioning stages are discussed in this section.

Modeling Issues

Limitations of Some Existing Hysteresis Models of PEAs

While the modeling of creep and vibration dynamics of PEAs have been relatively well address, the existing models of hysteresis, especially the rate-independent hysteresis models reviewed in Section 3.1.2, still possess issues, as follows. Though being highly effective in representing both the major hysteresis loops and the minor hysteresis loops, it is noted in the literature [40,53-56,58] that the Preisach hysteresis model cannot at the same time represent the initial ascending curve when the PEA is subject to nonnegative voltage inputs. This is due to the fact that the Preisach hysteresis operator has only two saturation values and without a separate initial section that connects the origin to the hysteresis loop (**Figure 2**(a)). The PI hysteresis models and the MRC hysteresis model do not suffer from this problem due to the existence of initial sections in their respective hysteresis operator besides the hysteresis loop (Figures 3(a) and 4(a)(2)). However, to limit the computational effort, the number of hysteresis operators used in PI or MRC hysteresis models in practice is usually very small (typically around 10) compared to the Preisach hysteresis model (infinity).

As a result, the PI and MRC models tend to be less accurate than the Preisach hysteresis model in representing the hysteresis loops, especially in the cases of the small minor loops. So, given a tractable amount of computational effort, there is a contradiction between the capability of representing all hysteresis phenomena, including one-sided hysteresis (the PI and MRC hysteresis models), and the accuracy (Preisach hysteresis models) in the existing rate-independent hysteresis models. This remains an issue to be solved in future research.

Applications of Different Actuator Structures

The applications of PEA-driven positioning systems with structures different from the conventional ones (those can be treated as a mass fixed to one end of the PEA) also pose modeling problems. A typical example is the PEAdriven stick-slip actuator shown in **Figure 12**, in which an end-effector is supported and guided by a movable platform that is driven by a PEA, and it works as follows.

During the course of slow expansion of the PEA, the end-effector moves along with the platform. If the PEA suddenly contracts, the end-effector slides on the platform because the force due to inertia becomes larger than the friction between the end-effector and the movable platform. As a result, the end-effector moves a step, Δs, with respect to its original position. Such steps can be accumulated to achieve a theoretically unlimited displacement (actually limited by the size of the moveable platform).

The modeling of the dynamics of the endeffector displacement has not been well addressed in the literature. Chang and Li [142] developed a model for the PEA-driven stick-slip actuator without considering the dynamics of the PEA while the friction involved was modeled as Coulomb friction, which is over simplified.

By taking into account presliding friction, models were developed and reported in more recent studies on stickslip actuators, e.g. [143]. However, issues including the influence of PEA nonlinearities and the end-effector mass on the performance of the PEA-driven stick-slip actuator remain to be addressed.

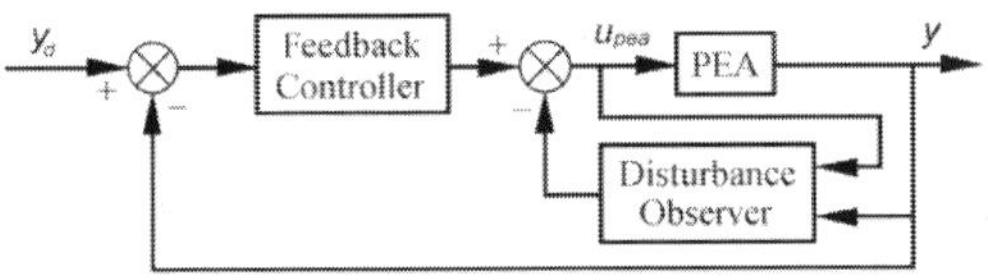

Figure 11. Augmenting the feedback control loop with a disturbance observer.

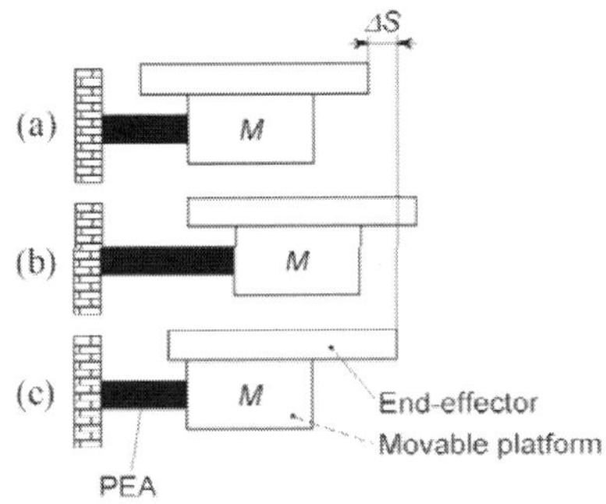

Figure 12. Actuation sequence of a PEA-driven stick-slip actuator; (a) Start; (b) Slow expansion and (c) Fast contraction.

Control Issues

Controller Design

As can be seen from Section 4.2, sliding mode control has been one of the most promising techniques for PEA control applications due to their capability to largely suppress hysteresis and creep in practice. However, there are two remaining problems. One problem of implementing the sliding mode control techniques on PEAs is chattering. This is a kind of high frequency vibration in the plant output displacement induced by the discontinuous control signal generated by a sliding mode controller, and it can induce wear, noises, and even resonance and instability in the plant. Chattering is especially severe in plants with fast responses, such as PEAs, because they cannot filter out the high frequency component in the switching control signal. Another problem is that the bounds of the matched uncertainties need to be known for designing the discontinuous control signal, but such bounds are usually difficult to determine in practice while conservatively setting the bounds to large values only leads to more severe chattering [110]. Existing research efforts have been focused on solving the chattering problem. For example, a method referred to as the sliding mode with boundary layer control was developed by approximating the discontinuous control signal with a continuous one, but with the cost of introducing nonzero steady state error, and chattering is still significant if the bounds of the matched uncertainties are large [110]. Higher order sliding surfaces are also used for chattering suppression, such as the integral sliding mode technique in [92,144] and the PID sliding surface in [115]. However, the second problem remains to be solved.

State Estimation

Many advanced control techniques reviewed in Section 4, such as sliding mode control [114], require state feedback. Since usually not all of the states of a PEA model are measurable in application, state observers must be used. However, the problem of estimating the states of PEA models has not been well addressed in the literature, as discussed below.

Currently, many kinds of state observers are applicable to PEA control and they can be classified into two categories: non-model-based and model-based. Non-modelbased filters/differentiators, such as the low-pass filter plus ideal differentiator, the α-β filter [145], the highgain differentiator [146], the integral-chain differentiator [146], and the sliding mode differentiator based on the super-twisting algorithm [147] (given that the states needed are the derivatives of the measured plant output), usually generate large phase lags if the desired level of noise suppression is enforced (for the first four aforementioned methods) or excessive chattering in noisy systems (for the last method). Compared to non-modelbased filters/differentiators, model-based observers, e.g., the extended Kalman filter [148], the unscented Kalman filter [149,150], and the high-gain observer [114], can generate more accurate estimations if the imperfection of the model can be ignored. However, with the presence of uncertainties such as the effects of hysteresis, creep, and external load on a PEA, however, the performances of the model-based observers degrade. For improvement, in the cases that the system uncertainties can be treated as a lumped unknown input to the system model, a kind of model-based observers should be used to estimate the system states even with the presence of the unknown input, which is referred to as unknown input observers (UIOs) in the literature. Many UIOs have been reported in the past three decades, including the full-order UIO [151], the reduced-order UIO [152], the UIO designed based on projection operator [153], and the sliding mode-based observers (SMOs) [110,153-155].

Applications of these UIOs require that the observer matching condition be satisfied [156,157], which states that the rank of the product of the output matrix and the unknown input matrix in the state space model of the system must be equal to that of the unknown input matrix [110]. However, existing PEA models, such as the one in [92], do not meet this condition. Attempts to relax the observer matching condition have been reported [156,157], but the resultant UIOs were very complicated. As such, a UIO with a simpler structure and the capability of relaxing the observer matching condition still needs to be developed in future research for the use in PEA tracking control.

Modeling and Control of Multi-Degree-of-Freedom (Multi-DOF) PEA-Driven Positioning Systems

All of the above discussions have been focused on single-DOF PEA-driven positioning systems. However, there are modeling and control issues uniquely related to multi-DOF PEA-driven positioning systems and they are briefly discussed in this section.The multi-DOF PEA-driven positioning systems reported in the literature can be divided in to two categories: series mechanisms and parallel mechanisms. Among which, multi-DOF PEA-driven series mechanisms [16] and the xyz-type parallel mechanisms (translations along x-, yand z-axis) such as those in [16,158-160] can usually be treated as a series of individual 1-DOF PEAdriven mechanisms without significant coupling between the moving axes, as such their modeling and control can be well address with the techniques applied to the single-DOF systems.The remaining multi-DOF PEA-driven parallel mechanisms whose moving axes are significantly coupled can be further divided into fully-actuated systems, i.e. the number of actuation is the same as the number of DOF [17,114,161-167], and over-actuated systems, i.e. the number of actuation is the more than the number of DOF [20,112].For the fully-actuated systems, their models are usually established based on the geometric structure of the mechanism and mechanics. Since most of such mechanisms use flexure hinges instead of conventional hinges to avoid friction and backlash [17,114,158,159-166], the modeling of the flexure hinges becomes an important part in modeling such mechanisms [114,160,166]. Two kinds of workflows can be adopted for modeling the multi-DOF PEA-driven positioning systems. For the first kind of workflows, a kinematic model is firstly constructed based on the geometric structure of the mechanism, and then either a phenomenological [114] or physical [160] dynamic model can be established based on the kinematic model. The second kind of workflows involves direct identification of the dynamic model of the mechanism without establishing the kinematic model [16].Many control schemes that have been applied to control single-DOF PEA-driven positioning systems can also be extended to control fully-actuated PEA-driven positioning systems, e.g. PID control [112,168], H_2 control [168], and sliding mode control [112] since such methods can be easily extended to fully-actuated multi-in putmulti-output cases.

Compared to the fully-actuated systems, the over actuated systems have certain advantages in the performance point of view, e.g. singularity elimination, dexterity improvement, and better load carrying ability [169]. However, the over-actuated structure poses difficulties in both modeling and control. Specifically, the PEAs in the mechanism need to cooperate well or excessive internal forces will occur and may damage the mechanism, whilst the involvement of PEA nonlinearities makes such cooperation difficult to model and control. As a result, very few researches on the modeling and control of over-actuated PEA-driven parallel mechanism have been reported compared to its popularity in macro parallel mechanisms. The only example is [112], in which the modeling and control of a $x\theta_y\theta_z$ 3-DOF positioning system driven by 4 PEAs are concerned. Specifically, in [112] each PEA in the mechanism is modeled and controlled individually and uses only the output displacements of the corresponding PEA as feedback, while the inverse kinematic model of the mechanism is used to generate the desired displacement for each PEA according to the desired position of the end-effector. The cross-axis couplings are considered as disturbances. As such, the control problem is transformed into several 1-DOF control problems, which are much simpler. However, in this scheme the actual position of the end-effector is not closed-loop controlled, and the requirement of being able to measure the displacement of each PEA in the mechanism limits the generality of this scheme. As such, more general schemes for modeling and controlling over-actuated multi DOF PEA-driven positioning systems that can relax the aforementioned requirement are still to be developed in the future.

CONCLUSIONS

Jingyang Peng, Xiongbiao Chen (chattering) and the corresponding state estimators (the incompatibility between the observer matching condition and the existing PEA models); finally, 6) the modeling and control of some PEA-driven positioning systems with unconventional structures, e.g. PEA-driven stick-slip actuators and over-actuated multiDOF PEA-driven positioning systems, are largely untouched in the literature, thus remain to be addressed in future research.

REFERENCES

1. D. Croft, G. Shed and S. Devasia, "Creep, Hysteresis, and Vibration Compensation for Piezoactuators: Atomic Force Microscopy Application," Journal of Dynamic Systems, Measurement, and Control, Vol. 123, No. 1, 2001, pp. 35-43. doi:10.1115/1.1341197
2. Q. Zou, K. K. Leang, E. Sadoun, M. J. Reed and S. Devasia, "Control Issues in High-Speed AFM for Biological Applications: Collagen Imaging Example," Asian Journal of Control, Vol. 6, No. 2, 2004, pp. 164-178. doi:10.1111/j.1934-6093.2004.tb00195.x
3. R. Kassies, K. O. Van der Werf, A. Lenferink, C. N. Hunter, J. D. Olsen, V. Subramaniam and C. Otto, "Combined AFM and Confocal Fluorescence Microscope for Applications in Bio-Nanotechnology," Journal of Microscopy, Vol. 217, No. 1, 2005, pp. 109-116. doi:10.1111/j.0022-2720.2005.01428.x
4. H. Song, G. Vdovin, R. Fraanje, G. Schitter and M. Verhaegen, "Extracting Hysteresis from Nonlinear Measurement of Wavefront-Sensorless Adaptive Optics System," Optics Letters, Vol. 34, No. 1, 2009, pp. 61-63. doi:10.1364/OL.34.000061
5. W. Yang, S.-Y. Lee and B.-J. You, "A Piezoelectric Actuator with a Motion-Decoupling Amplifier for Optical Disk Drives," Smart Materials and Structures, Vol. 19, No. 6, 2010, pp. 065027.1-065027.10. doi:10.1088/0964-1726/19/6/065027
6. G. Stöppler and S. Douglas, "Adaptronic Gantry Machine Tool with Piezoelectric Actuator for Active Error Compensation of Structural Oscillations at the Tool Centre Point," Mechatronics, Vol. 18, No. 8, 2008, pp. 426-433. doi:10.1016/j.mechatronics.2008.03.002
7. S. R. Viswamurthy, A. K. Rao and R. Ganguli, "Dynamic Hysteresis of Piezoceramic Stack Actuators Used in Helicopter Vibration Control: Experiments and Simulations," Smart Materials and Structures, Vol. 16, No. 4, 2007, pp. 1109-1119. doi:10.1088/0964-1726/16/4/020
8. M. S. Senousy, F. X. Li, D. Mumford, M. Gadala and R. K. N. D. Rajapakse, "Thermo-Electro-Mechanical Performance of Piezoelectric Stack Actuators for Fuel Injector Applications," Journal of Intelligent Material Systems and Structures, Vol. 20, No. 4, 2009, pp. 387-399. doi:10.1177/1045389X08095030
9. J.-J. Wei, Z.-C. Qiu, J.-D. Han and Y.-C. Wang, "Experimental Comparison Research on Active Vibration Control for Flexible

Piezoelectric Manipulator Using Fuzzy Controller," Journal of Intelligent and Robotic Systems, Vol. 59, No. 1, 2010, pp. 31-56. doi:10.1007/s10846-009-9390-2

10. D. Shu, J. Maser, M. Holt, R. Winarski, C. Preissner, A. Smolyanitskiy, B. Lai, S. Vogt and G. B. Stephenson, "Optomechanical Design of a Hard X-Ray Nanoprobe Instrument with Nanometer-Scale Active Vibration Control," AIP Conference Proceedings, Vol. 879, 2007, pp. 1321-1324. doi.org/10.1063/1.2436307
11. H. Zhou and B. Henson, "Analysis of a Diamond-Shaped Mechanical Amplifier for a Piezo Actuator," The International Journal of Advanced Manufacturing Technology, Vol. 32, No. 1-2, 2007, pp. 1-7. doi:10.1007/s00170-005-0303-7
12. M. Muraokaa and S. Sanada, "Displacement Amplifier for Piezoelectric Actuator Based on Honeycomb Link Mechanism," Sensors and Actuators A: Physical, Vol. 157, No. 1, 2010, pp. 84-90. doi:10.1016/j.sna.2009.10.024
13. Y. Zhang, G. Liu and J. Hesselbach, "On Development of a Rotary-Linear Actuator Using Piezoelectric Translators," IEEE/ASME Transactions on Mechatronics, Vol. 11, No. 5, 2006, pp. 647-650. doi:10.1109/TMECH.2006.882998
14. J. Li, R. Sedaghati, J. Dargahi and D. Waechter, "Design and Development of a New Piezoelectric Linear Inchworm Actuator," Mechatronics, Vol. 15, No. 6, 2005, pp. 651-681. doi:10.1016/j.mechatronics.2005.02.002
15. R. Merry, R. Van de Molengraft and M. Steinbuch, "Modeling of a Walking Piezo Actuator," Sensors and Actuators A: Physical, Vol. 162, No. 1, 2010, pp. 51-60. doi:10.1016/j.sna.2010.05.033
16. R. Merry, M. Uyanik, R. Van de Molengraft, R. Koops, M. Van Veghel and M. Steinbuch, "Identification, Control and Hysteresis Compensation of a 3 DOF Metrological AFM," Asian Journal of Control, Vol. 11, No. 2, 2009, pp. 130-143. doi:10.1002/asjc.89
17. T.-F. Lu, D. C. Handley, Y. K. Yong and C. Eales, "A Three-DOF Compliant Micromotion Stage with Flexure Hinges," Industrial Robot: An International Journal, Vol. 31, No. 4, 2004, pp. 355-361. doi:10.1108/01439910410541873
18. S. W. Lee, K.-G. Ahn and J. Ni, "Development of a Piezoelectric Multi-Axis Stage Based on Stick-and-Clamping Actuation Technology,"

Smart Materials and Structures, Vol. 16, No. 6, 2007, pp. 2354-2367. doi:10.1088/0964-1726/16/6/040

19. S. Devasia, E. Eleftheriou and S. O. R. Moheimani, "A Survey of Control Issues in Nanopositioning," IEEE Transactions on Control Systems Technology, Vol. 15, No. 5, 2007. pp. 802-823. doi:10.1109/TCST.2007.903345
20. P. Instrumente, "The World of Nanopositioning and Micropositioning 2005/2006," Physik Instrumente, Karlsruhe, 2005.
21. I. Mayergoyz and G.Bertotti (Eds.), "The Science of Hysteresis," Vol. 3, Elsevier, St. Louis, 2005.
22. H. J. M. T. S. Adriaens, W. L. De Koning and R. Banning, "Modeling Piezoelectric Actuators," IEEE/ASME Transactions on Mechatronics, Vol. 5, No. 4, 2000, pp. 331- 341. doi:10.1109/3516.891044
23. J. A. Main and E. Garcia, "Piezoelectric Stack Actuators and Control System Design: Strategies and Pitfalls," Journal of Guidance, Control, and Dynamics, Vol. 20, No. 3, 1997, pp. 479-485. doi:10.2514/2.4066
24. T. Fett and G. Thun, "Determination of Room-Temperature Tensile Creep of PZT," Journal of Materials Science Letters, Vol. 17, No. 22, 1998, pp. 1929-1931. doi:10.1023/A:1006608509876
25. W. T. Ang, F. A. Garmón, P. Khosla and C. N. Riviere, "Modeling Rate-Dependent Hysteresis in Piezoelectric Actuator," 2003 IEEE/RSJ International Conference on Intelligent Robots and Systems, Las Vegas, 27 October-1 November 2003, pp. 1975-1980. doi:10.1109/IROS.2003.1248937
26. W. T. Ang, P. K. Khosla and C. N. Riviere, "Feedforward Controller with Inverse Rate-Dependent Model for Piezoelectric Actuators in Trajectory-Tracking Applications," IEEE/ASME Transactions on Mechatronics, Vol. 12, No. 2, 2007, pp. 134-142. doi:10.1109/TMECH.2007.892824
27. R. S. Robinson, "Interactive Computer Correction of Piezoelectric Creep in Scanning Tunneling Microscopy Images," Journal of Computer-Assisted Microscopy, Vol. 2, No. 1, 1996, pp. 53-58.
28. X. B. Chen, Q. Zhang, D. Kang and W. Zhang, "On the Dynamics of Piezoactuated Positioning Systems," Review of Scientific Instruments, Vol. 79, No. 11, 2008, pp. 116101.1-116101.3. doi:10.1063/1.2982238
29. W. Chen and C. S. Lynch, "A Micro-Electro-Mechanical Model for Polarization Switching of Ferroelectric Materials," Acta Materialia, Vol.

46, No. 15, 1998, pp. 5305- 5311. doi:10.1016/S1359-6454(98)00207-9

30. W. Seemann, A. Arockiarajan and B. Delibas, "Modeling and Simulation of Piezoceramic Materials Using Micromechanical Approach," European Congress on Computaional Methods in Applied Sciences and Engineering 2004 (ECCOMAS 2004), Jyväskylä, 24-28 July 2004.
31. A. Arockiarajan, B. Delibas, A. Menzel and W. Seemann, "Studies on Nonlinear Electromechanical Behavior of Piezoelectric Materials Using Finite Element Modeling," IWPMA 2005, 2nd International Workshop on Piezoelectric Materials and Applications in Actuators, Heinz Nixdorf Institute, Paderborn, 22-25 May 2005.
32. B. Delibas, A. Arockiarajan and W. Seemann, "Intergranular Effects on Domain Switchings in Polycrystalline Piezoceramics," IWPMA 2005, 2nd International Workshop on Piezoelectric Materials and Applications in Actuators, Heinz Nixdorf Institute, Paderborn, 22-25 May 2005.
33. M. Kamlah and Q. Jiang, "A Constitutive Model for Ferroelectric PZT Ceramics under Uniaxial Loading," Smart Materials and Structures, Vol. 8, No. 4, 1999, pp. 441- 459. doi:10.1088/0964-1726/8/4/302
34. M. Kamlah and C. Tsakmakis, "Phenomenological Modeling of the Non-Linear Electro-Mechanical Coupling in Ferroelectrics," International Journal of Solids and Structures, Vol. 36, No. 5, 1999, pp. 669-695. doi:10.1016/S0020-7683(98)00040-7
35. M. Kamlah and U. Böhle, "Finite Element Analysis of Piezoceramic Components Taking into Account Ferroelectric Hysteresis Behavior," International Journal of Solids and Structures, Vol. 38, No. 4, 2001, pp. 605-633. doi:10.1016/S0020-7683(00)00055-X
36. C. M. Landis, J. Wang and J. Sheng, "Microelectromechanical Determination of the Possible Remanent Strain and Polarization States in Polycrystalline Ferroelectrics and the Implications for Phenomenological Constitutive Theories," Journal of Intelligent Material Systems and Structures, Vol. 15, No. 7, 2004, pp. 513-525. doi:10.1177/1045389X04041653
37. J. Schröder and H. Romanowski, "A Simple Coordinate Invariant Thermodynamic Consistent Model for Nonlinear Electro-Mechanical Coupled Ferroelectrica," European Congress on Computaional Methods in Applied Sciences and Engineering 2004 (ECCOMAS

2004), Jyvä- skylä, 24-28 July 2004.

38. J. Schröder and H. Romanowski, "A Thermodynamically Consistent Mesoscopic Model for Transversely Isotropic Ferroelectric Ceramics in a Coordinate-Invariant Setting," Achive of Applied Mechanics, Vol. 74, No. 11-12, 2005, pp. 863-877. doi:10.1007/s00419-005-0412-7
39. S. Klinkel, "A Thermodynamic Consistent 1D Model for Ferroelastic and Ferroelectric Hysteresis Effects in Piezoceramics," Communications in Numerical Methods in Engineering, Vol. 22, No. 7, 2006, pp. 727-739. doi:10.1002/cnm.845
40. T. Hegewald, B. Kaltenbacher, M. Kaltenbacher and R. Lerch, "Efficient Modeling of Ferroelectric Behavior for the Analysis of Piezoceramic Actuators," Journal of Intelligent Material Systems and Structures, Vol. 19, No. 10, 2008, pp. 1117-1129. doi:10.1177/1045389X07083608
41. IEEE, "ANSI/ IEEE Std. 176-1987: IEEE Standard on Piezoelectricity," The Institute of Electrical and Electronics Engineers, New York, 1988.
42. P. G. Harper, "Kinematic Theory of Piezoelectric Hysteresis," Journal of Applied Physics, Vol. 52, No. 11, 1981, pp. 6851-6855. doi:10.1063/1.328677
43. D. C. Jiles and D. L. Atherton, "Theory of the Magnetisation Process in Ferromagnets and Its Application to the Magnetomechanical Effect," Journal of Physics D: Applied Physics, Vol. 17, No. 6, 1984, pp. 1265-1281. doi:10.1088/0022-3727/17/6/023
44. R. C. Smith and C. L. Hom, "Domain Wall Model for Ferroelectric Hysteresis," Technical Report CRSC-TR99- 07, Center for Research in Scientific Computation, New York, 1999.
45. T. Hegewald, M. Kaltenbacher and R. Lerch, "Characterization and Modeling of Piezoelectric Stack Actuators," IWPMA 2005, 2nd International Workshop on Piezoelectric Materials and Applications in Actuators, Heinz Nixdorf Institute, Paderborn, 22-25 May 2005, pp. 99- 106.
46. E. Preisach, "On the Magnetic Aftereffect," Zeitschrift für Physik A Hadrons and Nuclei, Vol. 94, No. 5-6, 1935, pp. 277-302.
47. I. D. Mayergoyz, "Mathematical Models of Hysteresis," Physical Review Letters, Vol. 56, No. 15, 1986, pp. 1518- 1521. doi:10.1103/PhysRevLett.56.1518
48. I. D. Mayergoyz and G. Friedman, "Generalized Preisach model of Hysteresis," IEEE Transactions on Magnetics, Vol. 24, No. 1, 1988, pp.

212-217. doi:10.1109/20.43892

49. P. Hejda and T. Zelinka, "Generalized Preisach Model of Hysteresis—Theory and Experiment," Czechoslovak Journal of Physics, Vol. 40, No. 1, 1990, pp. 57-68. doi:10.1007/BF01598355
50. M.-J. Jang, C.-L. Chen and J.-R. Lee, "Modeling and Control of a Piezoelectric Actuator Driven System with Asymmetric Hysteresis," Journal of the Franklin Institute, Vol. 346, No. 1, 2009, pp. 17-32. doi:10.1016/j.jfranklin.2008.06.005
51. P. N. Sreeram, G. Salvady and N. G. Naganathan, "Hysteresis Prediction for a Piezoceramic Material System," The ASME Winter Annual Meeting, New Orleans, 28 November-3 December 3 1993, pp. 35-42.
52. D. C. Hughes and J. T. Wen, "Preisach Modeling and Compensation for Smart Material Hysteresis," Proceedings of SPIE, Vol. 2427, 1995, pp. 50-64.
53. P. Ge and M. Jouaneh, "Generalized Preisach Model for Hysteresis Nonlinearity of Piezoceramic Actuators," Precision Engineering, Vol. 20, No. 2, 1997, pp. 99-111. do:10.1016/S0141-6359(97)00014-7
54. H. Hu and R. B. Mrad, "On the Classical Preisach Model for Hysteresis in Piezoceramic Actuators," Mechatronics, Vol. 13, No. 2, 2002, pp. 85-94. doi:10.1016/S0957-4158(01)00043-5
55. G. Robert, D. Damjanovic and N. Setter, "Preisach Modeling of Piezoelectric Nonlinearity in Ferroelectric Ceramics," Journal of Applied Physics, Vol. 89, No. 9, 2001, pp. 5067-5074. doi:10.1063/1.1359166
56. G. Song, J. Zhao, X. Zhou and J. A. De Abreu-Garcia, "Tracking Control of a Piezoceramic Actuator with Hysteresis Compensation Using Inverse Preisach Model," IEEE/ASME Transactions on Mechatronics, Vol. 10, No. 2, 2005, pp. 198-209. doi:10.1109/TMECH.2005.844708
57. X. Zhao and Y. Tan, "Neural Network based Identification of Preisach-Type Hysteresis in Piezoelectric Actuator Using Hysteretic Operator," Sensors and Actuators A: Physical, Vol. 126, No. 2, 2006, pp. 306-311. doi:10.1016/j.sna.2005.10.023
58. X. Zhao and Y. Tan, "Modeling Hysteresis in Piezo Actuator based on Neural Networks," Advances in Computation and Intelligence, Lecture Notes in Computer Science, Vol. 5370, 2008, pp. 290-296.
59. K. Kuhnen and H. Janocha, "Adaptive Inverse Control of Piezoelectric Actuators with Hysteresis Operators," Proceedings of European

Control Conference (ECC), Karsruhe, 31 August-3 September 1999, paper F 0291.

60. P. Krejci and K. Kuhnen, "Inverse Control of Systems with Hysteresis and Creep," IEE Proceedings—Control Theory and Applications, Vol. 148, No. 3, 2001, pp. 185- 192. doi:10.1049/ip-cta:20010375
61. K. Kuhnen and H. Janocha, "Complex Hysteresis Modeling of a Broad Class of Hysteretic Nonlinearities," Proceedings of the 8th international conference on New Actuators, Bremen, 10-12 June 2002, pp. 688-691.
62. K. Kuhnen and F. Previdi, "Modeling, Identification and Compensation of Complex Hysteretic Nonlinearities: A Modified Prandtl-Ishlinskii Spproach," European Journal of Control, Vol. 9, No. 4, 2003, pp. 407-421. doi:10.3166/ejc.9.407-418
63. K. Kuhnen, "Modeling, Identification and Compensation of Complex Hysteretic Nonlinearities and Log(t)-Type Creep Dynamics," Control and Intelligent System, Vol. 33, No. 2, 2005, pp. 134-147. doi:10.2316/Journal.201.2005.2.201-1420
64. M. Goldfarb and N. Celanovic, "Modeling Piezoelectric Stack Actuators for Control of Micromanipulation," IEEE Control Systems Magazine, Vol. 17, No. 3, 1997, pp. 69- 79. doi:10.1109/37.588158
65. S.-H. Lee, T. J. Royston and G. Friedman, "Modeling and Compensation of Hysteresis in Piezoceramic Transducers for Vibration Control," Journal of Intelligent Material Systems and Structures, Vol. 11, No. 10, 2000, pp. 781- 790. doi:10.1106/GQLJ-JGEU-MHG1-7JDF
66. T.-J. Yeh, S.-W. Lu and T.-Y. Wu, "Modeling and Identification of Hysteresis in Piezoelectric Actuators," Journal of Dynamic Systems, Measurement, and Control, Vol. 128, No. 2, 2006, pp. 189-196. doi:10.1115/1.2192819
67. T.-J. Yeh, R.-F. Hung and S.-W. Lu, "An Integrated Physical Model that Characterizes Creep and Hysteresis in Piezoelectric Actuators," Simulation Modelling Practice and Theory, Vol. 16, No. 1, 2008, pp. 93-110. doi:10.1016/j.simpat.2007.11.005
68. R. B. Mrad and H. Hu, "A Model for Voltage to Displacement Dynamics in Piezoceramic Actuators Subject to Dynamic Voltage Excitations," IEEE/ASME Transactions on Mechatronics, Vol. 7, No. 4, 2002, pp. 479-489. doi:10.1109/TMECH.2002.802724
69. D. Song and C. J. Li, "Modeling of Piezo Actuator's Nonlinear and Frequency Dependent Dynamics," Mechatronics, Vol. 9, No. 4, 1999,

pp. 391-410. doi:10.1016/S0957-4158(99)00005-7

70. Y. Yu, Z. Xiaob, N. G. Naganathan and R. V. Dukkipati, "Dynamic Preisach Modeling of Hysteresis for the Piezoceramic Actuator System," Mechanism and Machine Theory, Vol. 37, No. 1, 2002, pp. 75-89. doi:10.1016/S0094-114X(01)00065-9
71. X. Tan, R. Venkataraman and P. S. Krishnaprasad, "Control of Hysteresis: Theory and Experimental Results," Pentagon Technical Report A687934, 2001.
72. H. Hu and R. Ben Mrad, "A Discrete-Time Compensation Algorithm for Hysteresis in Piezoceramic Actuators," Mechanical Systems and Signal Processing, Vol. 18, No. 1, 2004, pp. 169-185. doi:10.1016/S0888-3270(03)00021-9
73. S. Bashash and N. Jalili, "A Polynomial-Based Linear Mapping Strategy for Feedforward Compensation of Hysteresis in Piezoelectric Actuators," Journal of Dynamic Systems, Measurement, and Control, Vol. 130, No. 3, 2008, pp. 031008.1-031008.10. doi:10.1115/1.2907372
74. R. Venkataraman and P. S. Krishnaprasad, "Novel Algorithm for the Inversion of the Preisach Operator Smart Structures and Materials," Proceedings of SPIE 3984, Smart Structures and Materials 2000: Mathematics and Control in Smart Structures, 2000, pp. 404-414. doi:10.1117/12.388785
75. R. Venkataraman and P. S. Krishnaprasad, "Approximate Inversion of Hysteresis: Theory and Numerical Results," Proceedings of the 39th IEEE Conference on Decision and Control, Sydney, Australia, 12-15 December 2000, pp. 4448-4454. doi:10.1109/CDC.2001.914608
76. J. Y. Peng and X. B. Chen, "H_2-Optimal Digital Control of Piezoelectric Actuators," Proceedings of the 8th World Congress on Intelligent Control and Automation, WCICA- 2010, Jinan, 7-9 July 2010, pp. 3684-3690. doi:10.1109/WCICA.2010.5553921
77. U.-X. Tan, W. T. Latt, C. Y. Shee, C. N. Riviere and W. T. Ang, "Feedforward Controller of Ill-Conditioned Hysteresis Using Singularity-Free Prandtl-Ishlinskii Model," IEEE/ASME Transactions on Mechatronics, Vol. 14, No. 5, 2009, pp. 598-605. doi:10.1109/TMECH.2008.2009936
78. B. D. Coleman and M. L. Hodgdon, "A Constitutive Relation for Rate-Independent Hysteresis in Ferromagnetically Soft Materials," International Journal of Engineering Science, Vol. 24, No. 6, 1986, pp.

897-919. doi:10.1016/0020-7225(86)90023-6

79. R. Bouc, "Forced Vibrations of a Mechanical System with Hysteresis," Proceedings of the 4th Conference on Non-linear Oscillations, Prague, 5-9 September 1967, pp. 315.
80. Y.-K. Wen, "Method of Random Vibration of Hysteretic Systems," Journal of the Engineering Mechanics Division, Vol. 102, No. 2, 1976, pp. 249-263.
81. M. Ismail, F. Ikhouane and J. Rodellar, "The Hysteresis Bouc-Wen Model, a Survey," Archives of Computational Methods in Engineering, Vol. 16, No. 2, 2009, pp. 161- 188. doi:10.1007/s11831-009-9031-8
82. O. Gomis-Bellmunt, F. Ikhouane and D. MontesinosMiracle, "Control of Bouc-Wen Hysteretic Dystems: Application to a Piezoelectric Actuator," The 13th Power Electronics and Motion Control Conference, 2008. EPEPEMC 2008, Poznan, 1-3 September 2008, pp. 1670-1675. doi:10.1109/EPEPEMC.2008.4635507
83. M. Rakotondrabe, "Bouc-Wen Modeling and Inverse Multiplicative Structure to Compensate Hysteresis Nonlinearity in Piezoelectric Actuators," IEEE Transactions on Automation Science and Engineering, Vol. 8, No. 2, 2011, pp. 428-431. doi:0.1109/TASE.2010.2081979
84. D. H. Wang, W. Zhu and Q. Yang, "Linearization of Stack Piezoelectric Ceramic Actuators Based on BoucWen Model," Journal of Intelligent Material Systems and Structures, Vol. 22, No. 5, 2011, pp. 401-413. doi:10.1177/1045389X10386132
85. L. Deng and Y. Tan, "Modeling Hysteresis in Piezoelectric Actuators Using NARMAX Models," Sensors and Actuators A: Physical, Vol. 149, No. 1, 2009, pp. 106- 112. doi:10.1016/j.sna.2008.09.022
86. A. Badel, J. Qiu and T. Nakano, "A New Simple Asymmetric Hysteresis Operator and Its Application to Inverse Control of Piezoelectric Actuators," IEEE Transactions on Ultrasonics, Ferroelectrics and Frequency Control, Vol. 55, No. 5, 2008, pp. 1086-1094. doi:10.1109/TUFFC.2008.761
87. H. Jung, J. Y. Shim and D. Gweon, "New Open-Loop Actuating Method of Piezoelectric Actuators for Removing Hysteresis and Creep," Review of Scientific Instruments, Vol. 71, No. 9, 2000, pp. 3436-3440. doi:10.1063/1.1287627
88. K. Kuhnen and H. Janocha, "Compensation of Creep and Hysteresis Effects of Piezoelectric Actuators with Inverse Systems," 6th

International Conference on New Actuators, Actuator'98, Bremen, 17-19 June 1998, pp. 309-312.

89. H. Janocha and K. Kuhnen, "Real-Time Compensation of Hysteresis and Creep in Piezoelectric Actuators," Sensors Actuators A: Physical, Vol. 79, No. 2, 2000, pp. 83-89. doi:10.1016/S0924-4247(99)00215-0
90. J. S. Dewey, K. Leang and S. Devasia, "Experimental and Theoretical Results in Output-Trajectory Redesign for Flexible Structures," Journal of Dynamic Systems, Measurement, and Control, Vol. 120, No. 4, 1998, pp. 456-461. doi:10.1115/1.2801486
91. K. S. Abidi and A. Şabanoviç, "Sliding Mode Control for High-Precision Motion of a Piezostage," IEEE Transactions on Industrial Electronics, Vol. 54, No. 1, 2007, pp. 629-637. doi:10.1109/TIE.2006.885477
92. H. C. Liaw, B. Shirinzadeh and J. Smith, "Enhanced Sliding Mode Motion Tracking Control of Piezoelectric Actuators," Sensors and Actuators A: Physical, Vol. 138, No. 1, 2007, pp. 194-202. doi:10.1016/j.sna.2007.04.062
93. G. M. Clayton, S. Tien, A. J. Fleming, S. O. R. Moheimani and S. Devasia, "Inverse-Feedforward of ChargeControlled Piezopositioners," Mechatronics, Vol. 18, No. 5-6, 2008, pp. 273-281. doi:10.1016/j.mechatronics.2007.07.006
94. J. W. Li, X. B. Chen and W. J. Zhang, "A New Approach to Modeling System Dynamics—In the Case of a Piezoelectric Actuator with a Host System," IEEE/ASME Transactions on Mechatronics, Vol. 15, No. 3, 2010, pp. 371- 380. doi:10.1109/TMECH.2009.2026473
95. J. W. Li, X. B. Chen and W. J. Zhang, "Axiomatic-Design-Theory-Based Approach to Modeling Linear High Order System Dynamics," IEEE/ASME Transactions on Mechatronics, Vol. 16, No. 2, 2011, pp. 341-350. doi:10.1109/TMECH.2010.2043535
96. K. Kuhnen, H. Janocha, D. Thull and A. Kugi, "A New Drive Concept for High-Speed Positioning of Piezoelectric Actuators," Proceedings of the 10th International Conference on New Actuators, Bremen, 14-16 June 2006, pp. 82-85.
97. K. K. Leang and S. Devasia, "Feedback-Linearized Inverse Feedforward for Creep, Hysteresis, and Vibration Compensation in AFM Piezoactuators," IEEE Transactions on Control Systems Technology, Vol. 15, No. 5, 2007, pp. 927-935. doi:10.1109/TCST.2007.902956
98. X. Dang and Y. Tan, "An Inner Product-Based Dynamic Neural Network

Hysteresis Model for Piezoceramic Actuators," Sensors and Actuators A: Physical, Vol. 121, No. 2, 2005, pp. 535-542. doi:10.1016/j.sna.2005.04.003

99. C. Ru, L. Chen, B. Shao, W. Rong and L. Sun, "A Hysteresis Compensation Method of Piezoelectric Actuator: Model, Identification and Control," Control Engineering Practice, Vol. 17, No. 9, 2009, pp. 1107-1114. doi:10.1016/j.conengprac.2009.04.013
100. L. Sun, C. Ru, W. Rong, L. Chen and M. Kong, "Tracking Control of Piezoelectric Actuator Based on a New Mathematical Model," Journal of Micromechanics and Microengineering, Vol. 14, No. 11, 2004, pp. 1439-1444. doi:10.1088/0960-1317/14/11/001
101. C. Ru and L. Sun, "Hysteresis and Creep Compensation for Piezoelectric Actuator in Open-Loop Operation," Sensors and Actuators A: Physical, Vol. 122, No. 1, 2005, pp. 124-130. doi:10.1016/j.sna.2005.03.056
102. H. H. Najafabadi, S. M. Rezaei, S. S. Ghidary, M. Zareinejad, K. Razi and R. Seifabadi, "Hysteresis Compensation of Piezoelectric Actuators under Dynamic Load Condition," IEEE/RSJ International Conference on Intelligent Robots and Systems, 2007. IROS 2007, San Diego, 29 October-2 November 2007, pp. 1166-1171. doi:10.1109/IROS.2007.4399048
103. E. Kouno, "A Fast Response Piezoelectric Actuator for Servo Correction of Systematic Errors in Precision Machining," CIRP Annals-Manufacturing Technology, Vol. 33, No. 1, 1982, pp. 369-372. doi: 10.1016/S0007-8506(07)61444-9
104. J. F. Cuttino, A. C. Miller and D. E. Schinstock, "Performance Optimization of a Fast Tool Servo for Single-Point Diamond Turning Machines," IEEE/ASME Transactions on Mechatronics, Vol. 4, No. 2, 1999, pp. 169-179. doi:10.1109/3516.769543
105. J. Lin, H. Chiang and C. C. Lin, "Tuning PID Control Gains for Micro Piezo-Stage in using Grey Relational Analysis," 2008 International Conference on Machine Learning and Cybernetics, Kunming, 12-15 July 2008, pp. 3863-3868. doi: 10.1109/ICMLC.2008.4621078
106. H.-J. Shieh, Y.-J. Chiu and Y.-T. Chen, "Optimal PID Control System of a Piezoelectric Microospitioner," 2008 IEEE/SICE International Symposium on System Integration, Nagoya, 4 December 2008, pp. 1-5. doi:10.1109/SI.2008.4770417
107. D. Y. Abramovitch, S. Hoen and R. Workman, "SemiAutomatic

Tuning of PID Gains for Atomic Force Microscopes," Asian Journal of Control, Vol. 11, No. 2, 2009, pp. 188-195. doi:10.1002/asjc.95

108. K. K. Tan, T. H. Lee and H. X. Zhou, "Micro-Positioning of Linear-Piezoelectric Motors Based on a Learning Nonlinear PID Controller," IEEE/ASME Transactions on Mechatronics, Vol. 6, No. 4, 2001, pp. 428-436. doi:10.1109/3516.974856
109. W. S. Oates and R. C. Smith, "Nonlinear Control Design for a Piezoelectric-Driven Nanopositioning Stage," Pentagon Technical Report A591444, 2005.
110. C. Edwards and S. K. Spurgeon, "Sliding Mode Control: Theory and Applications," Taylor & Francis, Abingdon, 1998.
111. P.-K. Huang, P.-H. Shieh, F.-J. Lin and H.-J. Shieh, "Sliding-Mode Control for a Two-Dimensional PiezoPositioning Stage," Control Theory & Applications, IET, Vol. 1, No. 4, 2007, pp. 1104-1113. doi:0.1049/iet-cta:20060371
112. J.-C. Shen, W.-Y. Jywe1, C.-H. Liu, Y.-T. Jian and J. Yang, "Sliding-Mode Control of a Three-Degrees-ofFreedom Nanopositioner," Asian Journal of Control, Vol. 10, No. 3, 2008, pp. 267-276. doi:10.1002/asjc.33
113. J.-C. Shen, J.-C. Shen, H.-K. Chiang and Y.-L. Shu, "Precision Tracking Control of a Piezoelectric-Actuated System," Precision Engineering, Vol. 32, No. 2, 2008, pp. 71-78. doi:10.1016/j.precisioneng.2007.04.002
114. Q. Xu and Y. Li, "Dynamics Modeling and Sliding Mode Control of an XY Micropositioning Stage," The 9th International Symposium on Robot Control (SYROCO'09), Gifu, 9-12 September 2009, pp. 781-786.
115. Y. Li and Q. Xu, "Adaptive Sliding Mode Control with Perturbation Estimation and PID Sliding Surface for Motion Tracking of a Piezo-Driven Micromanipulator," IEEE Transactions on Control Systems Technology, Vol. 18, No. 4, 2010, pp. 798-810. doi:10.1109/TCST.2009.2028878
116. H. C. Liaw, B. Shirinzadeh and J. Smith, "Sliding-Mode Enhanced Adaptive Motion Tracking Control of Piezoelectric Actuation Systems for Micro/Nano Manipulation," IEEE Transactions on Control Systems Technology, Vol. 16, No. 4, 2008, pp. 826-833. doi:10.1109/TCST.2007.916301
117. X. Chen and T. Hisayama, "Adaptive Sliding-Mode Position Control

for Piezo-Actuated Stage," IEEE Transactions on Industrial Electronics, Vol. 55, No. 11, 2008, pp. 3927-3934. doi:10.1109/TIE.2008.926768

118. H. C. Liaw and B. Shirinzadeh, "Enhanced Sliding-Mode Constrained Motion Tracking Control of Piezo-Actuated Flexure-Based Mechanisms," IEEE/ASME International Conference on Advanced Intelligent Mechatronics, 2009. AIM 2009, Singapore, 14-17 July 2009 2009, pp. 1879- 1884. doi:10.1109/AIM.2009.5229763
119. G. Schitter, P. Menold, H. F. Knapp, F. Allgower and A. Stemmer, "High Performance Feedback for Fast Scanning Atomic Force Microscopes," Review of Scientific Instruments, Vol. 72, No. 8, 2001, pp. 3320-3327. doi:10.1063/1.1387253
120. M. S. Tsai and J. S. Chen, "Robust Tracking Control of a Piezoactuator Using a New Approximate Hysteresis Model," Journal of Dynamic Systems, Measurement, and Control, Vol. 125, No. 1, 2003, pp. 96-102. doi:10.1115/1.1540114
121. G. Schitter, A. Stemmer, and F. Allgöwer, "Robust TwoDegree-of-Freedom Control of an Atomic Force Microscope," Asian Journal of Control, Vol. 6, No. 2, 2004, pp. 156-163. doi: 10.1111/j.1934-6093.2004.tb00194.x
122. A. Sebastian and S. M. Salapaka, "Design Methodologies for Robust Nano-Positioning," IEEE Transactions on Control Systems Technology, Vol. 13, No. 6, 2005, pp. 868-876. doi:10.1109/TCST.2005.854336
123. Y. Okazaki, "A Micro-Positioning Tool Post Using a Piezoelectric Actuator for Diamond Turning Machines," Precision Engineering, Vol. 12, No. 3, 1990, pp. 151-156. doi:10.1016/0141-6359(90)90087-F
124. D. Croft, S. Stilson and S. Devasia, "Optimal Tracking of Piezo-Based Nanopositioners," Nanotechnology, Vol. 10, No. 2, 1999, pp. 201-208. doi:10.1088/0957-4484/10/2/316
125. C. J. Li, H. S. M. Beigi, S. Li and J. Liang, "Nonlinear Piezo-Actuator Control by Learning Self Tuning Regulator," Journal of Dynamic Systems, Measurement, and Control, Vol. 115, No. 4, 1993, pp. 720-723. doi:10.1115/1.2899203
126. H.-J. Shieh, F.-J. Lin, P.-K. Huang and L.-T. Teng, "Adaptive Tracking Control Solely Using Displacement Feedback for a Piezo-Positioning Mechanism," IEE Proceedings——Control Theory and Applications, Vol. 151, No. 5, 2004, pp. 653-660. doi:10.1049/ip-cta:20040795
127. X. Tan and J. S. Baras, "Adaptive Identification and Control of Hysteresis

in Smart Materials," IEEE Transactions on Automatic Control, Vol. 50, No. 6, 2005, pp. 827-839. doi:10.1109/TAC.2005.849215

128. H. C. Liaw and B. Shirinzadeh, "Enhanced Adaptive Motion Tracking Control of Piezo-Actuated FlexureBased Four-Bar Mechanisms for Micro/Nano Manipulation," Sensors and Actuators A: Physical, Vol. 147, No. 1, 2008, pp. 254-262. doi:10.1016/j.sna.2008.03.020

129. J.-H. Xu, "Neural Network Control of a Piezo Tool Positioner," Canadian Conference on Electrical and Computer Engineering, Vancouver, 14-17 September 1993, pp. 333-336. doi:0.1109/CCECE.1993.332324

130. F.-J. Lin, H.-J. Shieh and P.-K. Huang, "Adaptive Wavelet Neural Network Control with Hysteresis Estimation for Piezo-Positioning Mechanism," IEEE Transactions on Neural Networks, Vol. 17, No. 2, 2006, pp. 432-444. doi:10.1109/TNN.2005.863473

131. F.-J. Lin, H.-J. Shieh, P.-K. Huang and L.-T. Teng, "Adaptive Control with Hysteresis Estimation and Compensation Using RFNN for Piezo-Actuator," IEEE Transactions on Ultrasonics, Ferroelectrics and Frequency Control, Vol. 53, No. 9, 2006, pp. 1649-1661. doi:10.1109/TUFFC.2006.1678193

132. H. Numasato and M. Tomizuka, "Settling Control and Performance of a Dual-Actuator System for Hard Disk Drives," IEEE/ASME Transactions on Mechatronics, Vol. 8, No. 4, 2003, pp. 431-438. doi:10.1109/TMECH.2003.819999

133. K. Leang and S. Devasia, "Hysteresis, Creep, and Vibration Compensation for Piezoactuators: Feedback and Feedforward Control," Proceedings of the 2nd IFAC Conference on Mechatronic Systems, Berkley, 9-11 December 2002, pp. 283-289.

134. L. Y. Pao, J. A. Butterworth and D. Y. Abramovitch, "Combined Feedforward/Feedback Control of Atomic Force Microscopes," Proceedings of the 2007 American Control Conference, New York City, 9-13 July 2007, pp. 3509-3515. doi:10.1109/ACC.2007.4282338

135. D. Croft and S. Devasia, "Vibration Compensation for High Speed Scanning Tunneling Microscopy," Review of Scientific Instruments, Vol. 70, No. 12, 1999, pp. 4600- 4605. doi:10.1063/1.1150119

136. H. Hu, H. M. S. Georgiou and R. Ben-Mrad, "Enhancement of Tracking Ability in Piezoceramic Actuators Subject to Dynamic Excitation Conditions," IEEE/ASME Transactions on Mechatronics, Vol. 10, No. 2, 2005, pp. 230-239. doi:10.1109/TMECH.2005.844705

137. C.-J. Lin and S.-R. Yang, "Modeling of a Piezo-Actuated Positioning Stage Based on a Hysteresis Observer," Asian Journal of Control, Vol. 7, No. 1, 2005, pp. 73-80. doi:10.1111/j.1934-6093.2005.tb00230.x

138. S. S. Aphale, S. Devasia and S. O. Reza Moheimani, "High-Bandwidth Control of a Piezoelectric Nanopositioning Stage in the Presence of Plant Uncertainties," Nanotechnology, Vol. 19, No. 12, 2008, pp. 125503.1- 125503.9. doi:10.1088/0957-4484/19/12/125503

139. D. Croft and S. Devasia, "Vibration Compensation for High Speed Scanning Tunneling Microscopy," Review of Scientific Instruments, Vol. 70, No. 12, 1999, pp. 4600- 4605. doi:10.1063/1.1150119

140. S. Khan, M. Elitas, E. D. Kunt and A. Sabanovic, "Discrete Sliding Mode Control of Piezo Actuator in NanoScale Range," IEEE International Conference on Industrial Technology, 2006. ICIT 2006, Mumbai, 15-17 December 2006, pp. 1454-1459. doi:10.1109/ICIT.2006.372418

141. J. Yi, S. Chang and Y. Shen, "Disturbance-ObserverBased Hysteresis Compensation for Piezoelectric Actuators," IEEE/ASME Transactions on Mechatronics, Vol. 14, No. 4, 2009, pp. 4196-4201. doi:10.1109/TMECH.2009.2023986

142. S. Chang and S. Li, "A High Resolution Long Travel Friction-Drive Micropositioner with Programmable Step Size," Review of Scientific Instruments, Vol. 70, No. 6, 1999, pp. 2276-2782. doi:10.1063/1.1149794

143. A. Bergander and J. Breguet, "Performance Improvements for Stick-Slip Positioners," Proceedings of 2003 International Symposium on Micromechatronics and Human Science, Nagoya, 19-22 October 2003, pp. 59-66. doi:10.1109/MHS.2003.1249910

144. F. Castanos and L. Fridman, "Analysis and Design of Integral Sliding Manifolds for Systems with Unmatched Perturbations," IEEE Transactions on Automatic Control, Vol. 51, No. 5, 2006, pp. 853-858. doi:10.1109/TAC.2006.875008

145. J. H. Painter, D. Kerstetter and S. Jowers, "Reconciling Steady-State Kalman and Alpha-Beta Filter Design," IEEE Transactions on Aerospace and Electronic Systems, Vol. 26, No. 6, 1990, 1990, pp. 986-991. doi:10.1109/7.62250

146. X. Wang, "High-Order Integral-Chain Differentiator and Application to Acceleration Feedback," Submitted to Computer Science for Possible Publication, 2011.

147. A. Levant, "Robust Exact Differentiation via Sliding Mode Technique,"

Automatica, Vol. 34, No. 3, 1998, pp. 379-384. doi:10.1016/S0005-1098(97)00209-4

148. Q. Xu and Y. Li, "Sliding Mode Control of a PiezoDriven Micropositioning System Using Extended Kalman Filter," IEEE International Conference on Automation and Logistics (ICAL), 2010, Hong Kong and Macau, 16- 20 August 2010, pp. 427-432. doi:10.1109/ICAL.2010.5585322
149. J. L. Minase, T.-F. Lu and F. Wornle, "State Estimation of Nonlinear Piezoelectric Stack Actuator Hysteresis Model," Proceedings of SPIE, Vol. 6414, 2006, pp. 641403.1-641403.10.
150. J. Minase, T.-F. Lu, B. Cazzolato and S. Grainger, "Adaptive Identification of Hysteresis and Creep in Piezoelectric Stack Actuators," The International Journal of Advanced Manufacturing Technology, Vol. 46, No. 9-12, 2009, pp. 913-921. doi:10.1007/s00170-009-2033-8
151. F. Yang and R. W. Wilde, "Observers for Linear Systems with Unknown Inputs," IEEE Transactions on Automatic Control, Vol. 33, No. 7, 1988, pp. 667-681. doi:10.1109/9.1278
152. M. Hou and P. C. Muller, "Design of Observers for Linear Systems with Unknown Inputs," IEEE Transactions on Automatic Control, Vol. 37, No. 6, 1992, pp. 871-875. doi:10.1109/9.256351
153. S. Hui and S. Żak, "Observer Design for Systems with Unknown Inputs," International Journal of Applied Mathematics and Computer Science, Vol. 15, No. 4, 2005, pp. 431-546.
154. S. H. Zak and S. Hui, S. "Output Feedback Variable Structure Controllers and State Estimators for Uncertain/Nonlinear Dynamic Systems," IEE Proceedings D: Control Theory and Applications, Vol. 140, No. 1, 1993, pp. 41-50. doi:10.1049/ip-d.1993.0006
155. J.-L. Chang and T.-C. Wu, "Robust Disturbance Attenuation with Unknown Input Observer and Sliding Mode Controller," Electrical Engineering, Vol. 90, No. 7, 2008, pp. 493-502. doi:10.1007/s00202-008-0099-1
156. T. Floquet, C. Edwards and S. K. Spurgeon, "On Sliding Mode Observers for Systems with Unknown Inputs," International Journal of Adaptive Control and Signal Processing, Vol. 21, No. 8-9, 2007, pp. 638-656. doi:10.1002/acs.958
157. K. Kalsi, J. Lian, S. Hui and S. H. Zak, "Sliding-Mode Observers for Systems with Unknown Inputs: A HighGain Approach," Automatica,

Vol.46,No.2,2010,pp.347-353.doi:10.1016/j.automatica.2009.10.040

158. X. Liu, J. Jeong and J. Kim, "A Three Translational DoFS Parallel Cube-Manipulator," Robotica, Vol. 21, No. 6, 2003, pp. 645-653. doi:10.1017/S0263574703005198
159. Q. Xu, and Y. Li, "A Novel Design of a 3-PRC Translational Compliant Parallel Micromanipulator for Nanomanipulation," Robotica, Vol. 24, No. 4, 2006, pp. 527-528. doi:10.1017/S0263574705002559
160. Y. Yue, F. Gao, X. Zhao and Q. Jeffrey Ge, "Relationship among Input-Force, Payload, Stiffness and Displacement of a 3-DOF Perpendicular Parallel Micro-Manipulator," Mechanism and Machine Theory, Vol. 45, No. 5, 2010, pp. 756-771. doi:10.1016/j.mechmachtheory.2009.12.006
161. S. Awtar and A. Slocum, "Constraint-Based Design of Parallel Kinematic XY Flexure Mechanisms," Journal of Mechanical Design, Vol. 129, No. 8, 2007, pp. 816-830. doi:10.1115/1.2735342
162. Q. Yao, J. Dong and P. Ferreira, "A Novel ParallelKinematics Mechanisms for Integrated, Multi-Axis Nanopositioning: Part 1. Kinematics and Design for Fabrication," Precision Engineering, Vol. 32, No. 1, 2008, pp. 7- 19. doi:10.1016/j.precisioneng.2007.03.001
163. J. Dong, Q. Yao and P. Ferreira, "A Novel ParallelKinematics Mechanism for Integrated, Multi-Axis Nanopositioning: Part 2: Dynamics, Control and Performance Analysis," Precision Engineering, Vol. 32, No. 1, 2008, pp. 20-33. doi:10.1016/j.precisioneng.2007.03.002
164. J. Dong, S. Salapaka and P. Ferreira, "Robust Control of a Parallel-Kinematic Nanopositioner," Journal of Dynamic Systems, Measurement, and Control, Vol. 130, No. 4, 2008, pp. 041007.1-041007.15. doi:10.1115/1.2936861
165. Y. Tian, B. Shirinzadeh and D. Zhang, "A Flexure-Based Five-Bar Mechanism for Micro/Nano Manipulation," Sensors and Actuators A: Physical, Vol. 153, No. 1, 2009, pp. 96-104. doi: 10.1016/j.sna.2009.04.022
166. Y. Tian, B. Shirinzadeh and D. Zhang, "Design and Dynamics of a 3-DOF Flexure-Based Parallel Mechanism for Micro/Nano Manipulation," Microelectronic Engineering, Vol. 87, No. 2, 2010, pp. 230-241. doi:10.1016/j.mee.2009.08.001
167. K. Hu, J. Kim, J. Schmiedeler and C. Menq, "Design, Implementation, and Control of a Six-Axis Compliant Stage," Review of Scientific Instruments, Vol. 79, No. 2, 2008, pp. 025108.2-

025105.11. doi:10.1063/1.2841804

168. T. W. Seo, D. S. Kang and J. Kim, "Synthesis and Comparison of Fine Actuator Controllers for a 3-DOF Micro Parallel Positioning Platform," Proceedings of the 10th International Conference on Control, Automation, Robotics and Vision, Hanoi, 17-20 December 2008, pp. 1606- 1611. doi:10.1109/ICARCV.2008.4795765
169. M. Grotjahn, B. Heimann and H. Abdellatif, "Identification of Friction and Rigid-Body Dynamics of Parallel Kinematic Structures for Model-Based Control," Multibody System Dynamics, Vol. 11, No. 3, 2004, pp. 273-294. doi:10.1023/B:MUBO.0000029426.05860.c2

CHAPTER

15

Experimental Investigation of Photostrictive Materials for MEMS Application

Mosfequr Rahman[1], Mahbub Ahmed[2], Masud Nawaz[1], Gustavo Molina[1], Abdur Rahman[3]

[1]Georgia Southern University, Statesboro, GA, USA

[2]Southern Arkansas University, Magnolia, AR, USA

[3]Edinboro University, Edinboro, PA, USA

ABSTRACT

Micro-Electro-Mechanical-Systems (MEMS) is an emerging technology that integrates micromachined mechanical structures with integrated circuits (IC). During the last two decades, it has grown vastly as a prominent research area among scientists and engineers. Usually the main components of MEMS

Citation:- Rahman, M. , Ahmed, M. , Nawaz, M. , Molina, G. and Rahman, A. (2017), "Experimental Investigation of Photostrictive Materials for MEMS Application". Open Access Library Journal, 4, 1-23. doi: 10.4236/oalib.1103856.

are sensors, actuators, and structures with all in microscale. A photostrictive material has high potential to be used in a MEMS application to develop an advanced optical actuator. A strain is induced when a photostrictive material is illuminated by high intensity light, which is known as photostriction. This phenomenon can be generally described as the combined effect of the photovoltaic effect and the converge-piezoelectric effect. When an incident light strikes a photostrictive material, a photovoltage is produced and this photovoltage develops strain in the material. Lanthanum modified lead zirconate titanate (Pb, La) (Zr, Ti) O_3 **ceramic doped with WO_3**, called PLZT, is one of the photostrictive ceramics which has an advantageous use as a wireless remote control over traditional actuators. Traditional actuators require wire connections to transmit the control signal; these wires yield noise via external electromagnetic fields. Whereas, PLZT actuators can transmit the control signal without wires which can eliminate possible noise due to the external electromagnetic field. In this current research, the photostrictive effect of a thin PLZT film on a silicon wafer is investigated experimentally. The transverse deflection of the PLZT optical actuator cantilever beam has been measured for stationary continuous light as well as for pulses of light using an optical chopper at various light intensities and focused locations. Results indicate that transverse deflection increases with the increase of light intensity. Also, the maximum transverse deflection at the free end of the cantilever beam for stationary light has been found to be three times larger than that for the pulses of light.

Keywords: MEMS, Photostriction, PLZT, Actuators

INTRODUCTION

Photostriction is a phenomenon in which strain is induced in photostrictive materials by incident light. In principle, this effect arises from a superposition of the photovoltaic effect, i.e., the generation of large voltage from the irradiation of light, and the converse piezoelectric effect, i.e., expansion or contraction under the voltage applied. A "bulk photovoltaic effect" is noticeable in some ferroelectric (or non-centrosymmetric) materials, such as ferroelectric single crystals or polarized ferroelectric ceramics. Along with this photovoltage, the converse piezoelectric effect induces mechanical strain [1]. The combinations of these two effects make these ferroelectric materials candidates for wireless actuator applications, that is, actuators excited by incident light. Photostrictive ceramics are one promising material for micro-fabrication technologies, which have photovoltaic, piezoelectric, and pyroelectric properties. Photostrictive materials offer the potential for development of advanced actuators with many advantages over conventional electro-mechanical actuators, such as non-contact actuation, remote control, and immunity from electric/magnetic disturbances [2]. Many types of micro-fabrication technologies, such as device design, lithography, film formation, and structure and system integration, have been developed [3] . Traditional actuators, like shape memory alloy and piezoelectric materials, have drawbacks for hard-wire connections such as control signal contamination due to external electro-magnetic disturbances. Photostrictive material appears to have potential for less friction and less weight for use in micro-electro mechanical systems (MEMS) due to its ability to be used wirelessly or by remote control. Piezoelectric ceramics have been used commercially as actuators in various fields; it requires hard wiring to supply voltage for actuation. On the other hand, photostrictive actuators can directly convert photonic energy to mechanical motion; they need neither electric lead wires nor electric circuits. A photostrictive type of actuator uses pulses of light instead of electric voltages for actuation. One material that possesses a significant photostrictive behavior under illumination by high-energy light is lanthanum-modified lead zirconate titanate (Pb, La) (Zr, Ti) O_3 ceramic doped with WO_3, called PLZT. In PLZT ceramics, UV and visible light have been found to excite free electrons within the crystalline structure that have greater than band-gap voltages; this is characteristic of a photosensitive material [4]. Also, like a typical ferroelectric material, PLZT can be poled to a remnant saturation polarization state, and the polarization can be switched by the application of an electric field. Due to these photosensitive and ferroelectric responses, PLZT ceramics are often called photo-ferroelectric (PFE) materials.

Many researchers have investigated the characteristics of photostrictive materials and its possible applications for MEMS devices [3] [5] [6] [7]. In this current research, a PLZT optical actuator on a silicon wafer as a cantilever beam is experimentally investigated to understand its response on continuous illumination of light and pulses of light.

Motivation of the Research

Micro-Electro-Mechanical Systems or MEMS technology can be defined as miniaturized mechanical and electro-mechanical elements that are made using micro fabrication techniques. MEMS functional elements are miniaturized structures, sensors, actuators, and microelectronics; the most important elements are the micro-sensors and micro-actuators. Over the last several decades, MEMS researchers have invented large numbers of microsensors with many sensing modalities, including temperature, pressure, inertial forces, chemical species, magnetic fields, optics, etc. Most recently, MEMS researchers have developed optical switches and mirrors to redirect or modulate light beams [8]. The present research will experimentally investigate the feasibility of using photostrictive materials for switching applications in MEMS devices. Switching applications in MEMS are used to transfer data and control signals. A traditional transducer needs electrical wire connections, which may contaminate the control signal by external electro-magnetic field interference. Photostrictive materials, on the other hand, do not need an electrical wire to supply voltage for actuation; they require only external light to actuate and may be used for remote sensing.

Research Hypotheses

Transverse deflection of the PLZT optical actuator depends on the light intensity. The deflection of the PLZT optical actuator will vary for the continuous and pulses of lights. Fast response and significant deflection of the PLZT optical actuator under the illumination of light could be used in MEMS applications as a wireless control.

Research Objectives

Photostrictive actuators have the superiority of remote sensing capabilities and freedom from external electric/magnetic disturbances, unlike conventional actuators. Based on this inspiration from current research, a PLZT thin film placed on the top of a silicon wafer is used to support a systematic characterization of photostrictive properties to develop an actuator. The main goal of the research is to investigate the potential of the PLZT material as a wireless actuator to convert photonic energy to mechanical energy.

The objectives of the present research follow.

1) Investigate the photostrictive effect of a continuous high intensity near-ultraviolet (UV) light on a PLZT optical actuator cantilever beam, which includes:

- measuring transverse deflection of the photostrictive cantilever beam for UV light focused at three different locations on the beam;
- for each focused location on the cantilever beam, measuring the transverse deflection along the beam at three different light intensities: 250 mW/cm^2, 200 mW/cm^2 and 100 mW/cm^2; and

- observing the time response of the cantilever beam deflection at the free end of the cantilever beam at three different light intensities and focused locations.

2) Investigate the photostrictive effect of the pulses of high intensity UV light on a PLZT optical actuator cantilever beam using an optical chopper, which includes:

- measuring transverse deflection of the photostrictive actuator at three different light intensities: 250 mW/cm^2, 200 mW/cm^2 and 100 mW/cm^2; and
- observing the time response of the cantilever beam at the free end of the beam at three different light intensities.

3) Compare the transverse deflections along the optical actuator cantilever beam between continuous and pulses of light conditions at three light intensities.

Background of Photostrictive Actuators

Characteristics of the PLZT Ceramics

PLZT compositions are represented as $Pb_{1-x}La_x(Zr_yTi_z)_{1-x/4}O_3$ or PLZT(X/Y/Z), where X = 100x, Y = 100y and Z = 100z. Ferroelectric materials have a wide range of properties depending on the composition [3]. The photovoltaic effect was first discovered in $BaTiO_3$ material [9]. P.S. Brody first proposed photostrictive materials as an opto-mechanical actuator, which can be used for MEMS devices to convert optical energy to mechanical energy [10] [11] . When ultraviolet (UV) light falls on the materials, it absorbs photons. Absorbed photons energize electrons and holes to move in a non-random manner; as a result, charges started to flow in an external short circuit of PLZT material. Light illuminates from the top and along the thickness of the material in the H direction. The photocurrent of PLZT ceramics varies proportionally with light intensity. Photocurrent increases as the thickness of PLZT film decreases, up to a certain limit [12]. The maximum photocurrent is predicted at 33 μm thickness for PLZT (3/52/48) ceramic doped with 0.5% WO_3 [13]. If film thickness becomes ultra-thin, then the film does not have enough electrons to flow and photocurrent drops because of the size effect [12]. Photocurrent of PLZT materials also depends on the wavelength of illuminating light. The maximum photocurrent is observed at 365 nm wavelength of light, which is out of visible light wavelength range [14]. No significant difference in wavelength dependency has been found between impurity-doped or non-doped samples. However, the valence of impurity has a significant effect on the photovoltaic response. Impurity atoms are classified into three categories: acceptors, donors, and remainder. K^+, Na^+, Mg^{2+}, Ba^{2+}, Al^{3+} & Fe^{3+} atoms are in the acceptor group; Bi^{3+}, Y^{3+}, Nb^{5+}, Ta^{5+} & W^{6+} are in the donor group. Photostriction (x_{ph}) can be defined in terms of photovoltaic voltage (E_{ph}) and piezoelectric coefficient (d_{33}) as $x_{ph} = d_{33}E_{ph}$. Impurities-doping has an effect on the piezoelectric coefficient, which improves photostriction [4] . Steady photovoltage and photocurrent generate in non-centrosymmetric ferroelectric material when illumination light has a wavelength corresponding to the absorption edge of the materials [1] . Generated photovoltage is greater than the band-gap energy for some materials. In certain ferroelectric material, photovoltage is about 1 kV/cm and current is in the order of nA; with this photovoltage, mechanical strain is produced due to converse piezoelectric effect [4] [13] . Several models have been proposed to describe the photovoltaic effect [13] . Photovoltage depends on both impurities-doping as well as crystal asymmetry.

Literature Review on the Transverse Deflection of the PLZT Cantilever Beam with the Light Intensity

Tonokai et al. experimentally investigated PLZT on a 8 mm × 2 mm × 0.05 mm Ti wafer at 50 mW/cm², 150 mW/cm², 300 mW/cm² and 450 mW/cm² UV light intensities; deflections were found to be 5 μm, 11 μm, 18 μm and 25 μm respectively [15] . Liang et al. experimented on a single-layer PLZT at 3 mW/cm², 6 mW/cm², 9 mW/cm², 12 mW/cm², and displacements were reported to be 0.31 μm, 0.33 μm, 0.42 m and 0.57 μm respectively [16] . Liang et al. developed a bimorph with amplificatory plate to reach deflection 100 times that of a single layer. The bimorph is constructed by reversing the polarization direction of a pair of PLZT unilateral wafers. The upper PLZT wafer develops a positive charge and expands due to UV light irradiation, and the lower PLZT wafer develops a negative charge and contracts at the same time; as a result, the entire bimorph bends downwards. Thakoor et al. reported that two PLZT ceramics wafers bond together to make a bimorph. The size of the bimorph is 20 mm × 5 mm × 0.4 mm. The maximum deflection was observed 200 μm at the tip at 80 mW/cm² [17]. Fukuda et al. experimented on the two pieces of 21 mm × 5 mm × 0.2 mm PLZT, glued in the opposite direction of polarization; the maximum deflections were observed to be 61 μm, 58 μm and 37 μm at 200 mW/cm², 100 mW/cm² and 25 mW/cm² light intensity respectively [18].

Literature Review on the MEMS Applications

Photostrictive effect of PLZT is useful for MEMS applications because it can serve as a current and voltage source for optical sensors at the same time. However, it requires some improvements: 1) improvement of output current; and 2) improvement of response [19]. Using the electrical properties of PLZT, an optical motor was developed as an electro-mechanical device suitable for miniaturization [3]. The optical motor driving principle was described with stator pads and moving pads. The PLZT electrode is connected with stator pads. When light falls on a PLZT electrode, high voltage generates between the stator and moving pad; then the moving pad starts to move towards the stator side. The proposed optical motor rotor disk has four pads, and each stator disk has three pads. The rotor disk was fixed or aligned with the motor axis, which is on the bearing. When light falls on PLZT ceramics, high voltage generates in the stator, and the rotor is pulled into the stator in a clockwise direction from the top view. The optical motor could be rotated, but its speed would be very slow. Internet traffic has grown enormously with the development of broadband access; a PLZT optical actuator could play a significant role in high-speed data transfer systems [20] . PLZT actuators can be used as high speed optical switches with benefits including low-voltage, low power consumption, low-polarization independence, noise robustness and switching speed of less than a nano second [21] [22] . Furukawa et al. made a 1 × 8 PLZT optical switch whose switching speed is less than 2.5 ns [23]. Satol et al. developed an 8 × 8 channel switch comprised of a PLZT deflector, fiber array, and micro-lens and slab waveguide [7]. Switching speed was less than 1 μs. PLZT bimorphs could also be used as a photo driven relay, consisting of two ceramic plates bonded together in opposite polarization directions and a dummy PLZT plate placed adjacent to the bimorph. Switching could be controlled by alternately irradiating light on the bimorph and dummy plates [5]. A micro-walking machine was developed using photostrictive bimorphs. Two PLZT ceramic legs are fixed on a plastic board; light irradiates on the legs alternately and the micro-walking machine moves like an inchworm at a very low speed of 10 μm/min [24]. An optical gripper was developed using a PLZT bimorph actuator. There are two bimorph actuators and UV light is irradiated by optical fibers on each actuator independently using two mirrors. Displacement of the gripper at the tip was reported to be 100 μm. The gripper has strong anti-noise characteristics and could be used for micro-surgery [18]. A photocurrent generates due to irradiation of light on the photostrictive actuator; this property has been used as an image comparator. One image is stored as a sequence of spatially varying ferroelectric domains. The second image is projected on the first and generates a photocurrent representing the dot product of stored and projected images [25]. Photostrictive materials also can be used as solar tracking shutter and photo phone [16]. Photo generated deflection could be used to walk on the surface or to run a motor indirectly. This kind of wireless optical control of an advanced mobile vehicle would be used for hazardous, hard to reach locations, as long as it is in the range of sight. Scattering effect of PLZT has been used to construct display devices and optical shutters, optical information storage and processing [17] [26] .

Basic Photostriction Equations

When the high-energy illumination with certain wavelength irradiates the photostrictive optical actuator, a current opposite to the polarized direction is induced. This current generates a voltage between the two surface electrodes, called the photovoltaic effect. The induced photovoltaic voltage induces actuation strains due to the converse piezoelectric effect. The induced electric field $E_l(t)$ from the photovoltaic effect can be determined from [27] :

$$E_l(t) = E_l(t_{j-1}) + \left[E_s - E_l(t_{j-1})\right]\frac{\alpha}{a_s} I(t_j) e^{-\frac{\alpha}{a_s} I(t_j)\Delta t} \cdot \Delta t - E_l(t_{j-1})\beta e^{-\beta\Delta t} \cdot \Delta t \quad (1)$$

where E_s is the saturated photovoltaic field; $a_s = L_a/b_a$ (length/width) is the aspect ratio; α is optical actuator constant and β is a constant related to voltage leakage. The light intensity I(t) is a function of time, and $E_l(t_j)$ denotes the induced electric field at time t_j. While I(t) remains constant during the interval Δt, then the electric field at tj=tj−1+Δttj=tj−1+Δt is $E_l(t_j)$.

The actuator temperature can be defined in a difference equation [27]:

$$\theta(t_{j+1}) = \theta(t_j) + \Delta\theta = \theta(t_j) + \frac{\left[I(t_{j+1})P - \gamma\theta(t_j)\right]\Delta t}{H + \gamma\Delta t} \quad (2)$$

where P is the absorbed heat from the light illumination; H is the heat capacity of the actuator; g is the heat transfer rate from the actuator to the structure; $\theta(t_j)$ is the actuator body temperature at t_j; $I(t_j)$ is an illumination intensity at t_j; and Δθ is a temperature rise in a time interval Δt.

The temperature change can trigger the pyroelectric effect from which an additional voltage is generated. The electric field $E_q(t)$, contributed by pyroelectric effect, can be determined from [27] :

$$E_\theta(t) = \frac{P_n}{\varepsilon}\theta(t) \tag{3}$$

where P_n is the pyroelectric constant.

The total induced electric field E(t) including both the photovoltaic effect and the pyroelectric effect is [27] :

$$E(t) = E_l(t) + E_\theta(t) \tag{4}$$

The magnitude of the light induced strains can be written as [18] :

$$\bar{S}(t) = d_{33}\left[E_l(t) + E_\theta(t)\right] + \lambda\theta(t)/Y_a \tag{5}$$

where d_{33} is the piezoelectric-strain constant, λ is a thermal stress coefficient, and Y_a is the Young's modulus of the optical actuator.

The electric potential of the actuator is

$$E_Z(t) = E(t) \times \text{thickness} \tag{6}$$

The objective of the current research is to investigate the influence of continuous and pulses of light on the transverse deflection of PLZT optical actuator. How the transverse deflection of a cantilever beam depends on the light intensity (both continuous and pulses) will be investigated experimentally. If the intensity of the light and the focused locations of the beam affect the time response will also be investigated in the current research.

METHODOLOGY

Experimentation has been performed mainly using a high intensity light source, a good repeatability displacement measurement sensor, and a photostrictive optical actuator.

The goal of the experiment was to observe the transverse deflection of a PLZT thin film on a silicon wafer cantilever beam under the illumination of continuous and pulses of high intensity light.

Experiment Method

Experimental Set-Up Design and Development

The experimental test set-up was designed to measure the photostriction effect of a photostrictive optical actuator as shown in Figure 1.

The set-up consists of a high pressure short arc xenon lamp with lamp housing, power supply with igniter, hot mirror, band pass filter, optical chopper, photostrictive cantilever type optical actuator, and laser sensor head with controller. Based on the design, an actual experimental set-up was developed as shown in Figure 2.

A sample of PLZT (3/52/48) ceramic was doped with 0.5% WO_3 of 1 µm thin film deposited on top of a 550 µm silicon wafer as shown in Figure 3. PLZT material was deposited on the silicon wafer at the MEMS Laboratory of Georgia Institute of Technology using the spattering method. PLZT (3/52/48) means Pb concentration of 97%, La concentration of 3% with Zr/Ti ratio 52/48. (3/52/48) ceramic also can be represented as Pb0.97 La 0.03 (Zr0.52Ti 0.48) 0.99 O_3.

A 150 watt xenon short arc lamp was provided high intensity light for maximum photovoltaic effect on the optical actuator. The arc lamp was placed inside the Science Tech 201 - 100 series lamp housing. A model 500 - 200 series AC power supply and a model 500-IG igniter were used to turn on the short arc lamp. To produce pulses of high intensity light on an optical actuator the Scitech 300CD model an optical chopper was used. A Keyence LK-H087 laser sensor head with 0.1 µm repeatability was used to measure the transverse deflection of actuator beam.

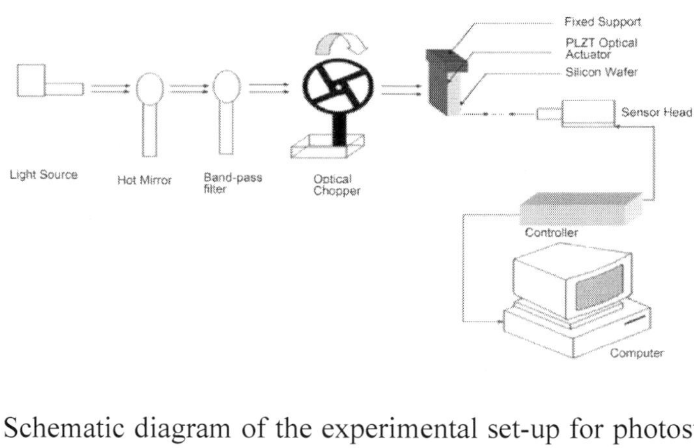

Figure 1. Schematic diagram of the experimental set-up for photostriction effect measurement.

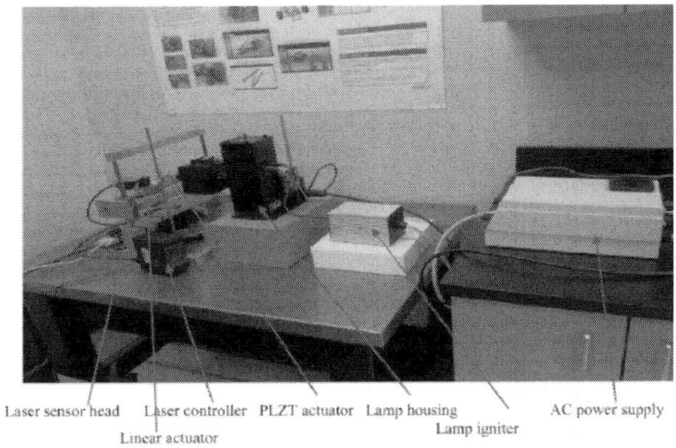

Figure 2. Experiment set-up for transverse deflection measurement of a cantilever beam.

Figure 3. PLZT sample manufactured at the MEMS laboratory.

Laser sensor data acquisition was controlled by the Keyence LK-G5001V model laser sensor controller. The beam deflection data was recorded in the computer through the laser sensor controller and data acquisition system. A heat absorbing glass was placed in front of the lamp housing, to filter high wavelength light that was generated by the arc lamp. In addition, a UV band-pass filter was placed behind a heat-absorbing filter, which allowed only 250 nm to 410 nm wavelengths of light to pass. This high intensity UV light went through a lamp housing window lens and was used to produce the photostriction effect on the optical actuator. A diamond-cutting blade was used to cut the sample into the desired shape (18 mm × 8 mm) for the experiment. PLZT thin film on silicon wafer sample was mounted as a cantilever beam on a stand. An LK-H087 laser displacement sensor was mounted on a linear actuator at the other side of the light source. The linear actuator can travel a minimum of 0.1 mm/step in the horizontal direction, thus allowing the laser displacement sensor head to move along the length of the sample cantilever beam. To irradiate pulses of light on the sample a variable frequency optical chopper was placed between the sample and the lamp housing window lens. Source light beam has a maximum light intensity of 250 mW/cm^2 at its focal point through the lamp housing window lens at 20 cm distance, which was measured by a UV meter. The UV meter sensor head was 12 mm in diameter, which measured the intensity on the average of sensor head area.

It was not experimentally possible for the whole cantilever beam to be exposed with equal light intensity. The light intensity starts to decrease as far as from the center of focal point on its perpendicular plane. Experiments were performed by focusing the light on the targeted cantilever beam at three different locations; these are the free end, middle, and fixed end as shown in Figure 4. Using the UV meter the light intensities were measured, at a distance of 20 cm, 25 cm, and 35 cm from the lamp housing window lens, were 250 mW/cm^2, 200 mW/cm^2, and 100 mW/cm^2 respectively.

Experimental Procedure

In conducting the experiment to produce photostriction effect and measuring the transverse deflection of the actuator, the following steps have been followed:

- A high-pressure short arc xenon lamp was placed properly inside the lamp housing with right polarity of cathode (−) and anode (+). This was used as a UV light source for the experiment.
- A power supply and a lamp igniter were connected to the lamp housing; and the lamp housing was placed on a vibration isolation table.
- A hot mirror lens and a 428 nm wavelength filter were placed after the lamp inside the housing filter slot. The source light first passed through the hot mirror then through the 428 nm wavelength filter.
- Output light was focused on the PLZT optical actuator cantilever beam at different distances from the output lens to get various light intensities on the targeted area.

- A frequency variable optical chopper was placed in between the light output lens and the PLZT cantilever beam to produce pulses of light on the targeted area as shown in Figure 5.

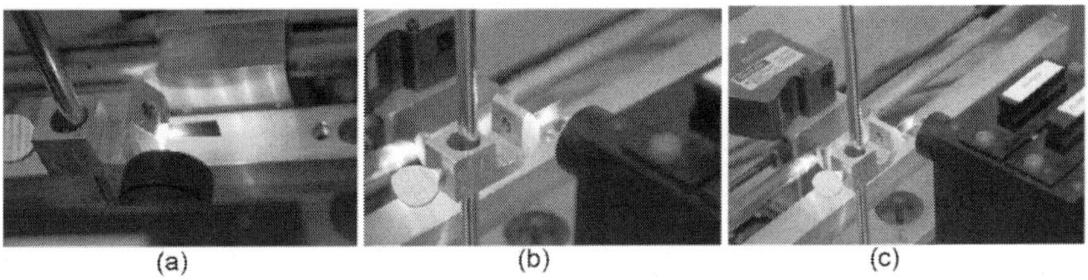

Figure 4. UV light was focused (a) At the fixed end; (b) At the middle; and (c) At the free end of the cantilever beam.

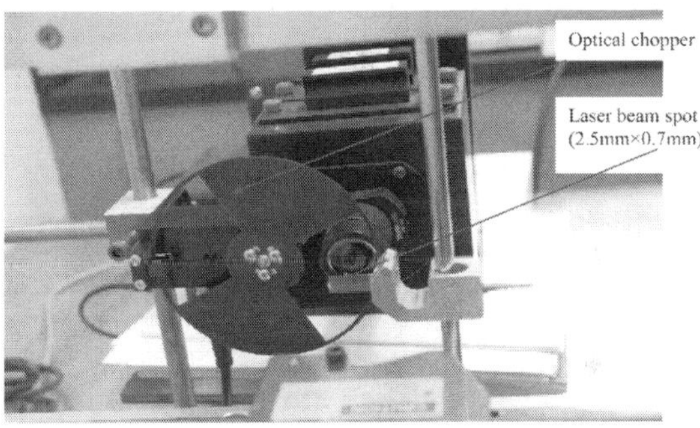

Figure 5. Optical chopper placement in the experiment set up to observe response for the pulses of light.

- The laser displacement sensor head was connected with the controller. The controller was powered by a 24V DC adapter. The laser sensor head was placed at the backside of the target that was the opposite side of the light source.
- The controller was connected to the computer by USB port to collect measured data on the personal computer (PC).
- Transverse deflection of the cantilever beam was measured by the laser displacement sensor along the length of the cantilever beam for every 1 mm distance from the free end. All measured data was recorded on the PC.
- Data was collected from the laser sensor at 0.1 sec intervals. An average of 20 - 30 seconds of collected data was recorded for analysis as shown in Figure 6.

- During the experiment, the photostrictive optical actuator cantilever beam was exposed for 5 minutes to the high intensity UV light then light was turned off and data was collected for 3 minutes.
- The average of three sets of data was plotted for the analysis.

RESULTS

Results of the Cantilever Beam for the Continuous Light

A photostrictive optical actuator was exposed under the uniform illumination of UV light for a certain period of time. Transverse deflection of the PLZT optical actuator cantilever beam is analyzed at various light intensities and various focused light positions. Comparison of transverse deflection is also presented.

Dependence of Transverse Deflection on Light Intensity at Specific Light Focused Position

Transverse deflections of the cantilever beam at 250 mW/cm², 200 mW/cm², and 100 mW/cm² light intensities were observed with light focused at different locations on the beam. Data plotted along the length of the cantilever beam are shown in Figures 7-9 respectively.

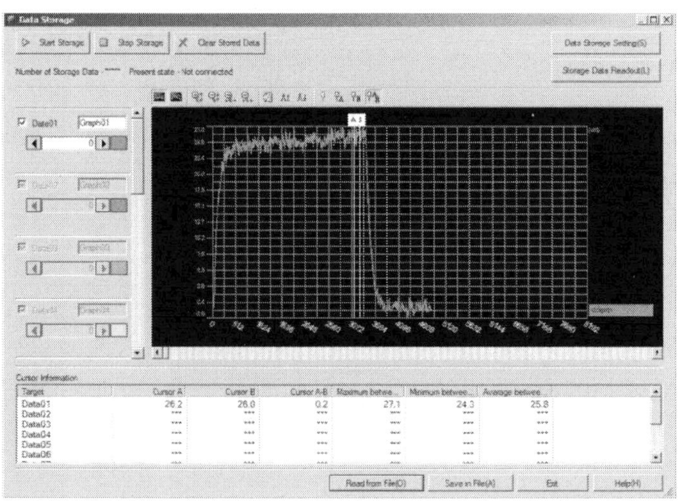

Figure 6. Controller output data collected in every 0.1 sec interval for 480 sec.

Deflections were measured every 1 mm distance along the length of cantilever beam. The negative deflection means the cantilever beam was deflected in the opposite side of the light source. From the figures, it can be seen that the transverse deflection near the fixed end of the beam was very small; it increased with the increase of distance from the fixed end. The maximum transverse deflection was observed at the free end of the beam. In addition, transverse deflection of the beam depends on the light intensity and it increases with the increase of light intensity at each specific focused location of the cantilever beam as shown in Figure 10.

Dependence of Transverse Deflection on Light Focused Position at Specific Light Intensity

Transverse deflection of the beam also depends on the locations of light focused on the beam. Deflection was maximized when the light was focused near the fixed end of the beam and it was minimized when the light was focused at the free end. Variation of transverse deflections due to light focused position variations were observed for three specific light intensities as shown in Figures 11-13. Deflections of the beam when the light was focused at the middle and at the free end were almost the same at the 250 mW/cm^2 as shown in Figure 11.

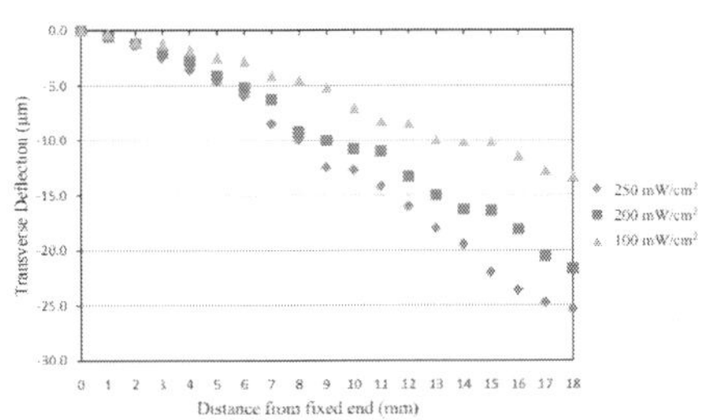

Figure 7. Comparison of transverse deflection for three different light intensities when light was focused at the fixed end of the cantilever beam.

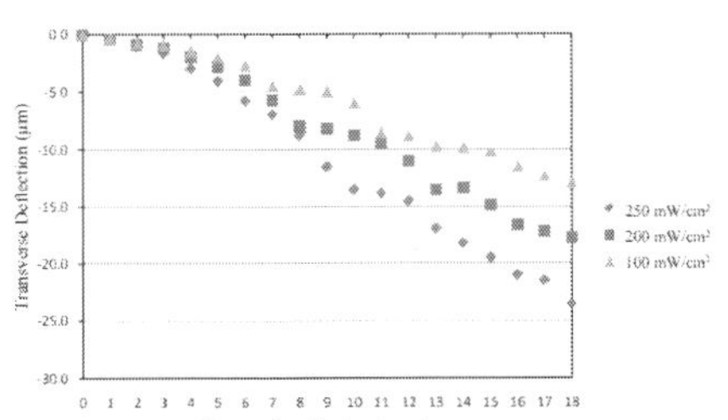

Figure 8. Comparison of transverse deflection for three different light intensities when light was focused at the middle of the cantilever beam.

At 200mW/cm^2, when light was focused at the middle, deflection of the beam was more than the deflection when the light was focused at the free end but less than the deflection when the light was focused at fixed end as shown in Figure 12.

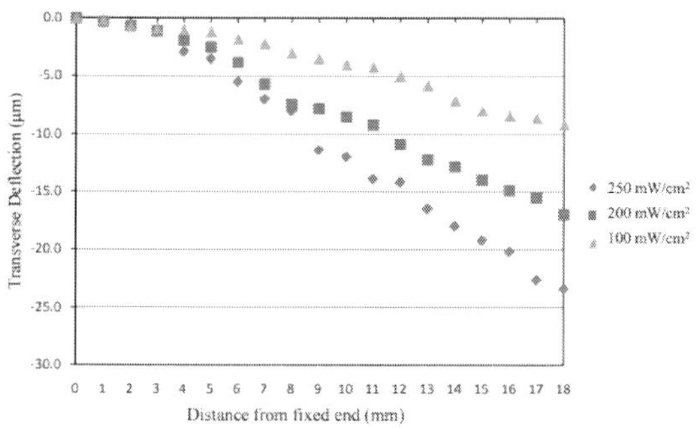

Figure 9. Comparison of transverse deflection for three different light intensities when light was focused at the free end of the cantilever beam.

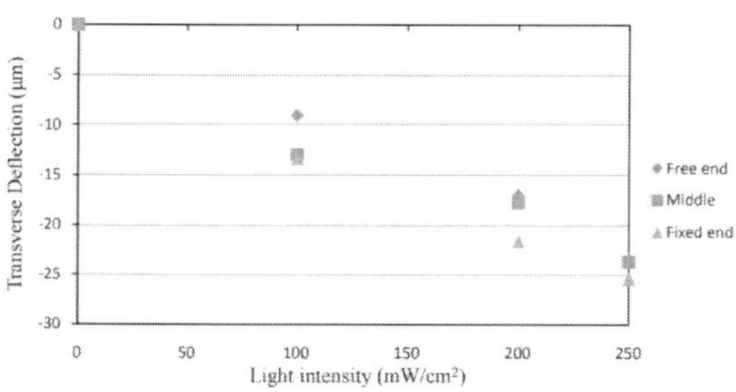

Figure 10. Transverse deflection at the free end of the beam with three different light intensities for three different focused locations on the cantilever beam.

When the gap between the light source and the targeted area of the beam increased the light beam diffused and allowed the cantilever beam to expose more light to the actuator surface; this happened especially when the light was focused at the middle position compare to the other two positions. At 100 mW/cm^2, the deflections were the same when the focused position of the light was at the middle and at the fixed end of the cantilever beam as shown in Figure 13. More diffused light fell on the beam surface when the light was focused at the middle of the beam compared to the other two positions. In addition, the light focused at the fixed end position allowed the beam to bend more near the fixed end, which ultimately resulted in these having the same beam deflection.

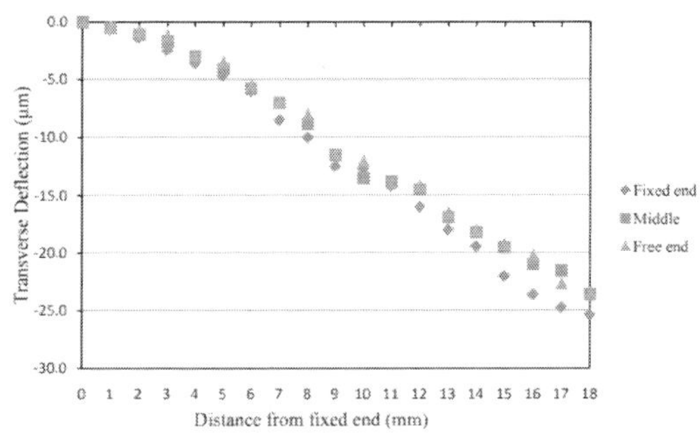

Figure 11. Variation of transverse deflection along the length of the cantilever beam for three different light focus positions on the beam at 250 mW/cm^2.

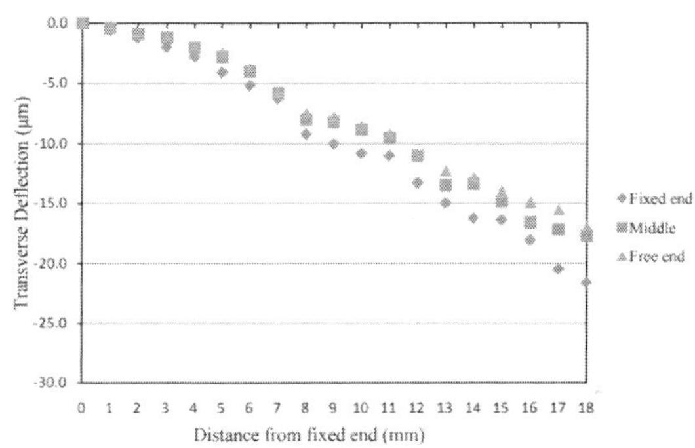

Figure 12. Variation of transverse deflection along the length of the cantilever beam for three different light focus positions on the beam at 200 mW/cm^2.

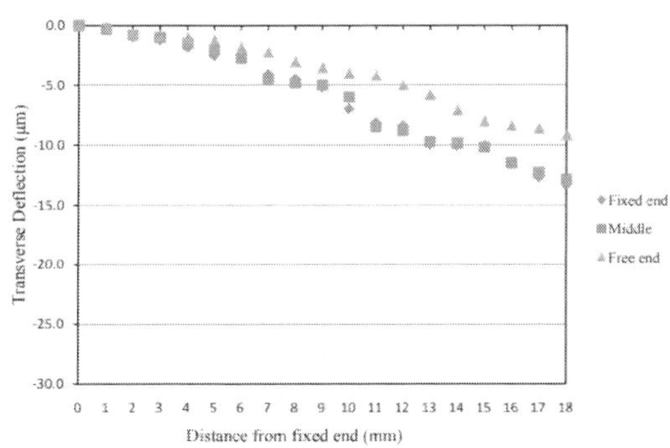

Figure 13. Variation of transverse deflection along the length of the cantilever beam for three different light focus positions on the beam at 100 mW/cm^2.

Time Response of the Cantilever Beam Deflection at the Free End for Three Different Focused Positions

High intensity UV light was focused at different locations on the cantilever beam. Deflection at the free end of the cantilever beam was observed to measure the time needed to reach the steady state condition as shown in Figures 14-16 respectively. Time response of the cantilever beam neither had any major influence on the light intensity nor on the focused locations of the actuator beam. Rising time was about 20 sec to reach 63% of the steady state maximum deflection. The cantilever beam was approaching the steady state after about 180 sec.

Hysteresis of the PLZT Cantilever Beam

The light was focused at the middle of the cantilever beam, and light was turned on and off for three times to observe the hysteresis in the actuator beam at a 200 mW/cm^2 light intensity. When the light turn on and off repeatedly, very small deflection on the cantilever beam was still observed when the light was still turned off as shown in Figure 17.

Results for the Pulses of Light

Tests were performed by placing an optical chopper in between the PLZT cantilever beam and the UV light source. The PLZT cantilever beam was exposed to illumination of pulses of light intermittently by the optical chopper. The two-slot blade optical chopper was run with a frequency of 0.5 Hz, which made the beam to be exposed to illuminations of light for 1 sec on and 1 sec off. The experiment was conducted at light intensities of 250 mW/cm^2, 200 mW/cm^2, and 100 mW/cm^2. In these pulses of light case, light was focused only at the middle of the cantilever beam.

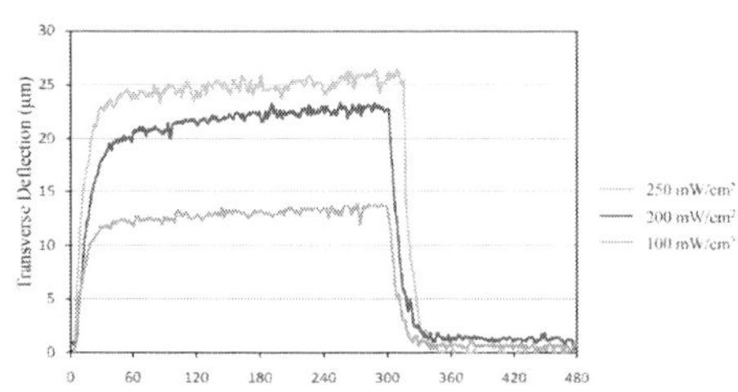

Figure 14. Comparison of time response on transverse deflection for three different light intensities when light was focused at the fixed end.

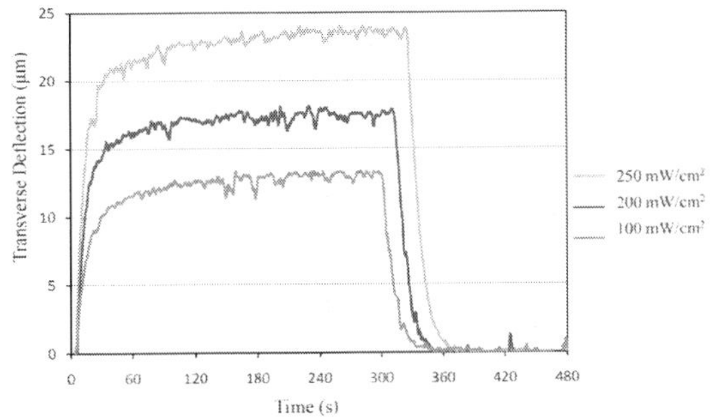

Figure 15. Comparison of time response on transverse deflection for three different light intensities when light was focused at the middle.

Comparison of Transverse Deflection with the Light Intensity

It has been observed that the magnitude of the corresponding transverse deflection along the length of the cantilever beam decreases in the pulses of light condition in comparison to the continuous light condition.

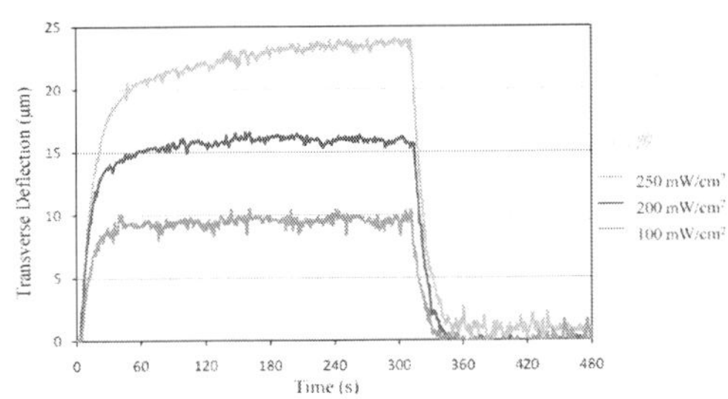

Figure 16. Comparison of time response on transverse deflection for three different light intensities when light was focused at the free end.

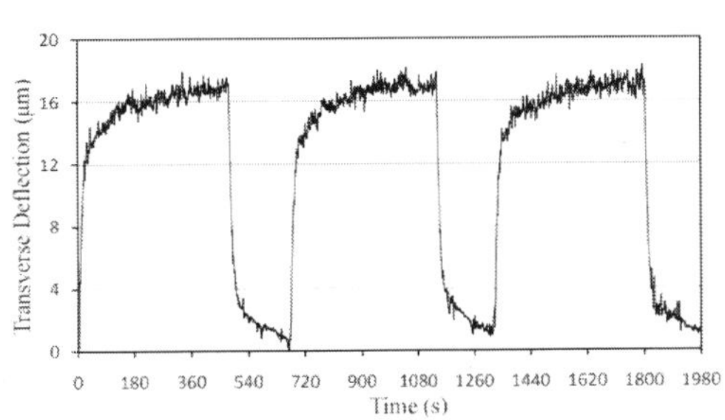

Figure 17. Hysteresis of the cantilever beam deflection at the free end.

Transverse deflection increases along the length of the actuator beam with the increase of distance from the fixed end as shown in Figure 18. It is also been observed that the deflection of the cantilever beam increases with increase of the light intensity as shown in Figure 19.

Time Response of the Cantilever Beam Deflection at the Free End for the Pulses of Light

The pulses of high intensity UV light were focused at the middle of the cantilever beam; deflection at the free end of the actuator beam was observed to measure the time needed to reach steady as shown in Figure 20. Time response of the cantilever beam has no major influence on the light intensities for the pulses of light condition.

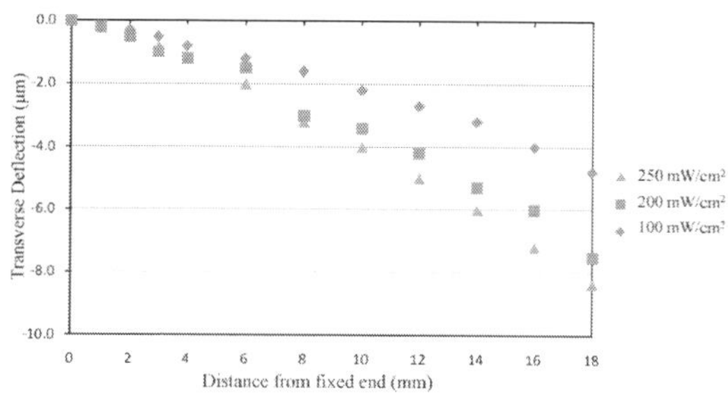

Figure 18. Transverse deflection along the length of the cantilever beam at three different light intensities for pulses of light condition.

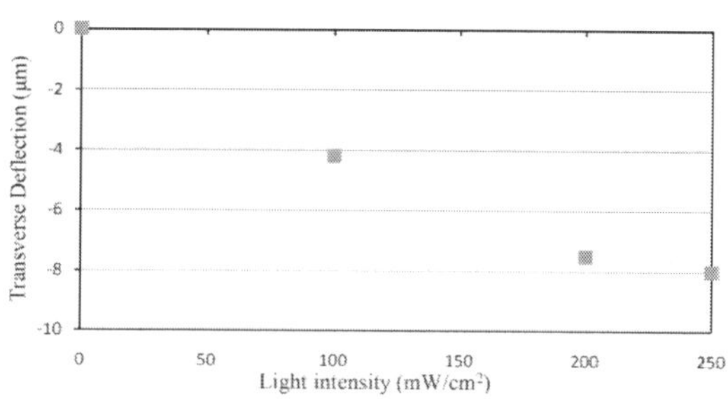

Figure 19. Transverse deflection depends on the light intensity in pulses of light condition.

Rising time was about 20 sec to reach 63% of the steady state maximum deflection. The deflection of the cantilever beam was approaching a steady state after about 180 sec.

Comparison of Deflection Results for the Continuous and Pulses of Lights

The transverse deflection along the length of the PLZT cantilever beam in a continuous light condition was compared to that of the pulses of light condition. Comparison was made between the continuous light and pulses of light conditions at three different light intensities when light was focused at the middle of the cantilever beam as shown in Figures 21-23. It has been noticed that, at each light intensity, the magnitude of the free end deflection of the beam for pulses of light was approximately 1/3rd of the free end deflection of continuous light.

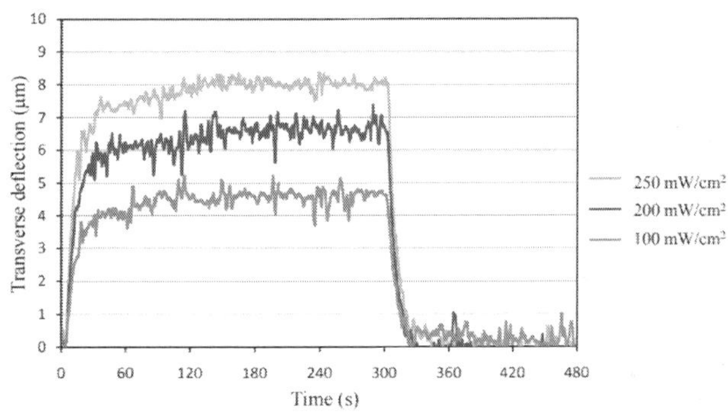

Figure 20. Comparison of time response on transverse deflection for three different light intensities in the pulses of light condition.

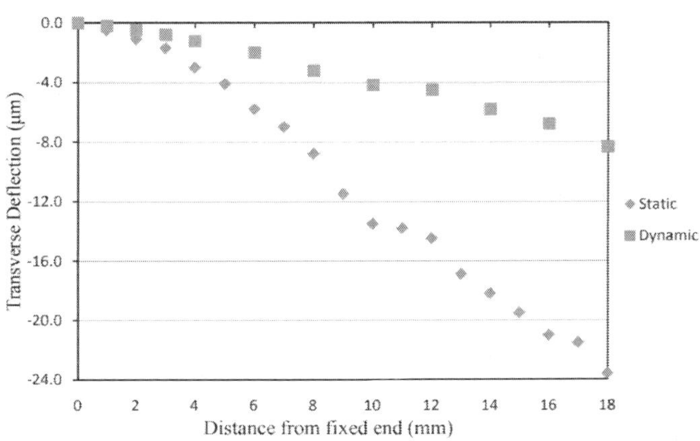

Figure 21. Comparison of transverse deflection between continuous and pulses of light conditions at 250 mW/cm^2 intensity.

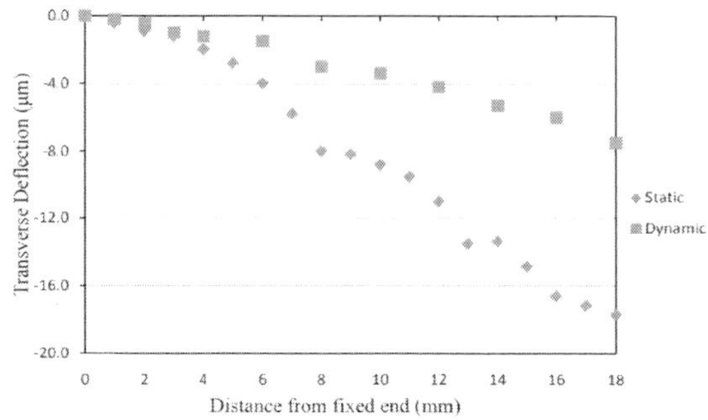

Figure 22. Comparison of transverse deflection between continuous and pulses of light conditions at 200 mW/cm^2 intensity.

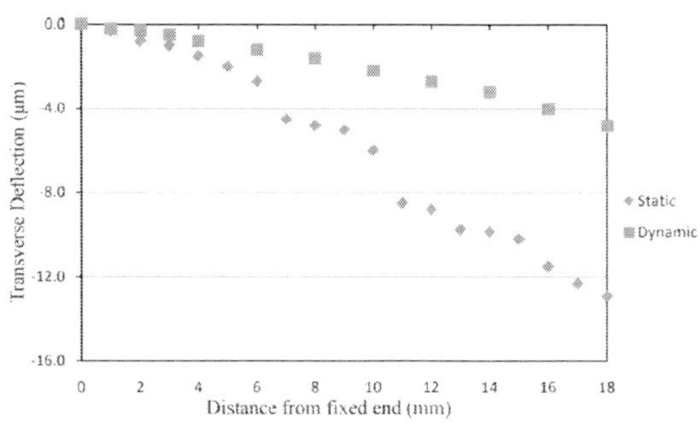

Figure 23. Comparison of transverse deflection between continuous and pulses of light conditions at 100 mW/cm^2 intensity.

In the pulses of light condition, the cantilever beam exposed to pulses of high intensity light, could not reach the maximum photostrictive effect in that amount of time that the continuous light condition could reach. That is why actuator beam deflection was less in the pulses of light condition than in continuous light condition.

CONCLUSIONS

From this experimental investigation, it has been observed that the continuous lights and the pulses of lights have a significant influence on the transverse deflection of PLZT optical actuator; less deflection was observed for the pulses of light compare to the continuous light. The PLZT optical actuator could be used in MEMS applications via a wireless control but the time response of the actuator observed was slow.

From the current study, analysis, and results of this research, the following conclusions can be made:

1) Experimental transverse deflection of a cantilever beam increases with the increase of light intensity for both continuous lights and pulses of lights.
2) Maximum tip deflection at the free end of the cantilever beam for pulses of light was about 1/3rd deflection of that for continuous light.
3) The light intensity and focused locations of the cantilever beam have no major influence on the time response. Rising time was observed to be about 20 sec to reach 63% of the steady state the maximum deflection for both continuous lights and pulses of light conditions. Deflection of the cantilever beam was approaching a steady state after about 180 sec for both continuous lights and pulses of lights.

REFERENCES

1. Fridkin, V.M. (1979) Photo Ferroelectrics. Springer Series in Solid State Sciences, Vol. 9, 85-113. https://doi.org/10.1007/978-3-642-81351-1_6
2. Rahman, M. (2005) A Finite Element Approach to Model and Analyze Photostrictive Optical Actuators. PhD Dissertation, University of Alabama, Tuscaloosa.
3. Kawaguchi, K., ichiki, M., Morikawa, Y. and Nakada, T. (2002) Electrical Properties of Photovoltaic PLZT and Its Application to Optical Motor. Ferroelectric, 273, 41-46.
4. Tanimura, M. and Uchino, K. (1988) Effect of Impurity Doping on Photostriction in Ferroelectric Ceramics. Sensors and Materials, 1, 47-56.
5. Sada, T., Inoue, M. and Uchino, K. (1987) Photostriction in PLZT Ceramics. Journal of the Ceramic Society of Japan, International Edition, 95, 499-504.
6. Okuym, M. (1998) Micro Sensors and Micro Actuators using Ferroelectric Thin Films. Proceedings of the International Symposium on Micromechatronics and Human Science, Nagoya, 29-34.
7. Satol, K., Sugama, A., Akahoshi, T., Kondo, M., Aoki, S. and Kurihara, K. (2007) Electro-Optic Effect of Lanthanum-Modified Lead Zirconate Titanate and Application to Optical Device. 16th IEEE International Symposium on Applications of Ferroelectrics, 827-828.
8. Huff, M.A. (2012) What Is Mems Technology? MEMS and Nanotechnology Exchange. https://www.mems-exchange.org/MEMS/what-is.html
9. Chynoweth, A.G. (1956) Surface Space-Charge Layers in Barium Titanate. Physics Review, 102, 705-714. https://doi.org/10.1103/PhysRev.102.705
10. Brody, P.S. (1975) High Voltage Photovoltaic Effect in Barium Titanate and Lead Titanate-Lead Zirconate Ceramics. Solid State Chemistry, 12, 193-200.
11. Brody, P.S. (1983) Optomechnical Bimorph Actuator. Ferroelectrics, 50, 27-33. https://doi.org/10.1080/00150198308014428
12. Qin, M., Kui, Y. and Yung, C.L. (2007) Photo Induced Current in (Pb0.97La0.03) (Zr0.52Ti0.48)O3 Thin Films of Different Thicknesses. Integrated Ferroelectrics, 88, 58-67. https://doi.org/10.1080/10584580601098647

13. Poosanaas, P., Dogan, A., Thakoor, S. and Uchino, K. (1998) Influence of Sample Thickness on the Performance of Photostrictive Ceramics. Journal of Applied Physics, 84, 1508-1512. https://doi.org/10.1063/1.368216
14. Yao, K., Gan, B.K., Chen, M. and Shannigrahi, S. (2005) Large Photo-Induced Voltage in a Ferroelectric Thin Film with In-Plane Polarization. Applied Physics Letters, 87, Article ID: 212906.
15. Tonokai, A., Takefumi, K., Daisuke, Y. and Koichi, S. (2010) PLZT Film Actuators Deposited by a Hydrothermal Method. IEEE International Conference on Robotics and Biomimetics, 1584-1589. https://doi.org/10.1109/ROBIO.2010.5723566
16. Liang, L., Shaoping, W. and Feng, C. (2006) Characteristics of Photo-Mechanical Device Using PLZT Wafer. Proceedings of the 2nd IEEE/ASME International Conference on Mechatronic and Embedded Systems and Applications, 1-4. https://doi.org/10.1109/MESA.2006.296973
17. Thakoor, S., John, M.M. and James, A.C. (1996) The Role of Piezoceramic Microactuation for Advanced Mobility. IEEE International Symposium on Application of Ferroelectrics, 1, 205-211.
18. Fukuda, T., Hattori, S., Arai, F., Matsuura, H., Hiramatsu, T., Ikeda, Y. and Maekawa, A. (1993) Characteristics of Optical Actuator-Servomechanisms using Bimorph Optical Piezoelectric Actuator. Proceedings of 1993 IEEE Robotics and Automation Conference, 618-623.
19. Ichiki, M., Furue, H., Kobayashi, T., Morikawa, Y., Nonaka, K., Maeda, R., Nakada, T. and Endo, C. (2005) Ultraviolet Sensor of Non-Bias Type; Layered Film Structure of Photovoltaic Ferroelectrics. The 13th International Conference on Solid-State Sensors, Actuators and Microsystems, 2, 2107-2110.
20. Mendinueta, J.M.D., Shinada, S., Furukawa, H. and Wada, N. (2017) Ultra-High-Capacity Optical Packet Switching Networks with Coherent Polarization Division Multiplexing QPSK/16QAM Modulation Formats. Photonics, 4, 27.
21. Nashimoto, K., Tanaka, N., LaBuda, M., Ritums, D., Dawley, J., Raj, M. and Kudzuma, D. (2005) High-Speed PLZT Optical Switches for Burst and Packet Switching. IEEE 2nd International Conference on Broadband Networks, 2, 1118-1123.

22. Nashimoto, K., Kudzuma, D. and Han, H. (2010) High-Speed Switching and Filtering using PLZT Waveguide Devices. 15th Opto Electronics and Communications Conference, 540-542.
23. Furukawa, H., Naoya, W., Hiroaki, H., Naganori, T., Keiichi, N. and Tetsuya, M. (2008) 640 Gbit/S/Port Optical Packet Switch Prototype with Optical Buffer using 1×8 PLZT Optical Switch and Parallel Pipeline Buffer Manager. 34th European Conference and Exhibition on Optical Communication, 1-2.
24. Uchino, K. (1989) Micro Walking Machine using Piezoelectric Actuators. Robotics and Mechatronics, 124, 44-47.
25. Martin, S.J., Butler, M.A. and Lan, C.E. (1988) Ferroelectric Optical Image Comparator using PLZT Thin Films. Electronics Letter, 24, 1486-1487. https://doi.org/10.1049/el:19881014
26. Thakoor, S., Poosanaas, P., Morookian, J.M., Yavrouian, A., Lowry, L., Marzwell, N.I., Nelson, J.G., Neurgaonkar, R.R. and Uchino, K. (1998) Optical Microactuation in Piezoceramics. Smart Structures and Materials Smart Electronics and MEMS. 5th Annual International Symposium on Smart Structures and Materials, San Diego, 20 July 1998.
27. Shih, H.R. and Tzou, H.S. (2000) Opto-Piezothermoelastic Constitutive Modeling of a New 2-D Photostrictive Composite Plate Actuator. ASME Control of Vibration and Noise: New Millennium, 61.

CHAPTER

16

Theory and Simulation Analysis of the Mode Shape and Normal Shape Actuators and Sensors

Jwo Ming Jou

Department of Mechanical Engineering, Cheng Shiu University, Kaohsiung City, Taiwan

ABSTRACT

In this paper, we will try to find a universal theoretical model and approximate solutions which can be applied to both mode shape and normal shape actuators and sensors, and which can be predicted the gain of the first three modes of the mode shape and normal shape actuators and sensors, finally through computer simulation analysis to validate. In order to prove the feasibility of the theory and as well as convenient to use on the electro-mechanical

Citation:- Jou, J. (2014), "Theory and Simulation Analysis of the Mode Shape and Normal Shape Actuators and Sensors". Open Journal of Acoustics, 4, 184-203. doi: 10.4236/oja.2014.44019.

engineering, we will try to simplify the three-dimension structure problem into an one-dimension structure problem. Furthermore we will design one kind of bimorph type piezoelectric cantilever beam, so that it can be used as with the actuator and sensor simultaneously, but also conducive to the theory and simulation analysis. As for the simulation analysis, we will use the ANSYS code.

Keywords: Mode Shape, Normal Shape, Actuator, Sensor, Gain, The First Three Modes, ANSYS Code

INTRODUCTION

Due to the piezoelectric materials (such as PZT or PVDF) have the advantages of fast response and high actuating force, so they are very suitable made for sensors and actuators to sense and control the vibration of the flexible structures, especially to sense or control the low frequency vibration, because the high-frequency vibration is very easy to be absorbed by the structure itself. So that the most piezoelectric sensors and actuators are used to sense and control the low frequency vibration of flexible structures. And since the design concept and theory of the modal sensors and actuators have been proposed by C.K., Lee [1] [2] , making them to sense and control the low frequency vibration of flexible structures are more rapid, sensitive, precise, accurate and efficient. In which the main design concepts and methods of the modal sensors and actuators are based on the coating of the effective electrode surface and changing of the polarization direction with specific modal or mode shape, shown as Figure 1. As for the theoretical model, it is entirely based on the mode shape functions or spatial distribution methods. It is successfully pulled out the transverse vibration displacement of the coupling multi-modal by the mode shape functions or spatial distribution methods. Next, it uses the coating methods or the functions of effective electrode surface to sense or control the low frequency vibration of the flexible structures. Since the relevant theoretical analysis, computer simulation and experimental eleven were uncovered or presented [3] - [9]. In addition, the theoretical model of modal or mode shape actuator is also applied to the different types of piezoelectric transformers, thereby to enhance their step-up ratio and conversion efficiency [10] [11] .

However, the theoretical model is only applicable to modal sensors and actuators, but it can't be applied on the normal shape sensors and actuators. If the theoretical model of modal sensors and actuators used in the normal shape sensors and actuators, we could not find the correct approximate solution. That is, so far, we still can't find one kind of universal equation to simultaneously describe or predict the behavioral of the mode shape and normal shape sensors and actuators, or to compare the gain or to distinguish the difference between the mode shape and normal shape sensors and actuators from past studies.

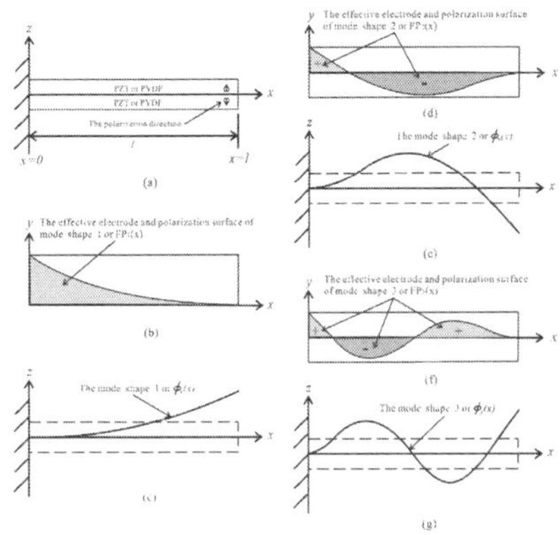

Figure 1. (a) The side view of the polarization direction of the bimorph piezoelectric cantilever beam; (b) The top view of the effective electrode surface of mode shape 1; (c) The side view of the mode shape 1; (d) The top view of the effective electrode surface of mode shape 2; (e) The side view of the mode shape 2; (f) The top view of the effective electrode surface of mode shaspe 3; (g) The side view of the mode shape 3.

In this paper, we will try from the original actuator and sensor equation, and electromechanical boundary conditions to find one kind of universal theoretical model and reasonable approximate solutions to compare the gain and difference between the mode shape and normal shape sensors and actuators. Furthermore we will also through the analysis results of computer simulation to verify the correctness and feasibility of the universal theoretical model and approximate solutions.

ACTUATOR EQUATION

In order to understand the differences between the mode shape and normal shape actuators and sensors, we just start from the one-dimension actuator equation of bimorph type piezoelectric cantilever beam. The main contribution of this equation comes from the bending effect, so the membrane effect can be completely ignored. Therefore, the equation can be simplified to as follows [12]:

$$\frac{\partial^2 M_1(x,t)}{\partial x^2} = \rho h \frac{\partial^2 w(x,t)}{\partial t^2}. \tag{1}$$

where the moment M_1 can be divided into two terms of the mechanical M_1^M and electrical moment M_1^E as follow:

$$M_1(x,t) = M_1^M(x,t) + M_1^E(x,t) = -D_{11}\frac{\partial^2 w(x,t)}{\partial x^2} + \Pi_3(t)FP(x). \tag{2}$$

As for the constants in the above Equations (1) and (2) can be defined as follow:

$$D_{11} = \frac{1}{3}\sum_{k=1}^{n}\left[(c_{11})_k\left(z_k^3 - z_{k-1}^3\right)\right]. \tag{3}$$

$$\rho h = \sum_{k=1}^{n}(\rho_k h_k). \tag{4}$$

$$h_k = z_k - z_{k-1}. \tag{5}$$

wherein above the symbols of c_{11}, D_{11}, h, w, t, x, z and ρ are defined as the Young's modulus, bending stiffness constant per unit width, thickness, transverse displacement, time, longitudinal coordinate, transverse coordinate and density.

Furthermore the electrical moment M_1^E of Equation (2) can be divided into two terms of time domain and space domain as follow:

$$\Pi_3(t) = 2\sum_{k=1}^{n}\left[e_{31}z^0 V_3(t)\right]_k = \Pi_3 e^{j\omega_i t} = 2\sum_{k=1}^{n}\left[z^0 e_{31} V_3 e^{j\omega_i t}\right]_k,\ i = 1,\ 2, \cdots, \infty,\ j = \sqrt{-1}. \tag{6}$$

And

$$\left[FP_n(x)\right]_{MS} = \mu_n \frac{\partial^2 \phi_n(x)}{\partial x^2} = \left(\frac{l}{\lambda_n}\right)^2 \frac{\partial^2 \phi_n(x)}{\partial x^2}. \tag{7}$$

wherein above the symbols of FP_n, l, $\omega_i(2\pi f_i)$, λ_n, μ_n and ϕ_n are defined as the function of effective electrode surface of the nth mode shape, length, natural angular velocity (natural frequency), mode shape eigenvalues, mode shape constant and mode shape function.

Let Equations (2)-(7) substituted into Equation (1), and after finishing, we can get a non-homogeneous partial differential equation of motion with transverse displacement of mode shape actuator as follows:

$$\left[\frac{\partial^4 w(x,t)}{\partial x^4}+\frac{\rho h}{D_{11}}\frac{\partial^2 w(x,t)}{\partial t^2}=\frac{\mu_n \Pi_3(t)}{D_{11}}\frac{\partial^4 \phi_n(x)}{\partial x^4}\right]_{MS}. \tag{8}$$

When the surface is uniformly coated and normal polarization with the bimorph type piezoelectric cantilever beam, that is, the function of effective electrode surface of Equation (7) can be redefined as:

$$\left[FP_n(x)\right]_{NS}=C_{NS}=\text{Constant}. \tag{9}$$

Then, the second derivative of electrical moment becomes zero.

$$\left[\frac{\partial^2 M_1^E(x,t)}{\partial x^2}=\Pi_3(t)\frac{\partial^2 FP(x)}{\partial x^2}=0\right]_{NS} \tag{10}$$

So for the normal shape actuator, Equation (8) can be simplified as:

$$\left[\frac{\partial^4 w(x,t)}{\partial x^4}+\frac{\rho h}{D_{11}}\frac{\partial^2 w(x,t)}{\partial t^2}=0\right]_{NS}. \tag{11}$$

In order for the mode shape and normal shape actuator equations can be applied to structures of different lengths, we can try to make Equation (8) and Equation (11) become the dimensionless equations:

$$\left[\frac{\partial^4 \overline{w}(\overline{x},t)}{\partial \overline{x}^4}+\frac{\rho h l^4}{D_{11}}\frac{\partial^2 \overline{w}(\overline{x},t)}{\partial t^2}=\frac{\overline{\mu}_n \Pi_3(t)}{D_{11}}\frac{\partial^4 \overline{\phi}_n(\overline{x})}{\partial \overline{x}^4}\right]_{MS}. \tag{12}$$

And

$$\left[\frac{\partial^4 \overline{w}(\overline{x},t)}{\partial \overline{x}^4}+\frac{\rho h l^4}{D_{11}}\frac{\partial^2 \overline{w}(\overline{x},t)}{\partial t^2}=0\right]_{NS}. \tag{13}$$

where the dimensionless mode shape constant can be redefined as:

$$\overline{\mu}_n=\frac{1}{\lambda_n^2}. \tag{14}$$

where the solution form of Equation (12) and Equation (13) can be expressed as:

$$\left[\overline{w}(\overline{x},t)=\overline{w}_h(\overline{x},t)+\overline{w}_p(\overline{x},t)\right]_{MS}. \tag{15}$$

And

$$\left[\overline{w}(\overline{x},t)=\overline{w}_h(\overline{x},t)\right]_{NS}. \tag{16}$$

Furthermore let the homogeneous solution of Equation (16) and Equation (17) are assumed to be as follows:

$$\overline{w}_h(\overline{x},t)=\overline{W}_h(\overline{x})e^{j\omega_i t}. \tag{17}$$

As for the particular solution of Equation (15) is assumed to be as follows:

$$\overline{w}_p(\overline{x},t)=A_n\overline{\phi}_n(\overline{x})e^{j\omega_i t}. \tag{18}$$

where the symbols of A_n and $\overline{W}_h$ **are defined as the undetermined coefficient and dimensionless homogeneous transverse displacement.**

We can take Equation (17) into Equation (12) and Equation (13), then after finishing, we can get a dimensionless homogeneous four-order ordinary differential equation, as:

$$\frac{d^4\overline{W}_h(\overline{x})}{dx^4}-\overline{\lambda}_i^4\overline{W}_h(\overline{x})=0,\ i=1,2,\cdots,\infty. \tag{19}$$

where the dimensionless natural eigenvalue of mode shape is defined as:

$$\overline{\lambda}_i^4=(l\lambda_i)^4=\frac{\rho h l^4\omega_i^2}{D_{11}}. \tag{20}$$

And the ith **natural resonance frequency of piezoelectric beam multilayer or mode shape or normal shape actuator can be derived by Equation (20), as:**

$$f_i=\frac{\lambda_i^2}{2\pi l^2}\sqrt{\frac{D_{11}}{\rho h}}. \tag{21}$$

We can further get a dimensionless homogeneous solution from Equation (19)

$$\overline{W}_h(\overline{x},t)=C_1\sinh\overline{\lambda}_i\overline{x}+C_2\cosh\overline{\lambda}_i\overline{x}+C_3\sin\overline{\lambda}_i\overline{x}+C_4\cos\overline{\lambda}_i\overline{x}. \tag{22}$$

In addition, we can take Equation (18) into Equation (12), so get the undetermined coefficient as:

$$A_n\left[\frac{\partial^4\bar{\phi}_n(\bar{x})}{\partial\bar{x}^4}-\bar{\lambda}_i^4\bar{\phi}_n(\bar{x})\right]=\frac{\bar{\mu}_n\Pi_3}{D_{11}}\frac{\partial^4\bar{\phi}_n(\bar{x})}{\partial\bar{x}^4}. \tag{23}$$

where the relationship of the fourth-order differential function and zero-order function is:

$$\frac{\partial^4\bar{\phi}_n(\bar{x})}{\partial\bar{x}^4}=\bar{\lambda}_n^4\bar{\phi}_n(\bar{x}). \tag{24}$$

So Equation (23) can be rewritten as:

$$A_n\left(\bar{\lambda}_n^4-\bar{\lambda}_i^4\right)\bar{\phi}_n(\bar{x})=\frac{\bar{\mu}_n\Pi_3}{D_{11}}\bar{\lambda}_n^4\bar{\phi}_n(\bar{x}). \tag{25}$$

Thus we can get the solution of the undetermined coefficient.

$$A_n=\frac{\bar{\lambda}_n^4}{\bar{\lambda}_n^4-\bar{\lambda}_i^4}\frac{\bar{\mu}_n\Pi_3}{D_{11}}=\frac{\omega_n^2}{\omega_n^2-\omega_i^2}\frac{\bar{\mu}_n\Pi_3}{D_{11}}. \tag{26}$$

where $\bar{\lambda}_n^4\neq\bar{\lambda}_i^4$ or $\omega_n^2\neq\omega_i^2$ **in normal or natural state.**

At this point, we can find the general solution of mode shape and normal shape actuator, as:

$$\bar{w}_{MS}(\bar{x},t)=\left[C_1\sinh\bar{\lambda}_i\bar{x}+C_2\cosh\bar{\lambda}_i\bar{x}+C_3\sin\bar{\lambda}_i\bar{x}+C_4\cos\bar{\lambda}_i\bar{x}+A_n\bar{\phi}_n(\bar{x})\right]e^{j\omega_i t}. \tag{27}$$

And

$$\bar{w}_{NS}(\bar{x},t)=\left[C_1\sinh\bar{\lambda}_i\bar{x}+C_2\cosh\bar{\lambda}_i\bar{x}+C_3\sin\bar{\lambda}_i\bar{x}+C_4\cos\bar{\lambda}_i\bar{x}\right]e^{j\omega_i t}. \tag{28}$$

Next, we can find the special solutions of the bimorph type piezoelectric cantilever beam from the electro- mechanical boundary conditions as follow:

$$\left[\bar{w}(0,t)=\bar{w}'(0,t)=0;\ \bar{w}''(1,t)=\left.\frac{M_1^E(\bar{x},t)}{D_{11}}\right|_{\bar{x}=1};\ \bar{w}'''(1,t)=\left.\frac{\partial M_1^E(\bar{x},t)}{D_{11}\partial\bar{x}}\right|_{\bar{x}=1}\right]_{MS}. \tag{29}$$

And

$$\left[\bar{w}(0,t)=\bar{w}'(0,t)=0;\ \bar{w}''(1,t)=\left.\frac{M_1^E(\bar{x},t)}{D_{11}}\right|_{\bar{x}=1};\ \bar{w}'''(1,t)=0\right]_{NS} \tag{30}$$

Or

$$\left[\bar{w}(0,t)=\bar{w}'(0,t)=0;\ \bar{w}''(1,t)=\left.\frac{\bar{\mu}_n\Pi_3(t)}{D_{11}}\frac{\partial^2\bar{\phi}_n(\bar{x})}{\partial\bar{x}^2}\right|_{\bar{x}=1};\ \bar{w}'''(1,t)=\left.\frac{\bar{\mu}_n\Pi_3(t)}{D_{11}}\frac{\partial^3\bar{\phi}_n(\bar{x})}{\partial\bar{x}^3}\right|_{\bar{x}=1}\right]_{MS}. \tag{31}$$

And

$$\left[\overline{w}(0,t)=\overline{w}'(0,t)=0;\ \overline{w}''(1,t)=\frac{C_{NS}\Pi_3(t)}{D_{11}};\ \overline{w}'''(1,t)=0\right]_{NS}. \tag{32}$$

where the dimensionless mode shape function of mode shape actuator under clamped-free boundary conditions as:

$$\overline{\phi}_n(\overline{x})=\cosh\overline{\lambda}_n\overline{x}-\cos\overline{\lambda}_n\overline{x}-\overline{\sigma}_n\left(\sinh\overline{\lambda}_n\overline{x}-\sin\overline{\lambda}_n\overline{x}\right). \tag{33}$$

And the dimensionless eigenvalues $\overline{\lambda}_n$ **and parameters** $\overline{\sigma}_n$ **of mode shape can be obtained from the following transcendental equation and formula:**

$$\cosh\overline{\lambda}_n\cos\overline{\lambda}_n+1=0. \tag{34}$$

And

$$\overline{\sigma}_n=\frac{\sinh\overline{\lambda}_n-\sin\overline{\lambda}_n}{\cosh\overline{\lambda}_n+\cos\overline{\lambda}_n}. \tag{35}$$

According to the electro-mechanical and clamped-free boundary conditions, we can determine the constants of Equation (27) and Equation (28), as follow:

$$C_1=-C_3=\frac{\left[B_1\left(\sin\overline{\lambda}_i-\sinh\overline{\lambda}_i\right)+B_2\left(\cosh\overline{\lambda}_i+\cos\overline{\lambda}_i\right)\right]}{2\left(1+\cosh\overline{\lambda}_i\cos\overline{\lambda}_i\right)}. \tag{36}$$

And

$$C_2=-C_4=\frac{\left[B_1\left(\cosh\overline{\lambda}_i+\cos\overline{\lambda}_i\right)-B_2\left(\sinh\overline{\lambda}_i+\sin\overline{\lambda}_i\right)\right]}{2\left(1+\cosh\overline{\lambda}_i\cos\overline{\lambda}_i\right)}. \tag{37}$$

where the constants of mode shape actuator are defined as

$$\left[B_1=\frac{\overline{\mu}_n\Pi_3}{D_{11}\overline{\lambda}_i^2}\left(1-\frac{\overline{\lambda}_n^4}{\overline{\lambda}_n^4-\overline{\lambda}_i^4}\right)\overline{\phi}_n''(1)\right]_{MS}. \tag{38}$$

And

$$\left[B_2=\frac{\overline{\mu}_n\Pi_3}{D_{11}\overline{\lambda}_i^3}\left(1-\frac{\overline{\lambda}_n^4}{\overline{\lambda}_n^4-\overline{\lambda}_i^4}\right)\overline{\phi}_n'''(1)\right]_{MS}. \tag{39}$$

where the second-order and third-order derivative mode shape function on the free boundary as follow:

$$\left[\overline{\phi}_n''(1)\right]_{MS}=\overline{\lambda}_n^2\left[\cosh\overline{\lambda}_n+\cos\overline{\lambda}_n-\overline{\sigma}_n\left(\sinh\overline{\lambda}_n+\sin\overline{\lambda}_n\right)\right]_{MS}. \tag{40}$$

And

$$\left[\bar{\phi}_n'''(1)\right]_{MS} = \bar{\lambda}_n^3\left[\sinh\bar{\lambda}_n - \sin\bar{\lambda}_n - \bar{\sigma}_n\left(\cosh\bar{\lambda}_n + \cos\bar{\lambda}_n\right)\right]_{MS}. \tag{41}$$

As for the constants of normal shape actuator were defined as

$$\left[B_1 = \frac{C_{NS}\Pi_3}{D_{11}\bar{\lambda}_i^2} = \frac{2\Pi_3}{D_{11}\bar{\lambda}_i^2}\right]_{NS}. \tag{42}$$

And

$$\left[B_2 = 0\right]_{NS}. \tag{43}$$

which let $C_{NS} = 2$ **for the function of the effective electrode surface of normal shape actuator.**

Until now, we can get a dimensionless general solution of the mode shape and normal shape actuator as follow:

$$\begin{aligned}
&\left[\bar{w}(\bar{x},t)\right]_{MS} = \left[\bar{W}(\bar{x})e^{j\omega_i t}\right]_{MS} \\
&= \frac{\left(1 - \frac{\bar{\lambda}_n^4}{\bar{\lambda}_n^4 - \bar{\lambda}_i^4}\right)\bar{\mu}_n\Pi_3 e^{j\omega_i t}}{2D_{11}\bar{\lambda}_i^2\left(1 + \cosh\bar{\lambda}_i\cos\bar{\lambda}_i\right)}\left\{\left(\sinh\bar{\lambda}_i\bar{x} - \sin\bar{\lambda}_i\bar{x}\right)\left[\bar{\phi}_n''(1)\left(\sin\bar{\lambda}_i - \sinh\bar{\lambda}_i\right) + \frac{\bar{\phi}_n'''(1)}{\bar{\lambda}_i}\left(\cosh\bar{\lambda}_i + \cos\bar{\lambda}_i\right)\right]\right. \\
&\left. + \left(\cosh\bar{\lambda}_i\bar{x} - \cos\bar{\lambda}_i\bar{x}\right)\left[\bar{\phi}_n''(1)\left(\cosh\bar{\lambda}_i + \cos\bar{\lambda}_i\right) - \frac{\bar{\phi}_n'''(1)}{\bar{\lambda}_i}\left(\sinh\bar{\lambda}_i + \sin\bar{\lambda}_i\right)\right]\right\} + \frac{\bar{\lambda}_n^4}{\bar{\lambda}_n^4 - \bar{\lambda}_i^4}\frac{\Pi_3 e^{j\omega_i t}}{\bar{\lambda}_n^2 D_{11}}\bar{\phi}_n(\bar{x})
\end{aligned} \tag{44}$$

And

$$\begin{aligned}
&\left[\bar{w}(\bar{x},t)\right]_{NS} = \left[\bar{W}(\bar{x})e^{j\omega_i t}\right]_{NS} \\
&= \frac{\Pi_3 e^{j\omega_i t}}{D_{11}\bar{\lambda}_i^2\left(1 + \cosh\bar{\lambda}_i\cos\bar{\lambda}_i\right)}\left[\left(\cosh\bar{\lambda}_i + \cos\bar{\lambda}_i\right)\left(\cosh\bar{\lambda}_i\bar{x} - \cos\bar{\lambda}_i\bar{x}\right) - \left(\sinh\bar{\lambda}_i - \sin\bar{\lambda}_i\right)\left(\sinh\bar{\lambda}_i\bar{x} - \sin\bar{\lambda}_i\bar{x}\right)\right]_{NS} \\
&= \frac{\Pi_3\left(\cosh\bar{\lambda}_i + \cos\bar{\lambda}_i\right)e^{j\omega_i t}}{D_{11}\bar{\lambda}_i^2\left(1 + \cosh\bar{\lambda}_i\cos\bar{\lambda}_i\right)}\left[\left(\cosh\bar{\lambda}_i\bar{x} - \cos\bar{\lambda}_i\bar{x}\right) - \frac{\sinh\bar{\lambda}_i - \sin\bar{\lambda}_i}{\cosh\bar{\lambda}_i + \cos\bar{\lambda}_i}\left(\sinh\bar{\lambda}_i\bar{x} - \sin\bar{\lambda}_i\bar{x}\right)\right]_{NS}
\end{aligned} \tag{45}$$

Due to the high frequency vibrations can easily be absorbed by the structure itself, so we are only interested in low-frequency vibration. And in order to understand the difference between the mode shape and normal shape actuators, we analyzed only for the first three resonant modes of the structures. And according to Equation (34) and Equation (35), we can get the eigenvalues and parameters of the first three modes, shown as Table 1. As the same time, we can get the dimensionless mode shape functions of the first three resonant modes by Equation (33) and Table 1, shown as Figure 2. And we can also get the dimensionless function of the effective electrode surface of first three modes when let Equation (33) take into Equation (7), shown as Figure 3. In addition, we can get a dimensionless transverse displacement of the mode shape and normal shape piezoelectric stator under steady state and the same driving conditions as shown in Figure 2.

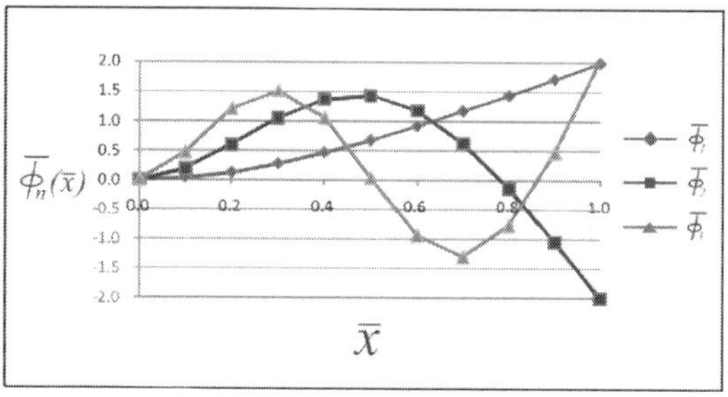

Figure 2. The dimensionless mode shape function of the first three modes of mode shape actuator relative to the unit length of structure when i = n.

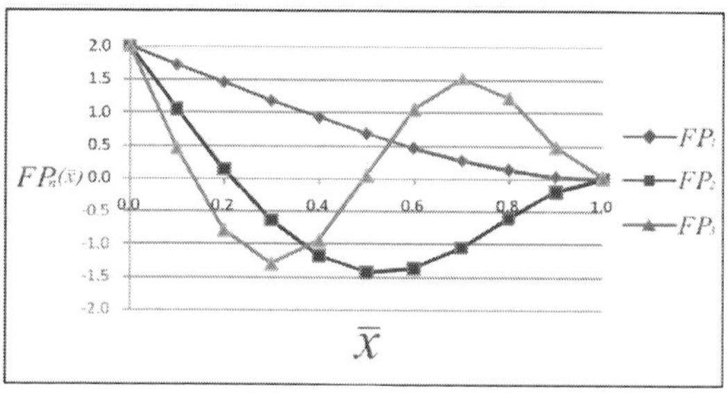

Figure 3. The dimensionless function of the effective electrode surface of the first three modes of mode shape actuator relative to the unit length of structure when i = n.

Table 1. The eigenvalues and parameters of the first three modes of mode shape actuator under clamped-free boundary conditions

Mode	λ_n	σ_n
1	1.8751	0.7341
2	4.6941	1.0185
3	7.8548	0.9992

Another according to Equation (40) and Equation (41), we can get the second and third derivative of mode shape function of the first three modes of mode shape actuator relative to the unit length of structure, shown as Figure 4 & Figure 5. And we can further that $\bar{\phi}''_n(1) = \bar{\phi}'''_n(1) = 0$ of the first three modes of mode shape actuator at free end. Therefore in, we can let Equation (44) further simplified as follows:

$$\left[\overline{w}(\overline{x},t)\right]_{MS} = \left[\overline{W}(\overline{x})e^{j\omega_i t}\right]_{MS} = \left[\frac{\overline{\lambda}_n^4}{\overline{\lambda}_n^4 - \overline{\lambda}_i^4}\frac{\Pi_3}{\overline{\lambda}_n^2 D_{11}}\overline{\phi}_n(\overline{x})e^{j\omega_i t}\right]_{MS}. \tag{46}$$

So far, Equation (46) of mode shape actuator appeared to be consistent with the particular solution of previously papers [1] [10] [11] .

Furthermore, we can make Equation (45) of normal shape actuator simplifies to

$$\left[\overline{w}(\overline{x},t)\right]_{MS} = \left[\overline{W}(\overline{x})e^{j\omega_i t}\right]_{MS} = \left[\frac{\overline{\lambda}_n^4}{\overline{\lambda}_n^4 - \overline{\lambda}_i^4}\frac{\Pi_3}{\overline{\lambda}_n^2 D_{11}}\overline{\phi}_n(\overline{x})e^{j\omega_i t}\right]_{MS}. \tag{47}$$

where the dimensionless mode shape function of normal shape actuator is defined as

$$\overline{\phi}_i(\overline{x}) = \cosh\overline{\lambda}_i\overline{x} - \cos\overline{\lambda}_i\overline{x} - \overline{\sigma}_i\left(\sinh\overline{\lambda}_i\overline{x} - \sin\overline{\lambda}_i\overline{x}\right);\ \text{where } i \neq n \text{ in normal state.} \tag{48}$$

And the dimensionless parameters $\overline{\sigma}_i$ **of normal shape actuator is defined as**

$$\overline{\sigma}_i = \frac{\sinh\overline{\lambda}_i - \sin\overline{\lambda}_i}{\cosh\overline{\lambda}_i + \cos\overline{\lambda}_i};\ \text{where } i \neq n \text{ in normal state.} \tag{49}$$

According to Equation (46) and Equation (47), we can get the ratio of dimensionless transverse displacement

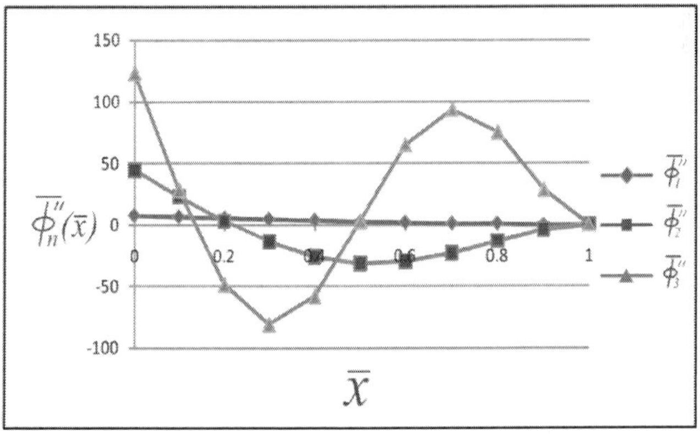

Figure 4. The second derivative of mode shape function of the first three modes of mode shape actuator relative to the unit length of structure when i = n.

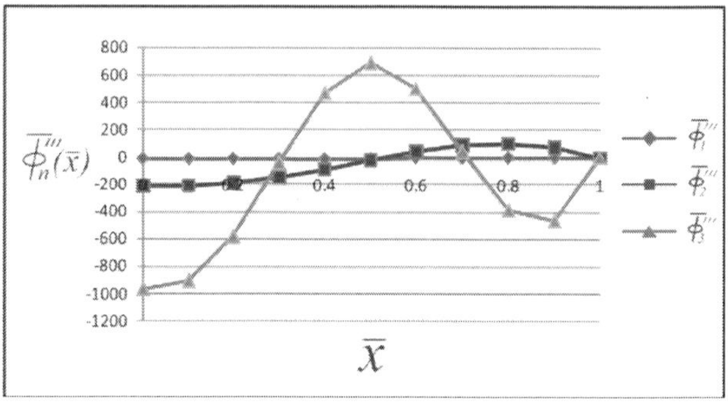

Figure 5. The third derivative of mode shape function of the first three modes of mode shape actuator relative to the unit length of structure when i = n.

of mode shape and normal shape actuator under conditions of steady state, constant driving voltage and the same bending stiffness constant per unit width as follows:

$$\left[\overline{W}(\overline{\lambda}_i)\right]_{MS}:\left[\overline{W}(\overline{\lambda}_i)\right]_{NS}=\left[\frac{1}{\overline{\lambda}_n^2}\frac{\overline{\lambda}_n^4}{\overline{\lambda}_n^4-\overline{\lambda}_i^4}\overline{\phi}_n(\overline{x})\right]_{MS}:\left[\frac{\left(\cosh\overline{\lambda}_i+\cos\overline{\lambda}_i\right)}{\overline{\lambda}_i^2\left(1+\cosh\overline{\lambda}_i\cos\overline{\lambda}_i\right)}\overline{\phi}_i(\overline{x})\right]_{NS}. \tag{50}$$

In order to facilitate understanding of the difference between mode shape and normal shape actuator, we can set the above ratio at the free end, that is, $\overline{\phi}_n(\overline{x})\big|_{\overline{x}=1}=\overline{\phi}_n(1)$ and $\overline{\phi}_i(\overline{x})\big|_{\overline{x}=1}=\overline{\phi}_i(1)$, shown as Figure 2. There fore Equation (50) can again be expressed as

$$\left[\overline{W}(\overline{\lambda}_i)\right]_{MS}:\left[\overline{W}(\overline{\lambda}_i)\right]_{NS}=\left[\frac{1}{\overline{\lambda}_n^2}\frac{\overline{\lambda}_n^4}{\overline{\lambda}_n^4-\overline{\lambda}_i^4}\overline{\phi}_n(1)\right]_{MS}:\left[\frac{\left(\cosh\overline{\lambda}_i+\cos\overline{\lambda}_i\right)}{\overline{\lambda}_i^2\left(1+\cosh\overline{\lambda}_i\cos\overline{\lambda}_i\right)}\overline{\phi}_i(1)\right]_{NS}. \tag{51}$$

SENSOR EQUATION

For the bimorph type piezoelectric cantilever beam under condition of constant electric potential, the sensor or current equation per unit length and width of the mode shape and normal shape sensor can be expressed as [12] :

$$I_p(\overline{x},t)=-e_{31}h\int_0^1\frac{\partial^3\overline{w}}{\partial\overline{x}^2\partial t}d\overline{x};\ p=MS\text{ or }NS. \tag{52}$$

And we can further take Equation (46) and Equation (47) into Equation (52) respectively, let Equation (52) be divided into two types of sensor or current equations as follow:

$$
\begin{aligned}
I_{MS}(\bar{x},t) &= -e_{31}h\int_0^1 \frac{\partial^3 \bar{w}_{MS}}{\partial \bar{x}^2 \partial t}\mathrm{d}\bar{x} = -e_{31}h\int_0^1 \frac{\partial^3\left[\frac{\bar{\lambda}_n^4}{\bar{\lambda}_n^4-\bar{\lambda}_i^4}\frac{\Pi_3\bar{\phi}_n(\bar{x})}{\bar{\lambda}_n^2 D_{11}}\mathrm{e}^{j\omega_i t}\right]_{MS}}{\partial \bar{x}^2 \partial t}\mathrm{d}\bar{x} \\
&= -\frac{j\omega_i e_{31}h\Pi_3 \mathrm{e}^{j\omega_i t}}{D_{11}}\frac{\bar{\lambda}_n^4}{\bar{\lambda}_n^4-\bar{\lambda}_i^4}\int_0^1 \frac{\partial^2\bar{\phi}_n(\bar{x})}{\bar{\lambda}_n^2\partial \bar{x}^2}\mathrm{d}\bar{x} \\
&= \frac{j\omega_i e_{31}h\Pi_3 \mathrm{e}^{j\omega_i t}}{D_{11}}\frac{1}{\bar{\lambda}_n}\left(\frac{\bar{\lambda}_n^4}{\bar{\lambda}_n^4-\bar{\lambda}_i^4}\right)\left[\bar{\phi}_n'(1)-\bar{\phi}_n'(0)\right]
\end{aligned}
\tag{53}
$$

And

$$
\begin{aligned}
I_{NS}(\bar{x},t) &= -e_{31}h\int_0^1 \frac{\partial^3 \bar{w}_{NS}}{\partial \bar{x}^2 \partial t}\mathrm{d}\bar{x} = -e_{31}h\int_0^1 \frac{\partial^3\left[\frac{\left(\cosh\bar{\lambda}_i+\cos\bar{\lambda}_i\right)\Pi_3\bar{\phi}_i(\bar{x})}{D_{11}\bar{\lambda}_i^2\left(1+\cosh\bar{\lambda}_i\cos\bar{\lambda}_i\right)}\mathrm{e}^{j\omega_i t}\right]_{NS}}{\partial \bar{x}^2 \partial t}\mathrm{d}\bar{x} \\
&= -\frac{j\omega_i e_{31}h\Pi_3\left(\cosh\bar{\lambda}_i+\cos\bar{\lambda}_i\right)\mathrm{e}^{j\omega_i t}}{D_{11}\left(1+\cosh\bar{\lambda}_i\cos\bar{\lambda}_i\right)}\int_0^1 \frac{\partial^2\bar{\phi}_i(\bar{x})}{\bar{\lambda}_i^2\partial \bar{x}^2}\mathrm{d}\bar{x} \\
&= -\frac{j\omega_i e_{31}h\Pi_3 \mathrm{e}^{j\omega_i t}}{D_{11}}\frac{1}{\bar{\lambda}_i}\left(\frac{\cosh\bar{\lambda}_i+\cos\bar{\lambda}_i}{1+\cosh\bar{\lambda}_i\cos\bar{\lambda}_i}\right)\left[\bar{\phi}_i'(1)-\bar{\phi}_i'(0)\right]
\end{aligned}
\tag{54}
$$

where the first derivative of dimensionless mode shape function of the first three modes can be obtained from clamped-free boundary conditions, or known from Figure 6. Therefore the ratio of dimensionless current of the mode shape and normal shape sensor dependent of eigenvalues under conditions of clamped-free, steady state, the same driving voltage and bending stiffness constant per unit width as follows:

$$
\bar{I}_{MS}(\bar{\lambda}_i):\bar{I}_{NS}(\bar{\lambda}_i) = \left[\frac{1}{\bar{\lambda}_n}\left(\frac{\bar{\lambda}_n^4}{\bar{\lambda}_n^4-\bar{\lambda}_i^4}\right)\bar{\phi}_n'(1)\right]_{MS} : \left[\frac{1}{\bar{\lambda}_i}\left(\frac{\cosh\bar{\lambda}_i+\cos\bar{\lambda}_i}{1+\cosh\bar{\lambda}_i\cos\bar{\lambda}_i}\right)\bar{\phi}_i'(1)\right]_{NS}.
\tag{55}
$$

where the first derivative of dimensionless mode shape function of the mode shape and normal shape sensor can be expressed as follow:

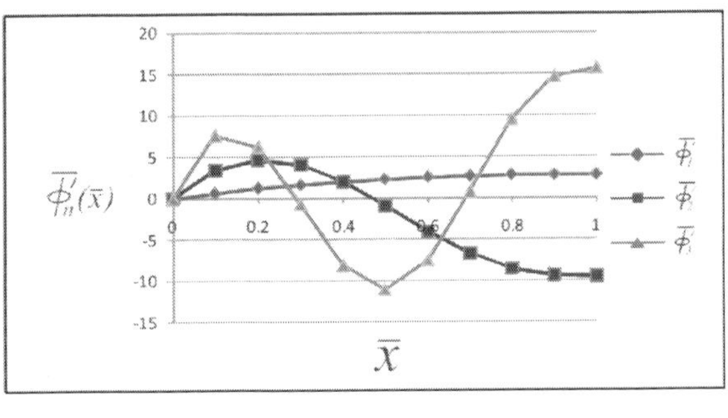

Figure 6. The first derivative of dimensionless mode shape function of the first three modes relative to the unit length of structure when i = n.

$$\left\{ \bar{\phi}_n'(1) = \bar{\lambda}_n \left[\sinh \bar{\lambda}_n + \sin \bar{\lambda}_n - \sigma_n (\cosh \bar{\lambda}_n - \cos \bar{\lambda}_n) \right];\ \bar{\phi}_n'(0) = 0 \right\}_{MS}. \tag{56}$$

And

$$\left\{ \bar{\phi}_i'(1) = \bar{\lambda}_i \left[\sinh \bar{\lambda}_i + \sin \bar{\lambda}_i - \sigma_i (\cosh \bar{\lambda}_i - \cos \bar{\lambda}_i) \right];\ \bar{\phi}_i'(0) = 0 \right\}_{NS}. \tag{57}$$

Since the current is proportional to voltage under condition of the same load R_L, So Equation (55) can also be expressed the ratio of dimensionless voltage of the mode shape and normal shape sensor as follows:

$$\bar{V}_{MS}(\bar{\lambda}_i) : \bar{V}_{NS}(\bar{\lambda}_i) = \left[\frac{1}{\bar{\lambda}_n} \left(\frac{\bar{\lambda}_n^4}{\bar{\lambda}_n^4 - \bar{\lambda}_i^4} \right) \bar{\phi}_n'(1) \right]_{MS} : \left[\frac{1}{\bar{\lambda}_i} \left(\frac{\cosh \bar{\lambda}_i + \cos \bar{\lambda}_i}{1 + \cosh \bar{\lambda}_i \cos \bar{\lambda}_i} \right) \bar{\phi}_i'(1) \right]_{NS}. \tag{58}$$

CASE STUDY: THEORY AND SIMULATION ANALYSIS

In order to understand the differences between mode shape and normal shape actuators and sensors by theory and simulation analysis, we specially design a series of one-dimension bimorph type piezoelectric cantilever beams, including the mode shape and normal shape actuators and sensors, shown as Figure 7.

Wherein the size and physical properties of the one-dimension bimorph type piezoelectric cantilever beam can refer to Table 2.

In addition, we will use the ANSYS code to simulate analysis the mode shape and normal shape actuators and sensors.

Where we have chosen the element type is the scalar tetragonal 98 of couple field.

Furthermore we will do the most sophisticated cutting of mesh, so that the analysis of the results can be quickly converged.

As for the analysis types, we were selected the modal, static and steady state.

Wherein the step of theoretical analysis is as follows:

(1-1) Using different frequency spacing $(\Delta f_i = 1\ \text{Hz},\ 10\ \text{Hz and } 100\ \text{Hz})$ to analyze the dimensionless transverse displacement and current or voltage of the first three modes of the mode shape and normal shape actuators and sensors from Equations (46) and (47) and Equations (52) and (53).

(1-2) Selecting the minimum frequency spacing $(\Delta f_i = 1\ \text{Hz})$ to analyze the ratio of the dimensionless transverse displacement and current or voltage and the mode shape function of the first three modes of the mode shape and normal shape actuators and sensors from Equation (50) and Equation (58).

As for the step of simulation analysis is as follows:

(2-1) Modeling of the mode shape and normal shape actuators and sensors respectively, including select element type, enter the physical properties, as well as coordinate system conversion, as shown Figure 8.

(2-2) Meshing of the mode shape and normal shape actuators and sensors respectively, including select the most sophisticated cutting of mesh or select the smart size 1, as shown Figure 9.

(2-3) Solving of the mode shape and normal shape actuators and sensors respectively, including setting boundary conditions of electro-mechanical, as shown Figure 10. All of which driving voltage on the effective electrode surface is 1.0 V.

(2-4) Post-processing of the mode shape and normal shape actuators and sensors respectively, includes processing the first three modes, the maximum deformation or electric potential, as shown Figures 11-13.

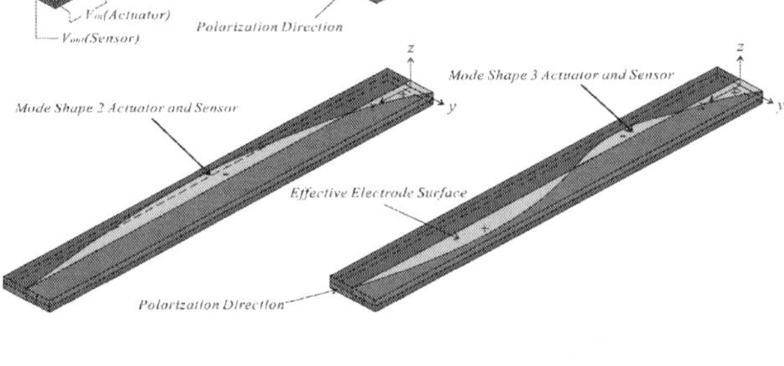

Figure 7. The schematic diagram of the mode shape and normal shape actuators and sensors.

Table 2. The size and physical properties of one-dimension bimorph type piezo-electric cantilever beam.

Physical Name	Physical Quantities
Size of Single Layer $(L \times W \times H)$	50 mm × 6 mm × 0.7 mm
Relative Permittivity (ε_{r})	1730
Piezoelectric Stress Constants (e_{31})	5.3 (N/V·m)
Young's Modulus (c_{11})	120 GPa
Density (ρ)	7600 (kg/m³)
Poisson Ratio (ν)	0.0

RESULTS AND DISCUSSION

According to the results of theory analysis, we found:

1) Under condition of the frequency spacing of 100 Hz, the maximum dimensionless transverse displacement of the second modal of the normal shape and mode shape actuators is smaller than the first and the third modals', as shown in Table 3 **and** Figure 14-(a-3). In addition, the maximum dimensionless voltage of the normal shape and mode shape sensors is proportional to the modal, as shown in Table 3 **and** Figure 14-(b-1).

2) Under condition of the frequency spacing of 10 Hz, the maximum dimensionless transverse displacement of the normal shape actuators is inversely proportional to the modal, as shown in Table 4 **and** Figure 14-(a-2). Furthermore, the maximum dimensionless transverse displacement of the second modal of the mode shape actuator is smaller than the first and the third modals', as shown in Table 4 **and** Figure 14-(a-2). As for the normal shape and mode shape sensors, the maximum dimensionless voltage is proportional to the modal un- der condition of the frequency spacing of 10 Hz, as shown in Table 4 **and** Figure 14-(b-2).

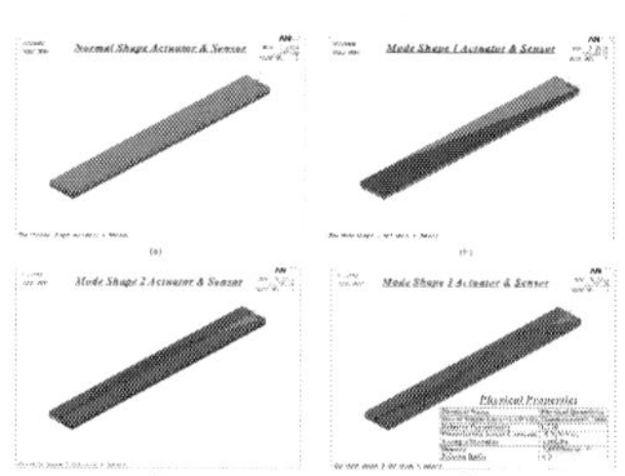

Figure 8. Modeling and setting physical properties of the mode shape and normal shape actuators and sensors. (a) The normal shape actuator and sensor; (b) The mode shape 1 actuator and sensor; (c) The mode shape 2 actuator and sensor; (d) The mode shape 3 actuator and sensor.

Table 3. The maximum dimensionless transverse displacement and voltage of the first three modes of the mode shape and normal shape actuators and sensors under condition of the frequency spacing of 100 Hz.

Mode	Actuators				Sensors			
	NSA (m/m)	MS1A (m/m)	MS2A (m/m)	MS3A (m/m)	NSS (V/V)	MS1S (V/V)	MS2S (V/V)	MS3S (V/V)
1	2.20	2.39	0.09	0.03	8.62	6.17	2.11	2.01
2	1.79	0.01	2.14	0.04	40.25	0.04	48.08	2.31
3	6.86	0.00	0.01	13.30	422.45	0.00	0.30	820.33

Table 4. The maximum dimensionless transverse displacement and voltage of the first three modes of the mode shape and normal shape actuators and sensors under condition of the frequency spacing of 10 Hz.

Mode	Actuators				Sensors			
	NSA (m/m)	MS1A (m/m)	MS2A (m/m)	MS3A (m/m)	NSS (V/V)	MS1S (V/V)	MS2S (V/V)	MS3S (V/V)
1	217.08	188.79	0.09	0.03	487.88	764.39	2.09	2.01
2	33.59	0.01	37.77	0.04	739.25	0.04	848.82	2.29
3	21.85	0.00	0.01	44.32	1348.82	0.00	0.30	2374.72

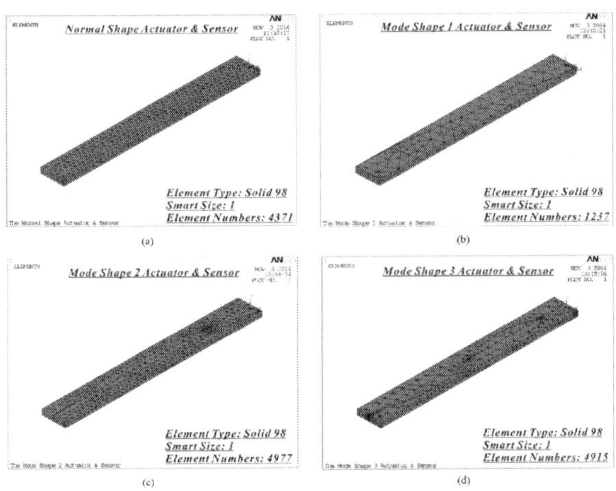

Figure 9. Meshing the mode shape and normal shape actuators and sensors. (a) The normal shape actuator and sensor; (b) The mode shape 1 actuator and sensor; (c) The mode shape 2 actuator and sensor; (d) The mode shape 3 actuator and sensor.

3) Under condition of the frequency spacing of 1 Hz, the maximum dimensionless transverse displacement of the second modal of the normal shape and mode shape actuators is larger than the first and the third modals', as shown in Table 5 **and** Figure 14-(a-1). As for the maximum dimensionless voltage of the normal shape and mode shape is proportional to the modal under condition of the frequency spacing of 1 Hz, as shown in Table 5 **and** Figure 14-(b-1).
4) Overall, in addition to the modal 1 or the first modal and under condition of the frequency spacing of 100 Hz, the gain or ratio of the second and third modal of the mode shape actuators and sensors are better than normal shape actuators and sensors, as shown in Table 6.
5) In terms of the resonance frequency, the approximate solutions through theoretical derivation are consistent with the analysis results by computer simulations, as shown in Table 7.
6) Under conditions of the simulation analysis of static, steady state and modal types, the first modal of the normal shape and mode shape actuators and sensors are the same gain or ratio, as shown in Table 8, Table 9 **and** Figure 15.
7) Furthermore, the gain or ratio of the second and third modal of the mode shape actuators and sensors are better than the normal shape actuators and sensors, as shown in Table 8, Table 9 **and** Figure 15.

CONCLUSION

According to the results of the theory and simulation analysis, on the whole, we found the first modal of the normal shape and mode shape actuators and sensors are the same gain.

Figure 10. Solving Process_ Setting boundary conditions of electro-mechanical the mode shape and normal shape actuators and sensors. (a) The normal shape actuator and

sensor; (b) The mode shape 1 actuator and sensor; (c) The mode shape 2 actuator and sensor; (d) The mode shape 3 actuator and sensor.

Table 5. The maximum dimensionless transverse displacement and voltage of the first three modes of the mode shape and normal shape actuators and sensors under condition of the minimum frequency spacing (**Δfi** = 1 Hz).

Mode	Actuators				Sensors			
	NSA (m/m)	MS1A (m/m)	MS2A (m/m)	MS3A (m/m)	NSS (V/V)	MS1S (V/V)	MS2S (V/V)	MS3S (V/V)
1	256.00	222.94	0.09	0.03	898.94	576.12	2.09	2.01
2	301.21	0.01	350.85	0.04	6637.87	0.04	7885.37	2.29
3	138.78	0.00	0.01	333.99	8562.97	0.00	0.30	20606.27

Table 6. The gain or ratio of dimensionless transverse displacement and voltage of the mode shape and normal shape actuators and sensors relative to the frequency under condition of different frequency spacing (**Δfi** = 1 Hz, 10 Hz & 100 Hz).

Frequency Spacing	Actuators			Sensors		
	MS1A. NS1A (m/m)	MS2A. NS2A (m/m)	MS3A. NS3A (m/m)	MS1S. NS1S (V/V)	MS2S. NS2S (V/V)	MS3S. NS3S (V/V)
1 Hz	0.87	1.16	2.40	0.64	1.19	2.41
10 Hz	0.87	1.12	2.03	0.64	1.15	2.03
100 Hz	1.08	1.20	1.94	0.72	1.20	1.94

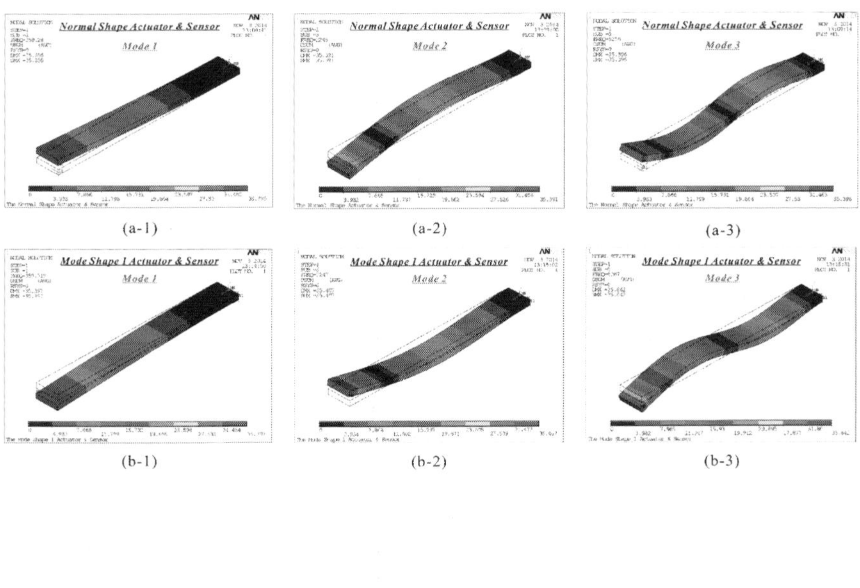

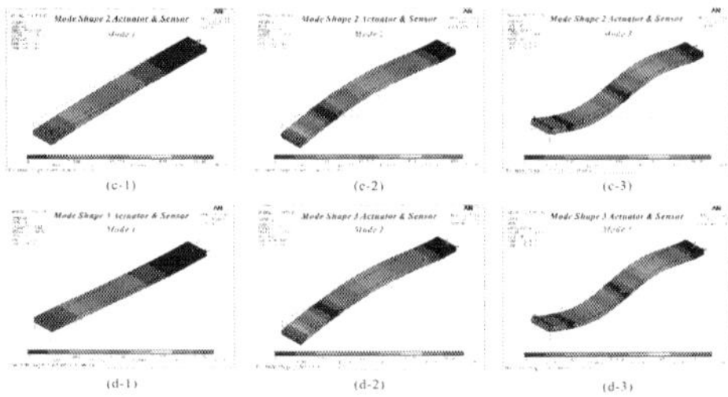

Figure 11. Post-processing the first three modes and the maximum deformation of the mode shape and normal shape actuators and sensors. (a) The first three modes of the normal shape actuator and sensor; (b) The first three modes of the mode shape 1 actuator and sensor; (c) The first three modes of the mode shape 2 actuator and sensor; (d) The first three modes of the mode shape 3 actuator and sensor (analysis type: modal).

Or in other words, the design concept of the first modal of mode shape actuator or sensor is not necessarily better than of the first modal of normal shape actuator or sensor, or even worse. However, the gain of the second and third modal of the mode shape actuators and sensors are better than the normal shape actuators and sensors under any operating states.

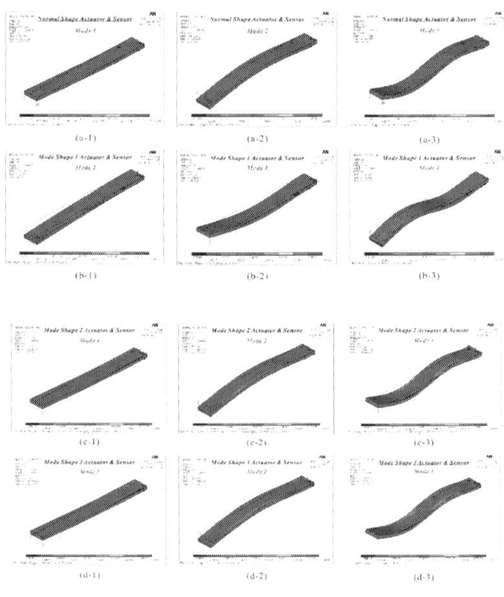

Figure 12. Post-processing the first three modes and the maximum electric potential of the mode shape and normal shape actuators and sensors. (a) The first

three modes of the normal shape actuator and sensor; (b) The first three modes of the mode shape 1 actuator and sensor; (c) The first three modes of the mode shape 2 actuator and sensor; (d) The first three modes of the mode shape 3 actuator and sensor (analysis type: modal).

Most importantly, we have to find a universal theoretical model and approximate solutions in this paper, which can predict the gain of the first three modes of normal shape and mode shape actuators and sensors, and through the analysis results of computer simulation to confirm.

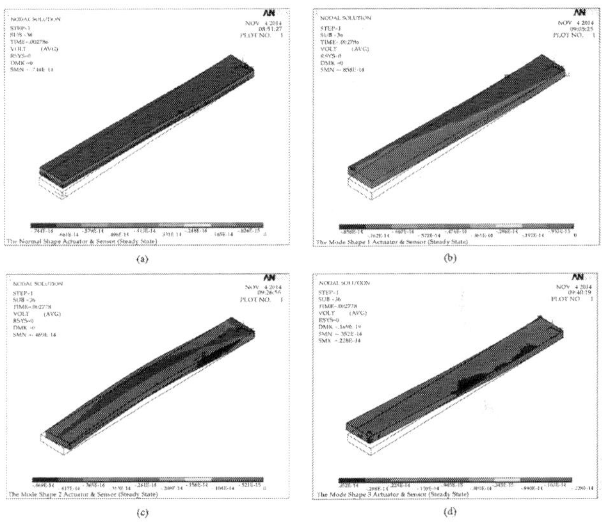

Figure 13. Post-processing the electric potential output state of the mode shape and normal shape actuators and sensors. (a) The normal shape actuator and sensor; (b) The mode shape 1 actuator and sensor; (c) The mode shape 2 actuator and sensor; (d) The mode shape 3 actuator and sensor (analysis type: steady state).

Table 7. The resonance frequency and the ratio of resonance frequency of the first three modes of the mode shape and normal shape actuators and sensors by theory and simulation analysis (analysis type: modal).

Mode	Theory (Hz)	N S (Hz)	M S 1 (Hz)	M S 2 (Hz)	M S 3 (Hz)	N S Theory	M S 1 Theory	M S 2 Theory	M S 3 Theory
1	359	359	359	360	360	1.000	1.000	1.003	1.003

2	2253	2245	2247	2247	2247	1.001	0.999	0.999	0.999
3	6308	6256	6267	6262	6264	0.992	0.994	0.993	0.993

Table 8. The maximum deformation and the ratio of maximum deformation of the mode shape and normal shape actuators by simulation analysis (analysis type: static & steady state).

Analysis Type	NSA (m)	MS1A (m)	MS2A (m)	MS3A (m)	MS1A. NSA (m/m)	MS2A. NSA (m/m)	MS3A. NSA (m/m)
Static & Steady State	5.18E−14	−1.52E−13	−9.57E−11	−1.74E−10	2.93	1846.88	3353.80

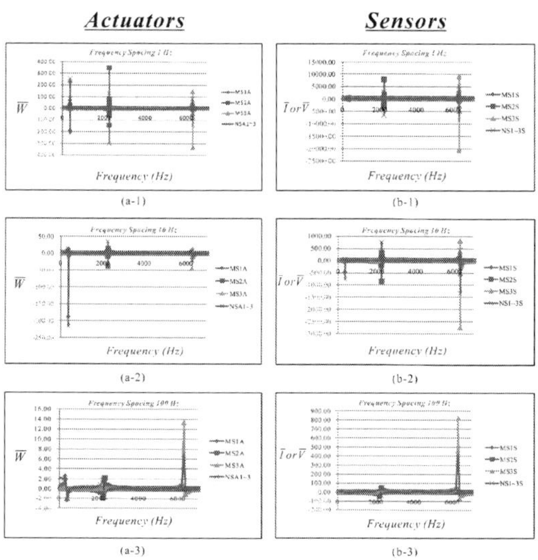

Figure 14. (a-1)-(a-3) The dimensionless transverse displacement $(\bar{W})$ of the mode shape and normal shape actuators relative to the frequency under condition of different frequency spacing; (b-1)-(b-3) The dimensionless current $(\bar{I})$ or voltage $(\bar{V})$ of the mode shape and normal shape sensors relative to the frequency under condition of different frequency spacing.

Table 9. The maximum sensing or output voltage and the ratio of maximum sensing or output voltage of the first three modes of the mode shape and normal shape sensors by simulation analysis (analysis type: modal).

Mode	NSS (V)	MS1S (V)	M S 2 S (V)	M S 3 S (V)	MS1S:NSS (V/V)	MS2S:NSS (V/V)	MS3S:NSS (V/V)
1	9	45	643	686	4.82	68.70	73.37
2	330	1506	25131	26925	4.56	76.13	81.56
3	2325	11946	194692	209525	5.14	83.75	90.13

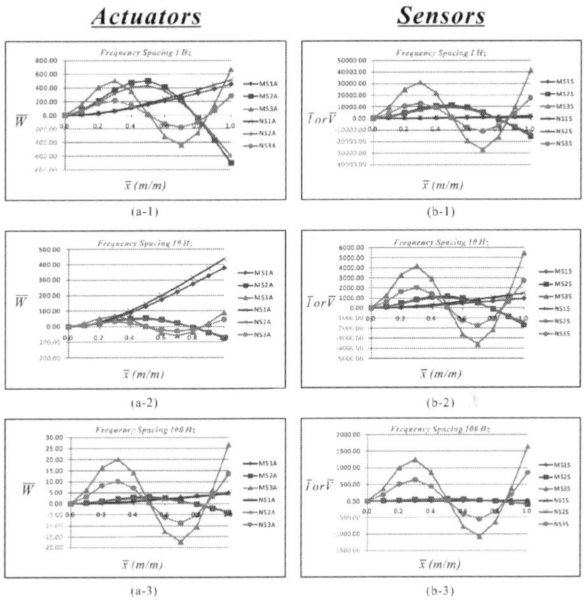

Figure 15. (a-1)-(a-3) The maximum dimensionless transverse displacement $(\overline{W})$ of the first three modes of the mode shape and normal shape actuators relative to the dimensionless length under condition of different frequency spacing; (b-1)-(b-3) The maximum dimensionless current $(\overline{I})$ or voltage $(\overline{V})$ of the first three modes of the mode shape and normal shape sensors relative to the dimensionless length under condition of different frequency spacing.

ACKNOWLEDGEMENTS

Jwo Ming Jou (Project No.: MOST103-2221-E-230-007).

REFERENCES

1. Lee, C.-K. (1987) Piezoelectric Laminates for Torsional and Bending Modal Control: Theory and Experiment. Ph.D. Dissertation, Cornell University, Ithaca.
2. Chiang, W.-W. and Lee, C.-K. (1989) Critical Active Damping Control of a Flexible Slender Plate Using a Distributed Modal Actuator and Sensor. American Control Conference, 700-706.
3. Vaz, A.F. and Farr, T.J. (1997) Experimental Validation of Analytical Models of Flexible Structures with Bonded Piezoelectric Film Actuators and Sensors. IEEE Instrumentation and Measurement Technology Conference, 2, 887-892.
4. Vat, A.F. (1998) Composite Modeling of Flexible Structures with Bonded Piezoelectric Film Actuators and Sensors. IEEE Transactions on Instrumentation and Measurement, 47, 513-520. http://dx.doi.org/10.1109/19.744200
5. Takagi, K., Nagase, K., Oshima, K. and Hayakawa, Y. (2001) Modal Filtering for Simply Supported Plate Using Equally Segmented Piezoelectric Film. American Control Conference, 2, 1593-1598.
6. Ballato, A. and Kim, Y.K. (2002) Displacements and Rotations of Practical Vibrational Modes of Piezoelectric Bimorph Cantilever Beams. Proceedings of IEEE Sensors, 2, 1294-1297.
7. Gou, X.K. and Tian, H.M. (2007) Active Vibration Control of a Cantilever Beam Using Bonded Piezoelectric Sensors and Actuators. 8th International Conference on Electronic Measurement and Instruments, 4, 85-88.
8. Rios-Gutierrez, M. and Silva-Navarro, G. (2009) Finite Element and Modal Modeling of a Cantilever Beam with Piezoelectric Patch Actuator for Vibration Absorption. 6th International Conference on Electrical Engineering Computing Science and Automatic Control, 1-6.
9. Sanchez-Rojas, J.L., Hernando, J., Donoso, A., Bellido, J.C., Gutierrez, L.M., Ababneh, A., Seidel, H. and Schmid, U. (2009) Piezoelectric Modal Sensors/Actuators Based on Microplates Applying Surface Electrode Patterning. Solid-State Sensors, Actuators and Microsystems Conference, 2346-2349.
10. Hsu, Y.-H., Lee, C.-K. and Hsiao, W.-H. (2005) Electrical and Mechanical Fully Coupled Theory and Experimental Verification

of Rosen-Type Piezoelectric Transformers. IEEE Ultrasonics, Ferroelectrics and Frequency Control, 52, 1829-1839. http://dx.doi.org/10.1109/TUFFC.2005.1561639

11. Huang, Y.-T., Wu, W.-J., Wang, Y.-C. and Lee, C.-K. (2007) Multilayer Modal Actuator-Based Piezoelectric Transformers. IEEE Ultrasonics Ferroelectrics and Frequency Control, 52, 359-367. http://dx.doi.org/10.1109/TUFFC.2007.249
12. Jou, J.M. (2003) Piezoelectricity Mechanics. Chuan Hwa Book Co. Ltd., Taipei.

CHAPTER

17

Experimental Study of the Response of Transonic Diffuser Flow to a Piezoceramic Actuator at Diffuser Throat

Minoru Yaga[1], Yusuke Uechi[2], Hiroaki Ozono[3], Masaaki Ishikawa[1], Isao Teruya[1]

[1]Department of Mechanical Systems, Faculty of Engineering, University of the Ryukyus, Okinawa, Japan

[2]Mie Metal Industry Co. Ltd., Mie, Japan

[3]Graduate School of Engineering and Science, University of the Ryukyus, Okinawa, Japan

ABSTRACT

An experimental study of the response of a piezoceramic actuator set at the throat to a transonic diffuser is carried out by measuring wall static pressure fluctuations and by visualizing the flow field using schlieren technique. The

Citation:- M. Yaga, Y. Uechi, H. Ozono, M. Ishikawa and I. Teruya, "Experimental Study of the Response of Transonic Diffuser Flow to a Piezoceramic Actuator at Diffuser Throat," Open Journal of Fluid Dynamics, Vol. 3 No. 2A, 2013, pp. 14-21. doi: 10.4236/ojfd.2013.32A003.

visualized flow fields are captured with a digital still camera and a digital high speed video camera. The piezo ceramic actuator is attached at the throat of the diffuser and driven by sinusoidal amplified voltage signals. The diffuser used in this experiment is circular arc half nozzle with the height h* and width w of 3 mm and 25 mm, respectively. The blockage factor of the piezoceramic actuator to the diffuser throat is 9.2% assuring the effect of change in the throat area rather than the boundary layer disturbances. The piezoceramic actuator is driven at the frequency of 100 Hz, 200 Hz, and 300 Hz and its amplitude is about 1 mm. It is found that the wall static pressure fluctuations and the behavior of the shock wave clearly correspond to the vibration of the piezo ceramic actuator for all the frequency ranges whereas the averaged shock position remains almost unchanged. All the results mentioned above suggest that driving the piezo ceramic actuator at the diffuser throat can be one of the promising techniques to control unsteady transonic diffuser flow.

Keywords: Compressible Flow; Shock Wave; Transonic Diffuser; Piezoceramic Actuator; Throat

INTRODUCTION

The unsteady flow field in a transonic diffuser has attracted a great deal of interest not only because of the practical industrial importance but also because of the complexity of the flow itself. It is well known that a shock wave in a transonic flow is basically unsteady due to the interaction between a shock wave and other flow phenomena, i.e., the shock wave oscillation is triggered and maintained by a local interaction between the shock foot and the boundary later developed along the wall surface or upstream and downstream propagating disturbances. Meier [1] reported that shock-induced separation causes the large-amplitude, unstable oscillation of a shock wave. In addition to the shock-induced separation, the effects of propagating disturbances toward the shock waves on the oscillations have been analytically, numerically and experimentally investigated [2,3], including forced oscillations by some functions or rotating rods as a forcing function to drive the shock waves. According to previous studies, several factors affect unsteady shock behavior in transonic diffusers. Therefore, in addition to revealing the causes of the shock oscillations, there have also been attempts to eliminate or reduce these unfavorable unsteady shock oscillations both for internal and external devices. It is quite difficult to deal with these types of oscillations unless the disturbances around the shock wave are attenuated or canceled by some techniques because the longitudinal and boundary layer associated disturbances cannot be separated.The approach to stabilize the unsteady flow fields is divided into two major controls, which are active control with jets [4,5] and passive control with a porous cavity [6] or vortex generators. These controls are expected to generate the opposite phase of signals or to retard separations of boundary layers. However, the causes of the oscillations include pressure disturbances in a core flow and unsteady phenomena associated with the shock boundary layer interaction, as mentioned above, the future trend for the oscillation controls might be a combination of passive and active controls.

Although the effect of the pressure disturbances in a core flow and disturbances associated with the boundary layer should be considered individually in order to clarify the factors affecting the oscillations, both effects cannot be divided because they interact with each other. Thus, in the present paper, we focus primarily on the mechanical disturbances by changing cross sectional area at the throat of a transonic diffuser, which is expected to affect the location of shock waves according to the isentropic flow relations between the throat area and local cross sectional area at a location of the shock wave.

EXPERIMENTAL APPARATUS AND PROCEDURE

Wind Tunnel Facility

Figures 1(a)-(c) show the wind tunnel, the detail of the test section, and the driving circuit of the piezoceramic actuator, respectively. The experimental apparatus consists of a 0.7 MPa compressor, a settling chamber, a regulator valve, and a circular arc transonic half diffuser in a blow down wind tunnel discharging to the atmosphere. The throat height of the diffuser and its radius are $h^* = 3$ mm and $R = 500$ mm, respectively. The span of the diffuser is 33 mm. The flow field in the diffuser is visualized by the schlieren technique and captured with a digital still camera or a digital high-speed video camera, which enables images of the flow field to be captured at a rate of 41,000 frames per second. Unsteady wall static pressure fluctuations are measured using semiconductor pressure sensors at the settling chamber, at the throat, and at positions $x = 20$, 30, 40, and 50 mm downstream of the throat. The signals from the pressure transducer are digitized at a sampling frequency of 10 kHz with 16-bit accuracy and are analyzed to evaluate the effect of the piezoceramic actuator on the diffuser flow fields. The resonance frequency of the transducers is 50 kHz assuring the accuracy of monitoring the pressure fluctuations. The total pressure at the settling chamber is monitored during the experiments by a computer, which controls the timing of the pressure measurements and flow visualizations.

Piezoceramic Actuator

The piezoceramic actuator is set at the diffuser throat, as shown in Figure 1(b). In order to achieve sufficient displacement by the actuator, a bimorph type piezoceramic actuator is adopted and driven by the circuit as shown Figure 1(c).

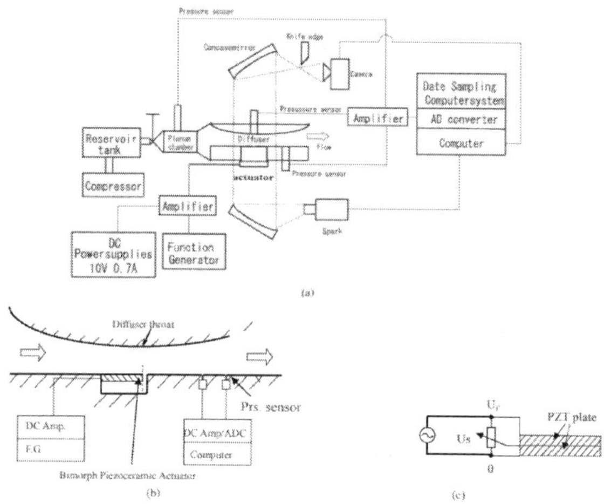

Figure 1. Experimental apparatus: (a) Wind tunnel and measurement system; (b) Detail of test section; (c) Driving circuit for piezoceramic actuator.

The length, width, and thickness of the piezoceramic actuator are 33 mm, 11 mm, and 0.8 mm, respectively. During the experiment, the piezoceramic actuator is connected to a DC type amplified voltage source and then bent in order to change the cross sectional area at the diffuser throat. The maximum input voltage from the signal source, displacement, and natural resonance frequency of the piezoceramic actuator are 60 Vp, 0.7 mm, and 400 Hz, respectively. Accordingly, the exciting input frequency to the piezoceramic actuator is limited to 300 Hz. From preliminary experiments, the displacement of the piezoceramic actuator is found to be sufficiently large compared to that with a layer type piezoceramic actuator. In this report, only sinusoidal signals are input into the DC amplifier to achieve simplicity of the frequency analyses.

EXPERIMENTAL RESULTS AND DISCUSSION

Results of Flow Visualization

Figures 2(a)-(d) show typical schlieren images of the flow fields and shock positions for the input frequency of f_p = 200 Hz. It is clear that some quite weak Mach waves are generated near the throat due to the effect of the moving piezoceramic actuator. At the same time, at a pressure ratio of p_0/p_b = 1.20, the starting shock wave appears to be very weak near x/h* = 6.0, which is followed by some weak shock waves due to the interaction between the shock wave and the boundary layers along the diffuser walls. Although the flow field downstream of the normal shock wave is supposed to be subsonic, Figure 2(a) and subsequent figures illustrate the occurrence of the shock wave downstream of the first normal shock wave.

The second shock wave or multiple shock waves are generated by the acceleration of the main flow due to the change in the effective cross sectional area of the diffuser and the disturbances approaching from downstream of the shock wave.

As reported in several previous studies, the shock wave near the throat is unstable, so that the shock wave observed in Figure 2(a) change its position, as will be discussed later herein. Figure 2(b) shows that the slight increase in the pressure ratio causes the downward displacement of the shock wave. Figure 2(c) illustrates the lambda-foot type shock wave on the upper and lower walls due to a typical moderate interaction between the shock wave and the boundary layer. Other shock waves downstream of the first shock position also correspond to the varying throat height.

Then, the distance between the throat and the averaged shock positions increases with the pressure ratio. Figures 2(c) and (d) show that the clear shock waves cause boundary layer separation followed by multiple shock waves downstream of the interaction region.

Shock Locations for Various Frequencies

According to Figure 2, the shock waves appear to move gradually in the previous section. Then, it is necessary to check the relation between rough locations and shock waves for various input frequencies of the piezoceramic actuator. Then, their positions are plotted with respect to the pressure ratio and are shown in Figure 3 as a parameter of the input frequencies of 0 Hz, 100 Hz, 200 Hz, and 300 Hz.

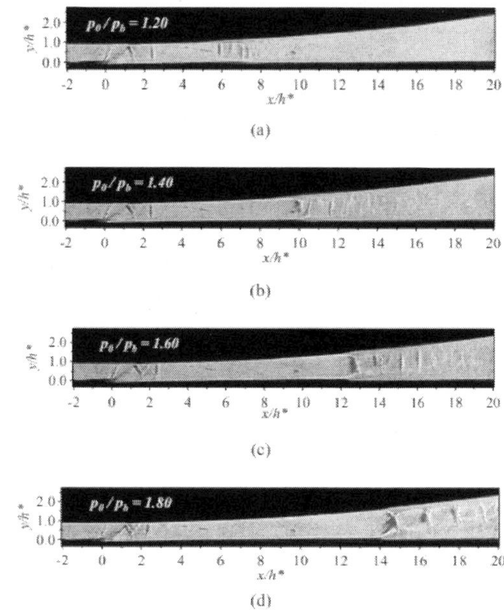

Figure 2. Schlieren images for frequency 200 Hz: (a) $p_0/p_b = 1.2$; (b) $p_0/p_b = 1.4$; (c) $p_0/p_b = 1.6$; (d) $p_0/p_b = 1.8$.

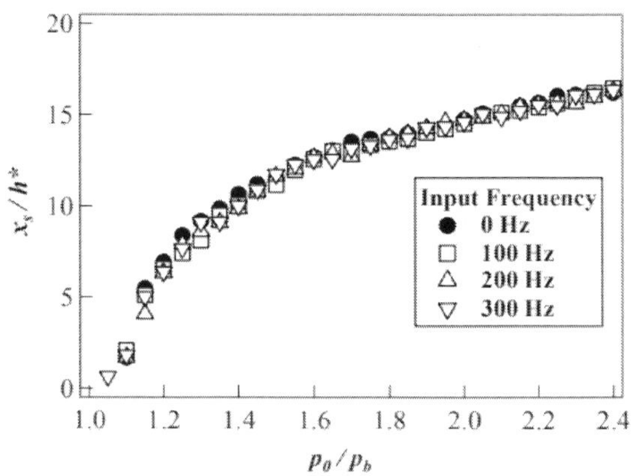

Figure 3. Relation between shock wave position and pressure ratio.

Figure 3 shows the relation between the positions of shock waves x_s divided by throat height h^* and the pressure ratio p_0/p_b the shock locations are measured in the schlieren images for various input frequencies of 0 Hz through 300 Hz. The shock wave is found to move monotonically downstream with the increase in the pressure ratio. However, the gradient of the shock displacement to the pressure ratio decreases with the pressure ratio because the shock-induced separation causes displacement delay. Considering the effect of the piezoceramic actuator, it is easily expected that the location of shock waves corresponds to the throat height because the Mach number just upstream of the shock wave depends on the ratio of the local cross sectional area to the throat area.

From the viewpoint of small-scale observation, the shock wave oscillates around its averaged position at the fixed frequency of the piezoceramic actuator, which will be discussed in the following section based on images captured by a high-speed video camera. The behavior of the shock wave is critical to the generation of large pressure fluctuations under the various pressure ratios.

Wall Static Pressure Fluctuation

Figures 4(a)-(d) show the time series pressure fluctuations at the throat x/h* = 0.0 and at x/h* = 10.0 for a frequency of 200 Hz. The wall static pressure pw divided by the back pressure pb at the throat exhibits a clear sinusoidal time history, which is considered to correspond to the behavior of the piezoceramic actuator.

On the other hand, the pressure fluctuation at x/h* = 10.0 indicates a severe large amplitude fluctuation. The large amplitude fluctuation suggests that the shock wave passes back and forth over the measurement position corresponding to the motion of the piezoceramic actuator. The increase in amplitude can be explained by Figure 3, which shows that the shock wave is located around x/h* = 10.0. Then, in case of a pressure ratio of 1.50, the shock wave is expected to stand slightly downstream of the pressure measurement position at x/h* = 10.0.

The influence of the shock wave could reach the pressure measurement position through the subsonic layer in the boundary layer which develops along the diffuser wall.

The pressure remains relatively low compared to that shown in Figure 4(a), which indicates that the shock wave stands further downstream than that for $p_0/p_b = 1.45$. However, the effect of the piezoceramic actuator can be still detected. Finally, in the case of $p_0/p_b = 1.55$, the shock wave moves far downstream of the pressure measurement position.

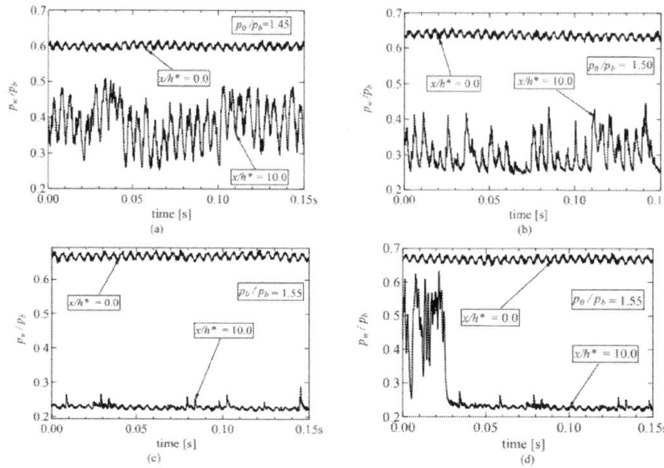

Figure 4. Wall static pressure fluctuations at throat and x/h* = 10.0 for input frequency 200Hz: (a) $p0/p^b = 1.45$; (b) $p^0/p^b= 1.50$; (c) $p^0/p^b = 1.55$; (d) $p^0/p^b= 1.55$.

This is also confirmed in Figure 3, which showing xs/h* = 10.0 at a pressure ratio of 1.55. Both the fluctuating and averaged pressure are much lower due to the downstream displacement of the pressure measurement position. These figures contain information about the shock position and the shock wave behavior. In other words, Figure 4(a) shows the severe pressure fluctuation at x/h* = 10.0, whereas the fluctuation at the throat does not. This means that the shock wave stands between the throat and the measurement position at x/h* = 10.0. As the pressure ratio increases, the fluctuation at x/h* = 10.0 gradually decreases the pressure fluctuation, achieving the smallest pressure fluctuation, as shown in Figure 4(c). The pressure fluctuation as shown Figure 4(c) denotes the supersonic state at x/h* = 10.0.

However, the sudden change in the average pressure was observed as shown in Figure 4(d) under the same condition, which is caused by the flow separation downstream of the shock wave. The increase in the pressure fluctuation indicates the shock wave displacement upstream due to the flow separation. However, sometime after the increase in pressure, the average value and fluctuation decrease, denoting the shock wave located downstream of the position at x/h* = 10.0. The pressure fluctuations at the throat are found to oscillate sinusoidally for all pressure ratios. Figure 4(d) also reveals that the wall pressure fluctuation at a pressure ratio of 1.55 suddenly changes its level at time t = 0.03 s, which suggests that the flow has just recovered from an unexpected separation downstream of the shock wave.

Root Mean Square of the Pressure Fluctuation

It is well known that wall static pressure fluctuations greatly decrease when a flow becomes supersonic.

In other words, monitoring the wall static pressure fluctuations clarifies whether the flow is supersonic or subsonic.

In the previous section, the pressure fluctuations also indicate the rough shock wave positions.

Then, for a more quantitative evaluation of the pressure fluctuations, the root mean square of pressure fluctuations is one of the indices of the shock wave position and its behavior.

Figure 5(a) shows that the root mean square *p*rms/*p*b for the input frequency of *fp* = 0 Hz at the throat shows almost no increase in *p*rms/*p*b with the increase in the pressure ratio *p*0/*p*b. On the other hand, activation occurs at the throat, the rms increases with the pressure ratio.

This indicates that the piezoceramic actuator always affects the pressure fluctuation at the throat. When the rms reaches approximately zero, the state of the measurement position is considered to be supersonic. This indicates that the hock wave is located downstream of the measurement position.

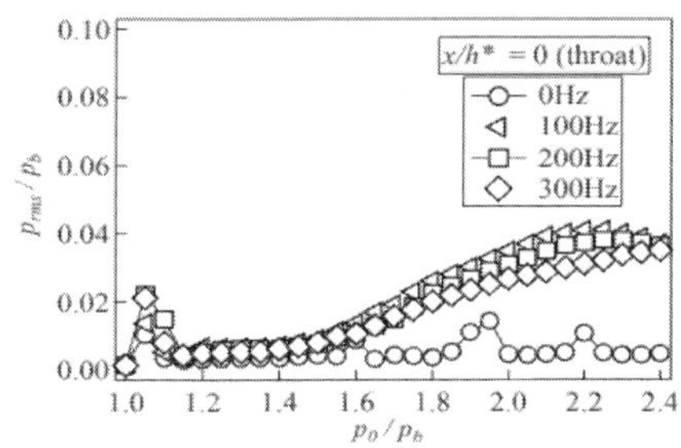

(a)

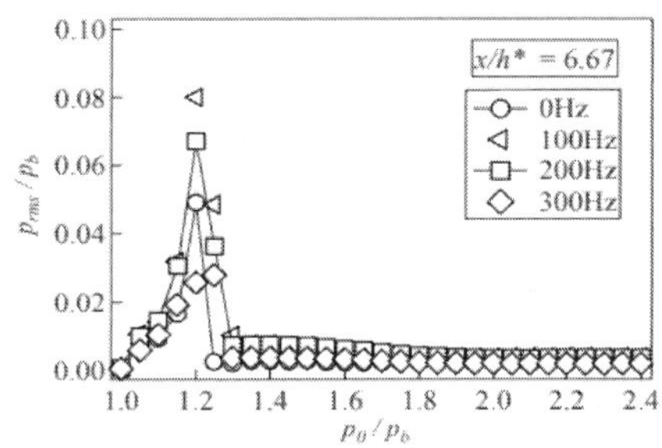

(b)

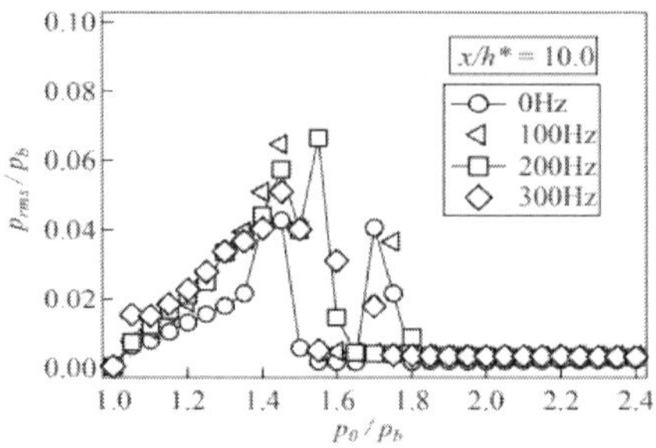

(c)

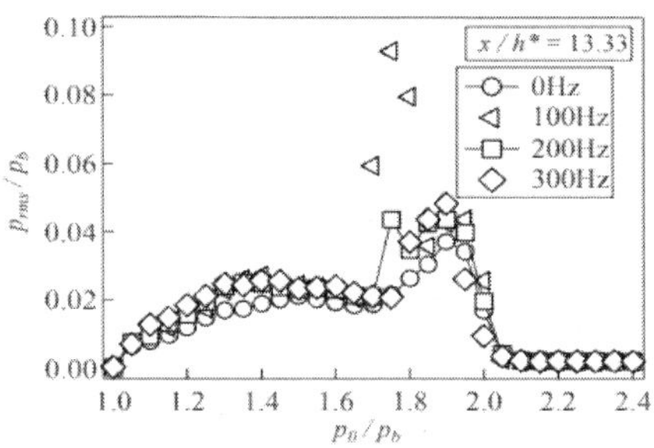

(d)

Figure 5. RMS of wall static pressure fluctuations: (a) x/h* = 0.0; (b) x/h* = 6.67; (c) x/h* = 10.0; (d) x/h* = 13.33.

In Figure 5(b), the clear peaks at a pressure ratio p^0/p^b of approximately 1.2 show that a single and relatively weak shock wave appears and moves downstream with the increase in the pressure ratio. The sudden drop in the rms indicates the completion of the process from the subsonic-to-supersonic transition due to the downstream displacement of the shock wave. This process is independent of the input frequencies of the piezoceramic actuator.

Figure 5(c) shows that the rms at a pressure ratio of approximately 1.7 sudden creases, which is caused by the unexpected separation of a boundary layer, as mentioned in the previous section. This also indicate the shock wave existence around the position of x/h* = 10.0. However, for a pressure ratio greater than 1.8, the rms for all input frequencies becomes approximately zero, which suggests that the flow is completely supersonic. Note that the source of the signal originates at the throat, and then even if the flow measurement position is supersonic, the signal from the throat can be detected. Figure 5(d) indicates the same variations that is all the rms for each input frequency suddenly decrease at the pressure ratio at 2.0 denoting the supersonic sta this postion.

FFT Analyses of Wall Pressure Fluctuation

The rms of the pressure fluctuation usually becomes large when the oscillating shock wave approaches the monitoring position. It is also important to examine the unsteady behavior of the shock wave in detail, especially in this case, in order to confirm the response of the flow field to the piezoceramic actuator.

One of the best ways to evaluate the effect of the actuation is to check the contributions of every frequency by means of FFT analysis of the pressure fluctuations at x/h* = 6.67. Figures 6(a)-(d) show the results of the FFT analysis of the pressure fluctuations for each pressure ratio. Figure 6(a) shows no dominant frequency and almost no frequency output for pressure ratios greater than $p^0/p^b = 1.25$, as deduced from the results for the rms shown in Figure 5(b). This indicates that the shock wave oscillates with no dominant frequency for all of the pressure ratios and that the state of the flow at x/h* = 6.67 becomes supersonic at pressure ratios greater than 1.25. Moreover, the levels of each spectrum are not so large because the shock wave at this position is not so strong.

On the other hand, Figures 6(b)-(d) clearly show the dominant frequencies for each pressure ratio, which corresponds to the input frequency to the piezoceramic actuator. Note that even in the supersonic state, the frequency of 200 Hz can be observed. These figures indicate the effect of the piezoceramic actuator on the flow fields. In addition to the results of the visualization, monitoring the wall static pressure fluctuation is a quantitative way to confirm the effect of the actuator.

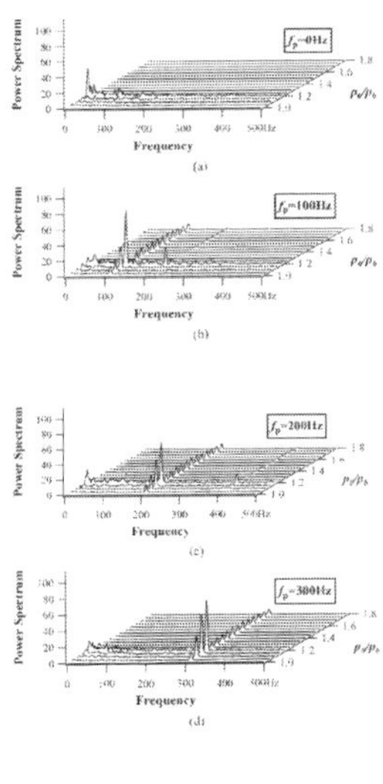

Figure 6. FFT analysis of wall static pressure fluctuations at *x/h** = 6.67: (a) *fp* = 0 Hz; (b) *fp* = 100 Hz; (c) *fp* = 200 Hz; (d) *fp* = 300 Hz.

Figures 7(a)-(d) show the FFT analysis of the pressure fluctuation at x/h* = 10.0 for the same input frequency as that in Figure 6. Figure 7(a) shows no dominant frequency but only a relatively low frequency, which has the same tendency as that in Figure 6(a)

Figures 7(b)-(d) show cleominant frequencies. This indicates the possibility of reducing the pressure fluctuation by controlling the shock position, which is expected to cancels the pressure fluctuation.

In the figures, a wide range of frequencies are observed at a pressure ratio of approximately 1.7 due to the unexpected shock-induced separation of the boundary layer.

All of the figures show a sudden decrease in the power spectrum at a pressure ratio of approximately 1.4. This implies that the shock wave completely passes the measurement position, resulting in a change in state from subsonic to supersonic.

FFT Analysis of Shock Positions

The wall static pressure fluctuations provide us with only local information on the shock wave behavior around the measurements positions. Accordingly, in order to clarify the entire flow field, the variations in unsteady shock positions are also important because the shock wave always contains all of the information on its upstream and downstream conditions regardless of its position.

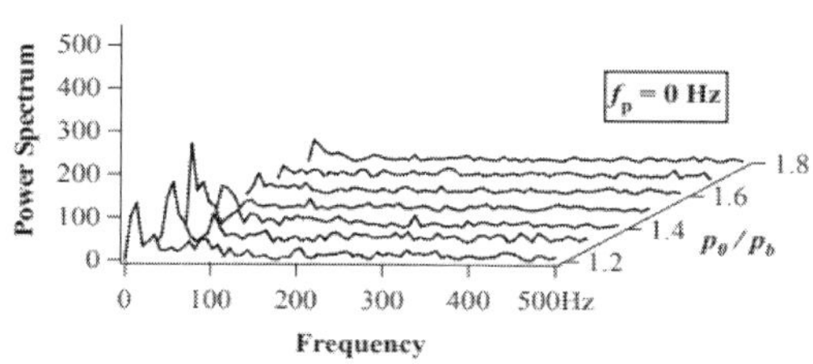

(a)

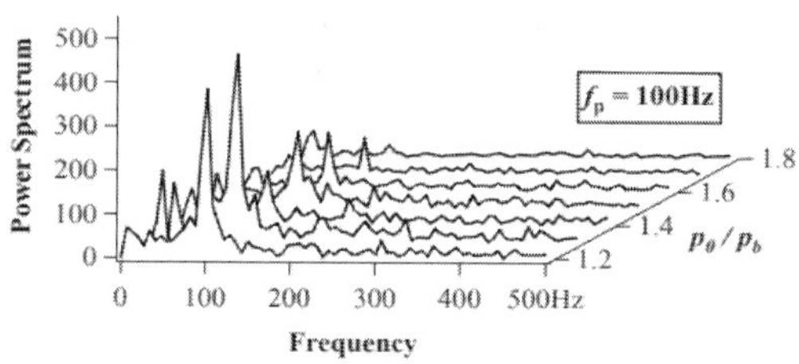

(b)

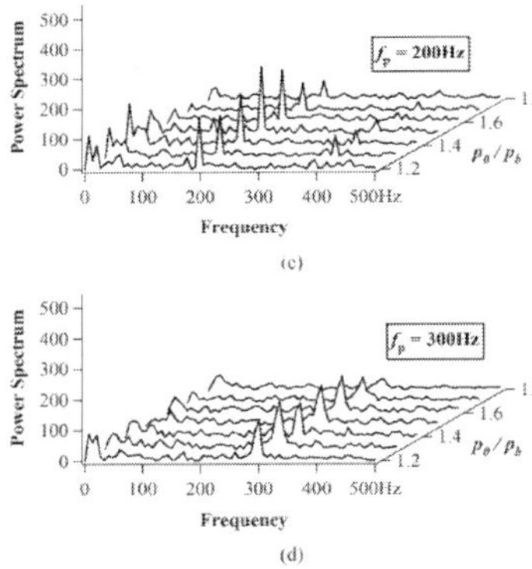

Figure 8. FFT analysis of Shock positions: (a) *fp* = 0 Hz; (b) *fp* = 100 Hz; (c) *fp* = 200 Hz; (d) *fp* = 300 Hz.

Figures 8(a)-(d) show the results of the FFT analyses of the shock positions under the influence of the piezoceramic actuator as well as with no control. In case of no control, there is no dominant frequency, as shown in Figure 8(a). The relatively low-frequency spectrum is observed as long as the pressure ratio is smaller than 1.4. This shows that the shock wave passes the measurement position monotonically as the pressure ratio increases. In contrast, for a shock wave frequency of 100 Hz, a dominant frequency is observed until a pressure ratio of 1.5. The peak value of the dominant frequency decreases as the pressure ratio increases. This decrease in the peak value of the dominant frequency is explained by the behavior of the shock wave, as shown in Figure 3. That is, the gradient of the shock wave displacement decreases as the pressure ratio increase as shown in Figure 3 due to the expansion rate of the cross sectional area of the diffuser.

This implies that shock wave located downstream is insensitive to the pressure fluctuation compared with that located upstream position. Then, the oscillation of the shock wave gradually decreases as it moves downstream. The diminishing of the dominant frequency for higher pressure ratios might be due to the relative increase in other frequencies.

On the other hand, when the higher frequency is applied to the throat, the clearly peaky dominant frequency, which is the exact same frequency as that input to the piezoceramic actuator, can be observed. Note also that the applied frequency remains, even for the higher pressure ratio, which suggests that the shock wave is always under the influence of the piezoceramic actuator. The tendencies observed in Figure 8(c) also appear in Figure 8(d). In other words, the input frequency of 300 Hz is the dominant frequency for all pressure ratios. Then, actuation at the throat is a promising method for controlling shock wave behavior.

CONCLUSIONS

A piezoceramic actuator is applied to the throat of a circular arc diffuser with various driving frequencies in order to clarify the response of the flow field and shock wave behaviors to the piezoceramic actuator. The piezoceramic actuator, which moves periodically as a reference signal, is considered to be a driving force for oscillation phenomena. The conclusions are summarized as follows:

1) The starting shock wave moves downstream monotonically with the increase in the wind tunnel pressure ratio, regardless of the input frequencies to the piezoceramic actuator.
2) The rms values of the wall pressure fluctuations decrease suddenly just after the shock wave completely passes over the measurement position for all driving frequencies.
3) The pressure fluctuations at the throat and downstream of the throat correspond to the exact same frequencies of the input frequencies to the piezoceramic actuator in both cases that the position is supersonic and subsonic state.
4) Shock wave behaviors are also confirmed to correspond to the piezoceramic actuator behaviors.

REFERENCES

1. G. E. A. Meier, "Shock Induced Flow oscillations in a Laval Nozzle," In: K. Oswatitsch and D. Rues, Eds., Symposium Transonicum II, Springer, Berlin, 1976, pp. 252-261.
2. T. Hsieh, T. J. Bogar and T. J. Coakley, "Numerical Simulation and Comparison with Experiment for SelfExcited Oscillations in a Diffuser Flow," AIAA Journal,
3. Vol. 25, No. 7, 1987, pp. 936-943. doi:10.2514/3.9725
4. R. Bur, R. Benay, A. Galli and P. Berthouze, "Experimental and Numerical Study of Forced Shock-Wave Oscillations," Aerospace Science and Technology, Vol. 10, No. 4, 2006, pp. 265-278. doi:10.1016/j.ast.2005.12.002
5. E. Benini, R. Biollo and R. Ponza, "Efficiency Enhancement in Transonic Compressor Rotor Blades Using Synthetic Jets: A numerical Investigation," Applied Energy Vol. 88, No. 3, 2011, pp. 953-962,
6. doi:10.1016/j.apenergy.2010.08.006
7. P. B. Salunkhe, J. Joseph and A. M. Pradeep, "Active Feed-Back Control of Stall in a Axial Flow Fan under Dynamic Inflow Distortion," Experimental Thermal and Fluid Science, Vol. 35, No. 6, 2011, pp. 1135-1142. doi:10.1016/j.expthermflusci.2011.03.008
8. M. Yaga, T. Haga and K. Oyakawa, "Study on Passive Control in a Transonic Diffuser," AIAA Paper 2000-0902, 38th Aerospace Sciences Meetings & Exhibit, Reno, January 2000, pp. 1-11.

CHAPTER

18

An ARX-Based PID-Sliding Mode Control on Velocity Tracking Control of a Stick-Slip Piezoelectric-Driven Actuator

Y. Cao, X. B. Chen

Department of Mechanical Engineering, University of Saskatchewan, Saskatoon, Canada

ABSTRACT

Piezoelectric-driven stick slip actuators have been drawn more and more attention in the nano- positioning application due to the high accuracy and theoretical unlimited displacement. However, the hysteresis of piezoelectric actuator (PEA) and the nonlinear friction force between the end- effector and the stage make control of piezoelectric-driven stick slip actuator challenge. This paper presents the development of an autoregressive exogenous

Citation:- Cao, Y. and Chen, X. (2015), "An ARX-Based PID-Sliding Mode Control on Velocity Tracking Control of a Stick-Slip Pi-ezoelectric-Driven Actuator". Modern Mechanical Engineering, 5, 10-19. doi: 10.4236/mme.2015.51002.

(ARX)-based proportional-integral- derive (PID)-sliding mode control (SMC) for the velocity tracking control of the piezoelectric-driven stick slip actuator. Stability is guaranteed by rigorously choosing the appropriate PID parameters and the zero steady state error is achieved. To verify the effectiveness of the proposed method, experiments were carried out on a commercially-available piezoelectric-driven stick slip actuator. The tracking errors were compared with the traditional PID controller, illustrating that in spite of existing of modeling error, the ARX-based PID-SMC is able to better improve the velocity tracking performance of piezoelectric-driven stick slip actuator, compared with the traditional PID controller.

Keywords: Autoregresive Progress, Sliding Mode Control, Slick-Slip, Piezoelectric Actuator

INTRODUCTION

Piezoelectric-driven stick-slip actuators play important roles in nano-positioning applications due to their simple configuration, high accuracy and theoretically unlimited displacement [1] - [6]. The end-effector is supported and guided by a movable stage, which is connected to the base through flexible hinges and driven by a PEA, as shown in Figure 1. The end-effector moves together with the stage under slow expansion of the PEA. Sudden contraction of the PEA leads to the sliding motion of the end-effector since the force due to the friction limit between the end-effector and the movable stage. As a result, the end-effector moves a step ΔS, with respect to its original position. Such steps can be accumulated to achieve a theoretically unlimited displacement (actually limited by the size of the moveable stage).

The hysteresis of PEA and the friction dynamics of the end-effector make the control of stick-slip actuator challenge. An analog electronic circuit was developed in [7] for driving stick-slip piezoelectric linear actuators. The key design task for the amplifier is to provide a high-voltage asymmetric sawtooth-like signal and feed it into a capacitive load.

To compensate the hysteresis effect, combination of a charge control scheme with switch- ing is proposed and the experiments confirm the superiority of the proposed method over other existing techniques. However, the charge control requires expensive hardware, which increases the difficulty and cost of its implementation.

To avoid this, a voltage/frequency proportional controller was proposed in [8] for high stroke displacement of stick-slip micro-positioning stage.

The proposed controller is essentially a combination of the sign controller, the classical proportional control and frequency proportional control, and the experimental results showed its priority over the basic controller.

To compensate the steady state error, integral control was employed in addition to the aforementioned method [9] . In [10] ,

Proportional-Derivative (PD) controller was applied in the control of the stick slip piezoelectric actuators. Its application in the sample holder for Atomic

Force Microscope (AFM), the micro-Electrical Discharge Machining (EDM) machine, the micro-assembly system and the micro telemanipulation system for biological specimens was described.

All the research mentioned above related to the displacement tracking control of the piezoelectric-driven stick-slip actuators under step or ramp reference signal.

The velocity tracking control has not been reported yet. This work focuses on the development of control method for the velocity tracking control of the piezoelectric- driven stick-slip actuators.

PID/PI/PD controller has shown great potential in the control application for nano-positioners.

The challenge of PID/PI/PD control in the velocity tracking control of stick-slip piezoelectric-driven stage is maintaining the system stability in the presence of uncertainty and disturbance.

It also has an issue with low gain margin in high frequency applications [1]. Recently, sliding model control (SMC) has been drawing considerable attention in the control research community worldwide due to its ability to compensate for the system uncertainties and disturbance [11] -[14] .

To solve the chattering problem of the regular SMC, a PID-based SMC (PID-SMC) was developed in [15], in which the discontinuous "bang-bang" switching function is replaced by a PID regulator. Due to the integral effect of the PID regulator, zero steady state tracking error can be achieved.

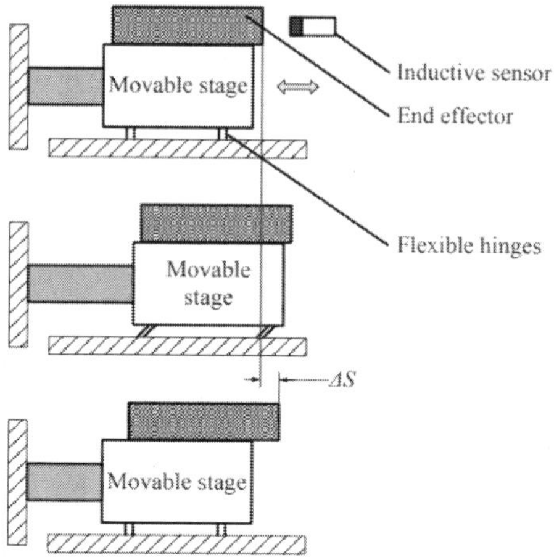

Figure 1. Working principle of the piezoelectric-driven stick- slip actuator.

As a state tracking control scheme, the PID-SMC developed in [15] has been shown effective in the control of the second order mechanical systems, in which the states can be readily estimated from the measured output and/or its derivatives. In some circumstances, the system state might not be readily or even be impossibly obtained due to system complexity. In such a case, the application of PID-SMC is challenged due to the lack of system state information. For this, an output tracking integrated PID-SMC is developed [16], in which the transfer function of the plant is considered to be a cascade of the numerator and the denominator. The numerator is compensated by an optimal inverse feed forward controller and as such the regular state tracking based SMC can be applied. The proposed method has been experimentally shown effective in the control of a commercially available one-degrees-of-freedom (DOF) piezoelectric actuator.

It is noted that if the dynamics of the plant can be represented by an auto-regressive (ARX) model without zeros, for example, $y(k)=\sum_{i=1}^{n}a_i y(k-i)+bu(k-1)$, where u(k) and y(k) are the system input and output at time instant k respectively, such a model will be readily transferred to a state space model with its state being the system output and the state tracking based PID-SMC is able to be directly applied for the dynamics compensation. Inspired by this, an ARX-based PID-SMC is developed in this paper and applied in the velocity control of a one-DOF stick-slip PEA. The effectiveness of the proposed method is experimentally verified and compared with the traditional PID controller introduced in [9].

ARX-BASED PID-SMC AND ITS DESIGN

The discrete nth order plant might be generally described by the following transfer function

$$G(z)=\frac{\sum_{j=0}^{n}\beta_i z^{-j}}{1+\sum_{i=1}^{n}\alpha_i z^{-i}}. \tag{1}$$

where $\alpha_i, \beta_i\ (i=1,2,\cdots,n)$ and β_0 are the parameters and the numerator and denominator are relatively prime polynomials. With the use of the system identification method, this plant model can be mathematically represented by an ARX model.

$$y(k)=\sum_{i=1}^{n}a_i y(k-i)+bu(k-1)+\varepsilon(k-1). \tag{2}$$

Equation (2) is identical to a discrete transfer function without zeros. The neglect of zeros definitely leads to model errors, which is considered to be disturbance in Equation (2).

The ARX model can be rewritten in a state space form

$$x(k+1)=Ax(k)+Bu(k)+D\varepsilon(k). \tag{3}$$

where $x(k)=\left[y(k-n)\ \cdots\ y(k-2)\ \ y(k-1)\right]^{\mathrm{T}}\in R^{n\times 1}$ is the state vector; $A=\begin{bmatrix}0_{(n-1)\times 1} & I_{(n-1)\times(n-1)}\\ -a_n & a\end{bmatrix}$ $(a=[-a_{n-1}\ \ -a_{n-2}\ \cdots\ -a_1])$, $B=\left[0_{1\times(n-1)}\ \ \beta_0\right]^{\mathrm{T}}$ and $D=\left[0_{1\times(n-1)}\ \ 1\right]^{\mathrm{T}}$ are the system matrixes of the discrete state space model. The state vector is represented in terms of the outputs in the past history, which suggest the state tracking is essentially the output tracking. Therefore, the state tracking SMC design method can be applied.

Denote the desired output vector to be $x_d(k)=\left[y_d(k-n)\ \cdots\ y_d(k-2)\ \ y_d(k-1)\right]^{\mathrm{T}}\in R^{n\times 1}$. The objective of SMC is to force the error state $e(k)=x(k)-x_d(k)$ to move to the sliding surface and then converge to zero. Equation (3) can be rewritten in terms of the dynamics of e(k),

$$e(k+1)=Ae(k)+Bv(k)+D\varepsilon(k)+Ax_d(k)-x_d(k+1). \tag{4}$$

Similar to the general SMC design approach introduced in [15], the input u (k) is divided into two parts $u_1(k)$ and $u_{SM}(k)$, i.e.

$$v_1(k)=-B^{\dagger}\left[Ax_d(k)-x_d(k+1)\right]. \tag{5}$$

It is noted that Equation (5) is non-casual if the future desired output is unknown. In such a case, $x_d(k+1)$ could be estimated by the trapezoid estimation

$$x_d(k+1) = 2x_d(k) - x_d(k-1). \tag{6}$$

Substituting Equation (5) into Equation (4) yields

$$e(k+1) = Ae(k) + Bu_{SM} + D\varepsilon(k) + (I - BB^\dagger)Ax_d(k) - (I - BB^\dagger)x_d(k+1). \tag{7}$$

For the system described by Equation (3), one has $(I - BB^\dagger)Ax_d(k) - (I - BB^\dagger)x_d(k+1) = 0$. Thus, Equation (7) is simplified as

$$e(k+1) = Ae(k) + Bu_{SM} + D\varepsilon(k). \tag{8}$$

For the sliding function that takes the following form of

$$s(k) = Se(k). \tag{9}$$

The control action can be considered consisting two parts, i.e.,

$$u_{SM}(k) = u_{eq}(k) + \Delta u(k). \tag{10}$$

where $u_{eq}(k) = -(SB)^{-1}SAe(k)$ is the equivalent control and $\Delta u(k)$ is the switching control.

Substituting Equation (10) into Equations (8) and (9) yields

$$e(k+1) = \left[I - B(SB)^{-1}S\right]Ae(k) + B\Delta u(k) + D\varepsilon(k). \tag{11}$$

$$s(k+1) = SB\Delta u(k) + SD\varepsilon(k). \tag{12}$$

For the convenience of following discussion, Equations (9) and (11) are rewritten as

$$e(k+1) = A_e e(k) + B\Delta u(k) + D\varepsilon(k), \quad s(k) = Se(k). \tag{13}$$

where $A_e = \left[I - B(SB)^{-1} S \right] A$. The system described by Equation (13) can be equivalent as a dynamic plant,

where $\Delta u(k)$ is the input and s(k) is the output. The switching control is to force the plant output to be zero. For this, PID controller can be employed to generate the switching control such that the equivalent control error s(k) is zero, which is given by

$$\Delta u(k) = -\left[Ps(k) + I\sum_{i=0}^{k} s(i)T + D\frac{s(k) - s(k-1)}{T} \right]. \quad (14)$$

where P, I and D are parameters of the discrete PID-based SMC; T is the sampling period.

Theorem 1: If the closed-loop system (13) is stable, the zero steady state error can be achieved [17] .

Theorem 2: There exist some P, I and D such that the closed-loop control system (13) is stable [16] .

EXPERIMENTS

To verify the effectiveness of the proposed method, experiments were carried out to control the motion velocity of a stick-slip piezoelectric-driven actuator (Figure 2). The PEA used to drive the stage generates displacement in a range of 15 μm with a resolution of 0.5 nm. An inductive sensor (SMU9000, Kaman, Windsor, CT) with a resolution of 10 nm was used to measure the displacement of the end-effector. Both the actuator and the sensor were connected to a host computer via an I/O board (PCI-DAS1602/16, Measurement Computing Corporation) and controlled via SIMULINK programs. The movable stage is driven by the PEA and the end-effector on the stage can slide horizontally on the stage via a pair of V grooves machined on them. Polished silicon wafers are attached to the V grooves as frictional surfaces to guarantee the uniformity of the frictional characteristics throughout the traveling range of the end-effector. All the displacements presented in this study were measured by the inductive sensor with a sampling interval of 0.05 ms.

Figure 3 shows the model configuration of the stick-slip piezoelectric-driven stage developed in [18] .

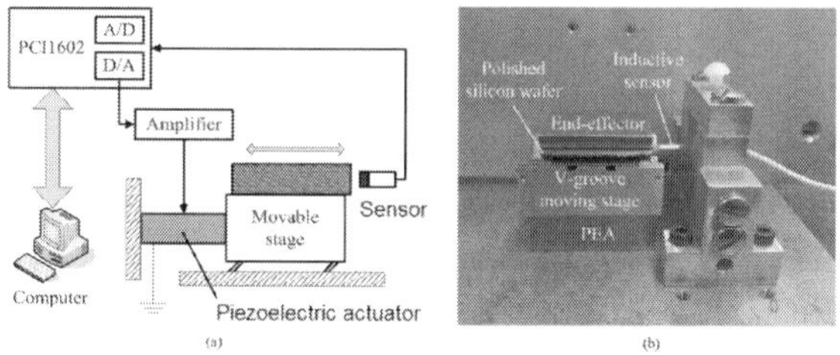

Figure 2. Experimental settings on the piezoelectric driven stage; (a) schematic and (b) picture.

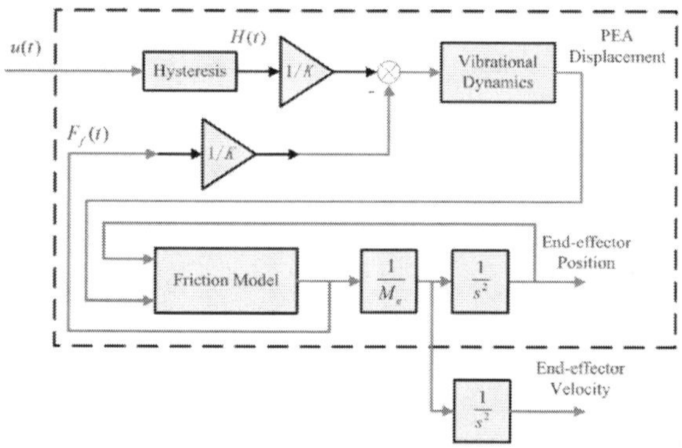

Figure 3. Model configuration of the piezoelectric-driven stick-slip actuator (Regenerated from [18]).

M_e is the mass of the end-effector, F_f is the friction force between the PEA stage and the end-effector. The hysteresis model H is cascaded with the vibration dynamics model. The output—PEA displacement is feed- forward to the friction model, which generates the friction force and the motion parameters of the end-effector.

To apply the ARX-based PID-SMC, the PEA and the end-effector are considered to be an integrated system with its input being the driven voltage u and its output being the motion velocity of the end-effector, as shown in Figure 3. The nonlinearities such as hysteresis and the friction are considered to be disturbances and then rejected by the proposed method. Experiments for parameter identification were first carried out to obtain the ARX model. Then, the ARX-based PID-SMC was applied to control the nano-postioning stage for tracking step reference signals. Finally, dynamic tracking control with the proposed methods was implemented and comparison to a PID controller was made.

Model Identification

To drive the end-effector, a saw-tooth generator is required, as seen in Figure 4. Each saw-tooth is 20 ms wide. The amplitude of the each saw-tooth is defined by the reference input voltage u(t). If u(t) < 0, a reverse saw- tooth signal is demanded to move the end-effector to the backward direction, as shown in Figure 5.

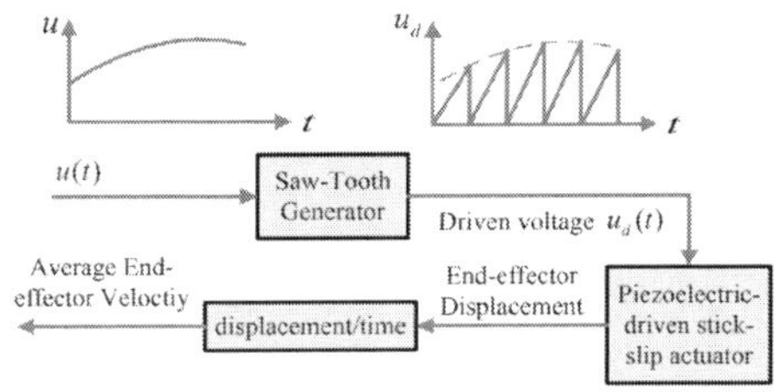

Figure 4. Saw-tooth generator for the stick-slip motion and the velocity calculator.

The driven voltage supplied to the PEA u_d can be represented in terms of the reference input voltage as

$$\text{if } u \geq 0,\ u_d(t) = u(t)\text{mod}(t,T)/T;$$
$$\text{if } u < 0,\ u_d(t) = -u(t) + u(t)\text{mod}(t,T)/T. \quad (15)$$

where T is the tooth width, mod represents the modulus after division.

The displacement of the end-effector is measured by the inductive sensor. The average velocity v is estimated by

$$v = \frac{S_{\text{saw-tooth}}}{T}. \quad (16)$$

where $S_{\text{saw-tooth}}$ is the displacement of the end-effector in one period of the saw-tooth signal.

The dynamics of the integral system can be regarded approximately as a second order system according to our previous research [19] [20]. As such, a second order ARX model was employed to describe the system dynamics, while the hysteresis exhibited by the actuator and the friction between the stage and the end-effector are considered as extra disturbances to the dynamic model. For the experiments, a 10 V step input reference voltage was provided to the PEA and the corresponding output velocity was estimated by Equation (23). With the recorded data, the parameters were identified by using the least square method, leading to

$$y(k) = 0.9985y(k-1) + 0.000972y(k-2) + 0.001458u(k-1). \tag{17}$$

Step Tracking

The ARX-based PID-SMC developed in this study was implemented in experiments to control the velocity of the piezoelectric-driven slick-slip actuator. The sliding surface was defined according to Equation (9), where $S = [m \quad 1]$ and m = −0.1. To minimize the rising time, the PID parameters in the PID regulator is adjusted to be P = 0.2, I = 3, D = 0. For comparison, the traditional PID controller was also applied, with its parameters adjusted by trails-and-errors to be P = 0.4, I = 4, D = 0.

Figure 6 shows the velocity step response of the piezoelectric-driven stick-slip actuator controlled by the developed control scheme and the traditional PID controller. Table 1 shows the comparison of the rising time and the overshoot.

It can be seen that the tracking errors approach to zero through the use of I component in the PID regulator. The velocity step response of the piezoelectric-driven stick-slip actuator controlled by the proposed method is faster than that controlled by the traditional PID controller. For example, when a 10 μm/s step reference input was provided, the rising time of the velocity step response controlled by the ARX-based PID-SMC is 0.13 s, 0.2 ms less than that controlled by the PID controller. Faster velocity response is also observed for the 20 μm/s step reference input. However, since the dynamics performance of the piezoelectric-driven stick-slip actuator varies with the amplitude of the input voltage, the model error increased when the 20 μm/s step reference input was provided to the actuator. As a result, the overshoot increased to 10% in this case.

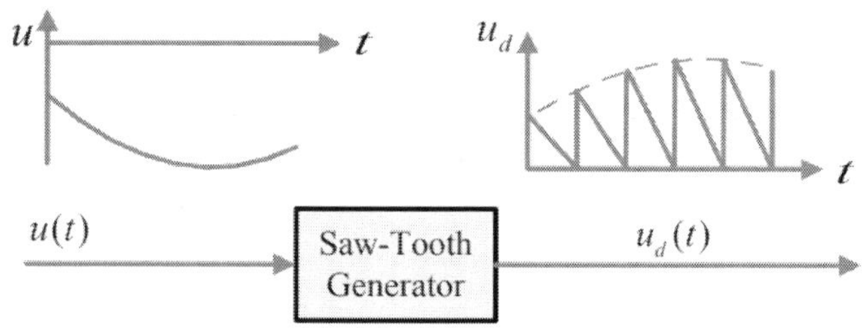

Figure 5. Reverse saw-tooth to driven the end-effector in the backward direction.

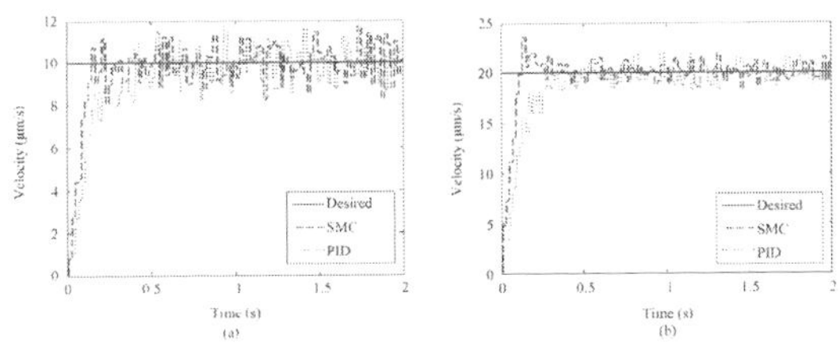

Figure 6. Velocity Step response of the piezoelectric-driven stick-slip actuator controlled by the proposed method and the traditional PID controller (a) 10 μm/s desired velocity; (b) 20 μm/s desired velocity.

Table 1. Comparison of the velocity step response between the ARX-based PID-SMC and the traditional PID controller.

Reference inputs	Controller	Rising time (s)	Overshoot (%)
10 μm/s step input	SMC	0.13	0
	PID	0.33	0
20 μm/s step input	SMC	0.11	10
	PID	0.21	0

To test the control performance of the proposed method, the same velocity step tracking experiments were also carried out with an increased mass of the end-effector. The same parameters is applied and compared with the traditional PID controller. Figure 7 shows the velocity step response of the piezoelectric-driven stick-slip actuator controlled by the developed control scheme and the traditional PID controller. Table 2 shows the comparison of the rising time and the overshoot. It can be concluded that the velocity step response of the piezoelectric-driven stick-slip actuator controlled by the proposed method is faster than that controlled by the traditional PID controller.

It is noted that the rising time of the velocity step response controlled by the ARX-based PID-SMC is 0.09 s. This is different from the result with an decreased mass of the end-effector, since the dynamics of the piezoelectric-driven stick-slip actuator also changes with the weight of the end-effector.

Sinusoidal Tracking

To further show the effectiveness of the proposed control method, sinusoidal tracking experiments with different frequencies were carried out on the piezoelectric-driven stick-slip actuator.

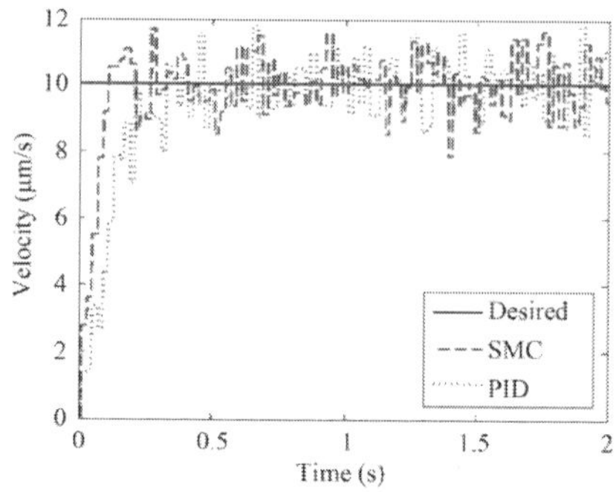

Figure 7. 10 μm/s Step response of the piezoelectric-driven stick-slip actuator controlled by the proposed method and the traditional PID controller.

Table 2. Comparison of the velocity step response between the ARX-based PID-SMC and the traditional PID controller with an increased mass of the end-effector.

Reference inputs	Controller	Rising time (s)	Overshoot (%)
10 μm/s step input	SMC	0.09	0
	PID	0.21	0

The same parameters were applied and the control performance was compared with the same PID controller. Figure 8 shows the sinusoidal tracking performance of the piezoelectric-driven stick-slip actuator controlled by the proposed method and the traditional PID controller.

Table 3 compares the tracking error evaluated in terms of the 2-norm of the difference between the desired output and the measured output. From Table 3, it can be seen that the developed control scheme performs better than the traditional PID controller. As frequency increases, the priority of applying the ARX-based PID-SMC is more obvious. For example, for a 0.25 Hz 10 μm/s sinusoidal reference input, the tracking error with ARX-based PID-SMC is 1.12 μm, which is 29.7% of that with the PID controller. While for a 1 Hz sinusoidal reference input, the improvement increases to 40.2%.

It is noted that the velocity tracking performance deteriorates when the motion direction of the end-effector changes. For example, for 10 μm/s 1 Hz sinusoidal reference velocity, the maximum tracking error in time interval

0.5 - 0.75 s and 1.5 - 1.75 s is 50% of the amplitude if controlled by the ARX-based PID-SMC. The maximum tracking error even reaches 100% of the reference amplitude of the traditional PID controller is applied. This might be due to the nonlinear friction force between the end-effector and moving stage when the velocity changes from positive to negative. Obviously, the ARX-based PID-SMC partially compensates the nonlinearity. Consideration of the nonlinear friction model and its integration in the controller design are required for further improvement on the velocity tracking performance of the piezoelectric-driven stick-slip actuator.

CONCLUSION

This paper presents the development of an ARX-based PID-SMC for the velocity control of the piezoelectric- driven stick-slip actuator. Specifically, by applying ARX model, the output-tracking problem is defined as one of state tracking, while the "bang-bang" switching control in SMC is replaced with PID-based one. With the developed control scheme, chattering or the state oscillation at a high frequency can be eliminated and also the zero steady state error can be achieved.

To verify the effectiveness of the developed control scheme, experiments were carried out on a piezoelectric-driven stick-slip actuator, whose dynamics was identified by experiments.

The results of velocity tracking performance with the proposed control scheme were compared to that with the traditional PID controller. It was shown that both the step response and the sinusoidal tracking performance can be further improved by applying the proposed method, although the ARX model error exists.

Table 3. Comparison of the velocity step response between the ARX-based PID-SMC and the traditional PID controller with an increased mass of the end-effector.

Reference inputs	Frequency (Hz)	Tracking error (μm/s)	
		ARX-based PID-SMC	PID
10 μm/s sinusoidal input	0.25	1.12	1.57
	0.5	2.13	2.89
	1	2.82	4.72

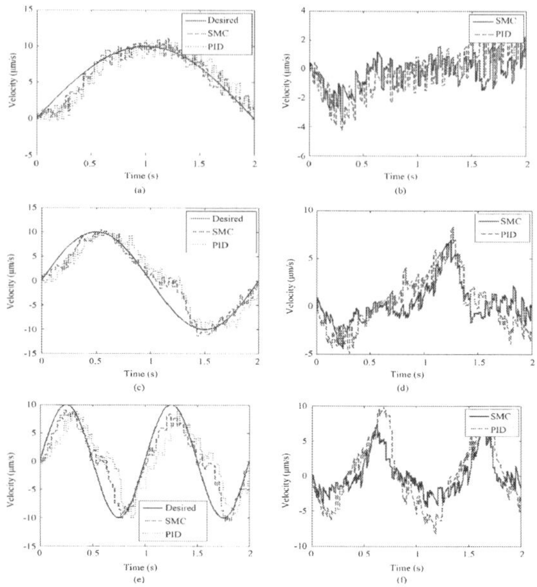

Figure 8. Comparison of sinusoidal tracking performance with different controllers: displacement and tracking error for (a) (b) 0.25 Hz frequency, (c) (d) 0.5 Hz frequency and (e) (f) 1 Hz frequency.

However, the control performance of the ARX-based PID-SMC depends on the accuracy of the identified dynamic model. This might be solved by employing a dynamic model with adaptive parameters and an adaptive ARX-based PID-SMC, which will be the focus of the future work. Furthermore, nonlinear friction model should be considered in the SMC design to further improve the velocity tracking performance of the piezoelectric-driven stick-slip actuator.

ACKNOWLEDGEMENTS

Y. Cao, X. B. Chen (CSC) and the Natural Sciences and Engineering Research Council (NSERC) of Canada is acknowledged.

REFERENCES

1. Devasia, S. and Moheimani, S.O.R. (2007) A Survey of Control Issues in Nano-Positioning. IEEE Transactions on Control System Technology, 15, 802-823. http://dx.doi.org/10.1109/TCST.2007.903345
2. Shiratori, H., Takizawa, M., Irie, Y., Hirata, S. and Aoyama, H. (2012) Development of the Miniature Hemispherical Tilt Stage Driven by Stick-Slip Motion Using Piezoelectric Actuators. Mechatronics-REM, Paris, 21-23 November.
3. Zhang, Q.S., Chen, X.B. Yang, Q. and Zhang, W.J. (2012) Development and Characterization of a Novel Piezoelectric-Driven Stick-Slip Actuator with Anisotropic-Friction Surfaces. International Journal of Advanced Manufacture Technology, 61, 1029-1034. http://dx.doi.org/10.1007/s00170-011-3771-y
4. Eigoli, A.K. and Vossoughi, G. (2010) Dynamic Modeling of Stick-Slip Motion in a Legged Piezoelectric Driven Microrobot. International Journal of Advanced Robotic Systems, 7, 201-208.
5. Kang, D. (2012) Modeling of the Piezoelectric-Driven Stick-Slip Actuators. Master Thesis, College of Graduate Studies and Research, University of Saskatchewan, Saskatoon.
6. Bergander, A. and Breguet, J.M. (1998) Performance Improvement for Stick-Slip Positioners. International Symposium on Micromechatronics and Human Science, 59-66.
7. Spiller, M. and Hurak, Z. (2011) Hybrid Charge Control for Stick-Slip Piezoelectric Actuators. Mechatronics, 21, 100- 108. http://dx.doi.org/10.1016/j.mechatronics.2010.09.002
8. Rakotondrabe, M., Haddab, Y. and Lutz, P. (2008) Voltage/Frequency Proportional Control of Stick-Slip Micropositioning Systems. IEEE Transactions on Control System Technology, 16, 1316-1322. http://dx.doi.org/10.1109/TCST.2008.917232
9. Rakotondrabe, M., Haddab, Y. and Lutz, P. (2009) Development of Modeling and Control of Mircro/Nano-Positioning 2-DOF Stick-Slip Device. IEEE Transactions on Mechatronics, 14.

10. Breguet, J.M. and Clavel, R. (1998) Stick and Slip Actuators: Design, Control, Performances and Applications. International Symposium on Micromechatronics and Human Sciences, Nagoya, 25-28 November 1998, 89-95.
11. Young, K.D., Utkin, V.I. and Özgüner, Ü. (1999) A Control Engineer's Guide to Sliding Mode Control. IEEE Transactions on Control System Technology, 17, 328-342. http://dx.doi.org/10.1109/87.761053
12. Edwards, C. and Spurgeon, S.K. (1998) Sliding Mode Control. Taylor & Francis Ltd, London.
13. Liawa, H.C., Shirinzadeh, B. and Smith, J. (2007) Enhanced Sliding Mode Motion Tracking Control of Piezoelectric Actuators. Sensors and Actuators, A138, 194-202. http://dx.doi.org/10.1016/j.sna.2007.04.062
14. Wang, S., Habibi, S., Burton, R. and Sampson, E. (2006) Sliding Mode Control of an Electrohydraulic Actuator System with Discontinuous Nonlinear Friction. Proceedings of the Institution of Mechanical Engineers, 222, Part I:J, Systems and Control Engineering.
15. Peng, J.Y. and Chen, X.B. (2014) Integrated PID-Based Sliding Mode State Estimation and Control for Piezoelectric Actuators. IEEE/ASME Transactions on Mechatronics, 19, 88-99. http://dx.doi.org/10.1109/TMECH.2012.2222428
16. Cao, Y. and Chen, X.B. (2014) Disturbance Observer Based Sliding Mode Control for a Three-DOF Nano-Positioning Stage. IEEE/ASME Transactions on Mechatronics, 19, 924-931. http://dx.doi.org/10.1109/TMECH.2013.2262802
17. Cao, Y. and Chen, X.B. (2012) Integrated Inversion-Feedforward and PID-Based Sliding Mode Control for Piezoelectric Actuators. American Control Conference, Montreal, 27-29 June 2012.
18. Peng, J.Y. and Chen, X.B. (2011) Modeling of Piezoelectric-Driven Stick-Slip Actuators. IEEE/ASME Transactions on Mechatronics, 16, 394-399. http://dx.doi.org/10.1109/TMECH.2010.2043849
19. Cao, Y. and Chen, X.B. (2012) A Novel Discrete ARMA-Based Model for Piezoelectric Actuator Hysteresis. IEEE/ ASME Transactions on Mechatronics, 17, 737-744. http://dx.doi.org/10.1109/TMECH.2011.2128339
20. Chen, X.B., Zhang, Q.S., Kang, D. and Zhang, W.J. (2008) On the Dynamics of Piezoelectric Positioning Systems. Review of Scientific Instrument, 79, Article ID: 116101. http://dx.doi.org/10.1063/1.2982238

INDEX

J

K

L

M

N

U

V

W

Z